# Effects of Processing on Food Composition, Nutritional Value and Sensory Quality

# Effects of Processing on Food Composition, Nutritional Value and Sensory Quality

Editors

**Mariusz Szymczak**
**Joanna Trafialek**

Basel • Beijing • Wuhan • Barcelona • Belgrade • Novi Sad • Cluj • Manchester

*Editors*

Mariusz Szymczak  
West Pomeranian University  
of Technology  
Szczecin, Poland

Joanna Trafialek  
Warsaw University of Life  
Sciences-SGGW  
Warsaw, Poland

*Editorial Office*  
MDPI  
St. Alban-Anlage 66  
4052 Basel, Switzerland

This is a reprint of articles from the Special Issue published online in the open access journal *Applied Sciences* (ISSN 2076-3417) (available at: https://www.mdpi.com/journal/applsci/special_issues/sensory_quality).

For citation purposes, cite each article independently as indicated on the article page online and as indicated below:

Lastname, A.A.; Lastname, B.B. Article Title. *Journal Name* **Year**, *Volume Number*, Page Range.

ISBN 978-3-0365-9951-9 (Hbk)  
ISBN 978-3-0365-9952-6 (PDF)  
doi.org/10.3390/books978-3-0365-9952-6

Cover image courtesy of Mariusz Szymczak

# Contents

# About the Editors

**Mariusz Szymczak**

Mariusz Szymczak is an Associate Professor at the Faculty of Food Sciences and Fisheries at the West Pomeranian University of Technology in Szczecin, Poland. He graduated with a Ph.D. in Food Science and Technology in 2004. He has 20 years of experience as a researcher in fish technology, especially focusing on raw materials with high protein contents. He collaborates with many industrial partners from the fish industry in the following R&D areas: fresh and frozen fish, salting, marinating, smoking, and surimi. In this research area, he has several patents and significant know-how, which are implemented in the industry. He is a board member of a number of journals and a reviewer for different papers in food science, biotechnology, and enzymes.

**Joanna Trafialek**

Joanna Trafialek is a food technologist and a professor at the University of Life Sciences in Warsaw. She has also been working as the head of postgraduate studies in the "Food Safety and Food Quality Management Systems" program. Her main research topic is food safety. She has developed effective tools which allow for the precise assessment of food safety in practice. Prof. Joanna Trafiałek is a co-author of numerous publications published in internationally recognized journals and a co-author of several chapters in books connected with gastronomy, food hygiene, and food safety management systems as well. She has significant professional insight into the food industry as an auditor, a food technologist, and a specialist in food safety. She used to work as a project coordinator of the project entitled "Transnational Quality Education for Organic Food Safety" (Akronim SAFE-ORGfood).

 *applied sciences*

*Article*

# Insights on the Potential of Carob Powder (*Ceratonia siliqua* L.) to Improve the Physico-Chemical, Biochemical and Nutritional Properties of Wheat Durum Pasta

Mirabela Ioana Lupu [1], Cristina Maria Canja [1,*], Vasile Padureanu [1], Adriana Boieriu [1], Alina Maier [1], Carmen Badarau [1], Cristina Padureanu [1], Catalin Croitoru [2], Ersilia Alexa [3] and Mariana-Atena Poiana [3,*]

[1] Faculty of Food and Tourism, Transilvania University of Brasov, Castelului 148, 500014 Brasov, Romania
[2] Faculty of Materials Science and Engineering, Transilvania University of Brasov, Eroilor 29 Blvd., 500039 Brasov, Romania
[3] Faculty of Food Engineering, University of Life Sciences "King Mihai I" from Timisoara, Calea Aradului 119, 300645 Timisoara, Romania
* Correspondence: canja.c@unitbv.ro (C.M.C.); marianapoiana@usab-tm.ro (M.-A.P.);
Tel.: +40-723-246-331 (C.M.C.); +40-726-239-838 (M.-A.P.)

**Abstract:** The aim of this research was to improve the physical-chemical properties and processability of wheat durum pasta while adding supplementary nutritional benefits. This was accomplished by incorporating carob powder into the conventional wheat pasta recipe. The study investigated the properties of pasta made with different proportions of carob powder (2%, 4%, 6% *w/w*) and evaluated its nutritional profile, texture, dough rheological properties and the content of bioactive compounds such as phenolic compounds. The physical and chemical properties (total treatable acidity, moisture content, and protein content), compression resistance, rheological properties of the dough and sensory analysis were also analyzed. Results showed that incorporating up to 4% carob powder improved the sensory and functional properties of the pasta. Additionally, the study found that the pasta contained phenolic compounds such as Gallic, rosmarinic, rutin and protocatechuic acids, ferulic, coumaric, caffeic acid, resveratrol and quercetin, and increasing the percentage of carob powder improved the polyphenolic content. The study concluded that it is possible to create innovative value-added pasta formulas using carob powder. Thus, the information revealed by this study has the potential to expand the portfolio of functional pasta formulations on the food market.

**Keywords:** carob powder; pasta; bioactive compounds; pasta properties; compression resistance; sensory analysis

**Citation:** Lupu, M.I.; Canja, C.M.; Padureanu, V.; Boieriu, A.; Maier, A.; Badarau, C.; Padureanu, C.; Croitoru, C.; Alexa, E.; Poiana, M.-A. Insights on the Potential of Carob Powder (*Ceratonia siliqua* L.) to Improve the Physico-Chemical, Biochemical and Nutritional Properties of Wheat Durum Pasta. *Appl. Sci.* **2023**, *13*, 3788. https://doi.org/10.3390/app13063788

Academic Editors: Joanna Trafialek and Mariusz Szymczak

Received: 5 February 2023
Revised: 13 March 2023
Accepted: 14 March 2023
Published: 16 March 2023



## 1. Introduction

Pasta represents one of the most consumed food products throughout the world, being distinguished by easiness and quickness of preparation [1]. It is an important source of carbohydrates (mainly starch), it contains proteins, a variable number of fibers and a very low concentration of fat [2].

The last report of the International Pasta Organization (IPO) indicated that 14.5 million tons of pasta were produced worldwide in 2018 [3] and there is an expected growth rate of 2.7% annually until 2025 [4]. Pasta is a popular food product in Europe, with high rates of consumption being reported in the Western region, in Italy, Spain and France; the Balkan region, in Greece; and in the Southern region; as well as in Chile, Argentina, Peru and Venezuela, according to a study conducted by Statista Global Consumer Survey [5] There is an increasing demand for food products with functional properties. These dietary items, also referred to as functional foods, can be enhanced, enriched or fortified.

To increase the nutritional value of the pasta, different studies have been carried out by adding different cereals (e.g., barley and pigmented cereals) [6,7], maize bran, grape marc and brewers spent grain flour [8], legume and pseudocereal flours [9–11] and ingredients from various origins, such as oregano and carrot leaf meal [12].

Plant phytochemicals and phenolic antioxidants, such as those found in fruits, vegetables, herbs and spices, have been identified as active ingredients for use in functional foods [4]. This aspect launched the opportunity to create new types of pasta, involving novel ingredients [13].

Therefore, in our study, we wanted to highlight the influence and benefits of carob powder on the quality of pasta. Carob, scientifically denominated *Ceratonia siliqua* L., is an ancient crop, cultivated in countries located in the Mediterranean area. The plant belongs to the Leguminosae (Fabaceae) family, being used by humans for nourishment since early times, according to archeologists. Carob powder can be used as an additive for enhancing the aromatic profile of products, as a coloring thickener agent, as a sweetener and as an anticoeliac agent [14] in the pharmaceutical industry. The nutritional value of carob was considered relevant due to its high content of dietary fiber and phenolic compounds [15]. Several studies have investigated the effect of the addition of carob flour in bread and bakery products, revealing the functional profile of the products [16–19].

Due to its sweetness and flavor similar to chocolate, as well as its low price, the carob milled into flour is widely used in the Mediterranean region as a cocoa substitute for sweets, biscuits and processed drinks production [20–23]. Additionally, the advantage of using carob powder as a cocoa substitute is that it does not contain caffeine and theobromine [24]. In the last years, carob pods have gained considerable attention because of their high carbohydrate and mineral content: many high-value-added products, such as lactic acid, mannitol, citric acid and pullulans, were produced from carob fermentation [25]. In parallel, carob pods have been used as a resource for bioethanol production. In Lebanon, carob pulp is mainly used for the preparation of carob syrup or carob molasses denoted "dibs", which is consumed by the Lebanese population as a sweetener [26].

The idea behind this study focused on the fact that, to date, little information is available on the usefulness of carob as an unconventional ingredient for improving the physicochemical, biochemical and nutritional properties of durum wheat pasta. Only a few attempts to use carob flour as an alternative source of polyphenolic compounds to design new pasta formulations have been reported. In this respect, the studies by Biernacka et al. [27] and Seczyk et al. [28] are a valuable starting point for our research. In terms of dough processability and sensory properties of the pasta, a major influence is given by the level of ingredients incorporated and requires in-depth studies in order to optimize formulations. It has also been noted that the fortification of pasta with functional ingredients rich in polyphenolic compounds allows for increasing the content of bioactive compounds, but this effect may be limited to a different extent by several factors, including the binding of polyphenolic compounds with food matrix constituents as a result of interactions of these compounds with proteins and starch. [29].

Following the above information, the goal of this research was to highlight the beneficial properties of carob powder by using it in the production of composite flours in the proportions of 2%, 4% and 6% ($w/w$) to develop functional wheat durum pasta. For this purpose, the content of bioactive compounds such as individual phenolic compounds were studied, as well the physicochemical properties (total titratable acidity, moisture content, protein content), dough rheological properties and compression resistance, and the sensory analysis was conducted.

## 2. Materials and Methods

### 2.1. Raw Materials

The raw material used in the technological process of carob pasta is durum wheat semolina flour produced by Valse Mollen Denmark. Premium semolina is a grist of coarse, relatively uniformly sized particles (200–425 μm) of the endosperm with minimal fines (flour) and bran content. In the process of milling durum wheat to obtain flour of particle size 300–500 μm, wheat bran and germ are removed from the flour [30].

The commercial profile of semolina, according to the common European semolina specifications, included: moisture 15%, protein 12%, ash 0.88%, granulation: over 500 μm

(0%), over 355 μm (20%), over 250 μm (45%), over 180 μm (20%), under 180 μm (15%) [30]. Supplementary nutritional parameters were provided by the producer: 1.9% fat, 68% carbohydrates, 3.5% fibers and 0.01% salt.

Carob (*Ceratonia siliqua* L.) flour used for sample preparation produced by Sanovita Valcea was purchased from a local specialized shop and stored in a dry place until use. The nutritional profile offered by the producer was: proteins (5.1%), lipids (0.3% of which was saturated), carbohydrates (80.7%) and fibers (10.8%).

Iodized salt used in pasta production was purchased from a local market and tap water used was collected from the national water distribution. Eggs were achieved from a local farm, produced in an organic method.

### 2.2. Pasta Preparation

The recipes of pasta with the addition of carob included mixtures in which the share of the mass of carob powder was 2%, 4% and 6% (m/m) of the mass of the mixture (100 g). A control sample with no addition of carob powder was also prepared. Thus, four pasta formulations were prepared. The recipe of the control sample consists of 100 g wheat flour, 37.5 mL of water, 1 g salt and 21 g egg. In pasta formulas with added carob, the wheat flour was replaced by a mixture of 98 g wheat flour and 2 g carob powder, 96 g wheat flour and 4 g carob powder and 94 g wheat flour and 6 g carob powder. Pasta sample coding is released in Table 1.

**Table 1.** Sample coding.

| Type | Coding | Percentage (%) of Carob Flour Used in the Pasta Formulations |
| --- | --- | --- |
| Control sample | PM | 0 |
| Sample with 2% carob flour | P2CP | 2 |
| Sample with 4% carob flour | P4CP | 4 |
| Sample with 6% carob flour | P6CP | 6 |

The pasta manufacturing process is shown in Figure 1.

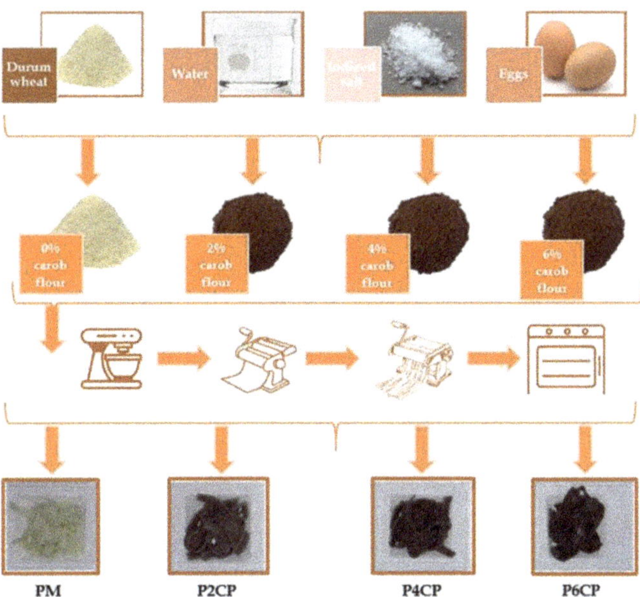


**Figure 1.** Pasta production method. PM: control sample; P2CP: sample with 2% (*w/w*) carob powder; P4CP: sample with 4% (*w/w*) carob powder; P6CP: sample with 6% (*w/w*) carob powder.

The raw materials and ingredients were dosed, weighted and scaled before being mixed at 250 rpm for 15 min in a laboratory mixer (Tefal Wizzo–QB307, Tefal, Rumilly, France) until a homogenous and firm dough was obtained. The dough was then divided into two 50 g pieces and laminated with a laminating machine (Laica PM2000, LAICA S.p.A., Barbarano Mossano (VI), Italy), pressed to eliminate air bubbles and excessive water deposit until the moisture content of the dough was 12 percent and the thickness of 2.5 mm was achieved. Then, the dough sheet was cut into tagliatelle shapes with a width of 4 mm and then dried by using a drying rack (KitchenAid—Pasta Drying Rack 5KPDR) and a laboratory drying oven at 50 °C (Deca PT-40) until pasta moisture reached the value of 12% moisture/100 g of pasta. The samples were carried out in triplicate for each type of flour.

## 2.3. Rheological Properties of the Dough

The biaxial stretching rheological properties of the dough at constant hydration were determined with a Chopin alveograph (Chopin Technologies, F Model Alveographe NG, Chopin, France). The method is based on the tensile strength of a dough sheet that is rested for 20 min and then subjected to a constant air pressure current until it forms a bubble and subsequently breaks. The air pressure inside the bubble is recorded until the bubble breaks and is then extrapolated graphically as a curve that reflects the dough's resistance to deformation [31].

The following parameters were recorded for each dough formulation: maximum overpressure P [mmH$_2$O], swelling index G (mm) and deformation energy W (J). The indicator L (mm) represents the dough extensibility (mm), estimating its handling properties. The P/L ratio indicates the ratio between the dough's toughness and its extensibility and is an important indicator along with W for characterizing flours for different bread and pastry products.

## 2.4. Extraction of Polyphenolic Compounds

The extraction of polyphenolic compounds was performed according to the method by Gaita et al. [29] by grinding the dry pasta samples and then performing in a hydroalcoholic medium with ethanol 45% (*v/v*). The solid ratio: solvent of 1:20 was stirred at a temperature of 25 °C for one hour, using the shaker Heidolph Promax 1020 (Heidolph Instruments GmbH & Co. KG, Schwabach, Germany). The extraction was filtered, and the obtained clear fractions were used for further analysis.

## 2.5. Determination of Total Phenolic Content and Polyphenolic Compounds Profile

Folin–Ciocalteu assay was used to determine the total phenolic content [32]. We used 0.5 mL from the extraction of each type of pasta treated with 1.25 mL of Folin–Ciocalteu reagent (Merck, Darmstadt, Germany) diluted 1:10 with water for this purpose. After incubating the sample for 5 min at room temperature, 1 mL Na$_2$CO$_3$ 60 g/L was added. A UV-VIS spectrophotometer was used to measure sample absorption at 750 nm after 30 min of incubation at 50 °C (Analytic Jena Specord 205). The calibration curve was created with Gallic acid as the standard at concentrations ranging from 5–250 g/mL.

The results of total phenolic content (TPC) were expressed in micrograms of Gallic acid equivalents (GAE) per g of investigated sample.

The profile of polyphenolic compounds was determined using high-performance liquid chromatography coupled with mass spectrometry (LC-MS) as described by Abdel-Hameed et al. [33]. The main polyphenols in carob flour pasta samples were determined using LC-MS with SPD-10A UV (Shimadzu) and LC-MS 2010 detectors, column EC 150/2 NUCLEODUR C18 Gravity SB 150 × 2 mm × 5 µm. The following were the chromatographic conditions: mobile stages A: water with formic acid at pH-3, B: acetonitrile with formic acid at pH-3, gradient program: 0.01–20 min 5% B, 20.01–50 min 5% B, 5–55 min 5% B, 55–60 min 5% B. Temperature 20 °C, mobile phase 0.2 mL/min The monitoring wavelengths were 280 nm and 320 nm. The curve calibration was drawn between 20 and 50 µg/mL. The results were given in µg/mL.

### 2.6. Determination of Total Titratable Acidity (TTA)

The samples of pasta were weighed to 15 g and chopped into small pieces for an easement of the homogenization process. The resulting samples were mixed with 100 mL of distilled water in a glass flask until homogenization. Meanwhile, the titration agent was formulated (Sodium Hydroxide 0.1 N Belle Chemical LLC, Billings, MT, USA). Three drops of phenolphthalein 1% in alcoholic solution were added and the titration started drop by drop, continuing until the indicator turned light pink and persisted for 30 s. The results were the average value of the two measurements. Total titratable acidity (acidity degrees/100 g product) was evaluated using the relationship presented in Equation (1):

$$\text{Total Titratable Acidity (acidity degrees/100 g products)} = \frac{V \times 0.1}{m} \times 100 \qquad (1)$$

### 2.7. Determination of Moisture Content

The moisture content of pasta samples was determined by using the gravimetric method reported by Rios et al. [34] and detailed in ISO 712:2009 [35]. Samples were weighed to 5 g using an analytical balance into a weighting vial and dried using a drying stove at 130 °C until reaching constant weight. Each sample was analyzed in triplicate and the results were expressed as average. The results were reported as weight percentage (%).

### 2.8. Determination of Compression Resistance

The crushing resistance of pasta was performed using a compression kit tester (Zwick-Roell Z005, Zwick, Ulm-Einsingen, Germany) [27]. Each sample was tested in triplicate and the results were expressed as averages. Measurements were carried out using cooked pasta and chilled until reaching the room temperature of 22.5 °C. Each sample was placed on the inferior machine pan (Zwich Z005) and compressed using a compression kit (2.5 mm thick) at a speed of 25 mm/min. The test was performed for determining the maximum force necessary for sample compression until the structure is smashed at 70% strain.

### 2.9. Determination of Protein and Fiber Content

The protein content of obtained pasta samples was evaluated by using the Kjeldahl mineralization method after nitrogen analysis using spectrophotometric analysis, according to the method reported by Heeger et al. [36] and Rios et al. [34]. The calibration was performed using ammonium chloride ($NH_4Cl$) and a conversion factor corresponding to the analyzed category was applied (5.7) to determine the protein content of the samples. The analysis was performed in triplicate and the results were expressed as the average of the three trials. The result was expressed as weight percentage (%). The fiber content was calculated taking into account the fiber content of raw materials and the recipe of pasta fabrication according to [37].

### 2.10. Sensory Evaluation

The overall acceptability of the pasta was evaluated by considering two profiles—visual profile and taste profile [28]. The visual profile included attributes such as color, appearance, attractiveness and overall acceptability, and the taste profile included flavor, taste, texture, consistency and aftertaste.

For performing the analysis, a 9-point hedonic scale was used. Each scale corresponded to a perception in order to facilitate data processing as follows: 1—"Dislike extremely"; 2—"Dislike very much"; 3—"Moderately dislike"; 4—" Slightly dislike"; 5—"Neither like nor dislike"; 6—" Slightly like"; 7—" Moderately like"; 8—"Like very much"; and 9—"Like extremely".

Sensory analysis was performed at the Faculty of Food and Tourism, Brasov, Romania. Panelists who declared suffering from different digestive problems were excluded from the study due to the high risk of aggravation of the diseases. While performing the analysis, the panelists were presented with a total of four samples of the product, as indicated in Table 1.

A number of 30 panelists aged 18–60 were requested to evaluate the samples. The participants in the study are persons with experience and competencies in sensory analysis, teachers and students who have studied and taught the mentioned subject. Panelists were asked to evaluate each sample for the assessed parameters. Panelists received the samples by turn in order to diminish the similarity and interdependence between samples with a 5-min rest before each sample tasting. Samples were coded accordingly to the codification mentioned above and each sample was distributed in glass containers at $20 \pm 1$ g and placed at a temperature in the range of 18–20 °C until serving. Panelists were served a glass of water as a palate cleanser. Consent for each panelist was required.

### 2.11. Statistical Analysis

Data analyses were performed using SPSS software (Statistical Package for Social Sciences Statistics, version 25.0.0, IBM, 2009, New York, NY, USA) and analyzed by ANOVA and Duncan's multiple range test (scored as significant if $p < 0.05$). The analysis was made in triplicate and the results were reported as mean value $\pm$ standard deviation.

## 3. Results and Discussion

### 3.1. Dough Rheological Properties

The rheological behavior of doughs and the quality attributes of final goods are greatly influenced by the water absorption capacity of flours, which varies among different flour sources [38].

The rheological parameters of pasta dough formulations are given in Table 2.

**Table 2.** Rheological data for the pasta dough formulations.

| Parameter | PM | P2CP | P4CP | P6CP |
|---|---|---|---|---|
| P (mmH$_2$O) | $170 \pm 0.15$ [a] | $103 \pm 0.09$ [b] | $101 \pm 0.03$ [b] | $93 \pm 0.01$ [c] |
| L (mm) | $22 \pm 0.05$ [a] | $33 \pm 0.12$ [b] | $33 \pm 0.09$ [b] | $27 \pm 0.06$ [c] |
| G (mm) | $10.4 \pm 0.11$ [a] | $12.8 \pm 0.13$ [b] | $12.8 \pm 0.07$ [b] | $11.6 \pm 0.05$ [c] |
| W ($10^{-4}$ J) | $169 \pm 0.25$ [a] | $130 \pm 0.28$ [b] | $125 \pm 0.21$ [c] | $100 \pm 0.14$ [d] |
| P/L | $7.73 \pm 0.13$ [a] | $3.12 \pm 0.09$ [b] | $3.06 \pm 0.18$ [b] | $3.44 \pm 0.05$ [b] |

PM: control sample; P2CP: sample with 2% ($w/w$) carob powder; P4CP: sample with 4% ($w/w$) carob powder; P6CP: sample with 6% ($w/w$) carob powder. P: the maximum overpressure; L: extensibility; G: swelling index; W: strain energy; P/L: indicates the ratio between the tenacity of the dough and its extensibility. Values followed by different letters differ significantly by ANOVA test ($p < 0.05$).

In the case of using 4% $w/w$ carob flour, it can be noticed that there was an improvement in the extensibility (L) of the dough and the swelling index (G) by up to 33%. The extensibility of the dough is influenced by the raw materials of the dough preparation. Dube et al. [39] reported that the extensibility of 100% wheat control dough progressively decreased from 156 mm to 77 mm for 40% sorghum wheat dough, while Sibanda et al. [40] also recorded similar results of a decrease in extensibility from 132 mm for 100% wheat control dough to 36 mm for 30%. The pasta supplemented with 10% carob fruit showed better texture parameters (less hard, less sticky and less adhesive) [41].

Figure 2 illustrates the variation of rheological parameters P and W with the amount of carob added to the pasta formulations.

Dough containing carob flour has the ability to retain gas and has a low capacity for extension without breaking. The dough with a more extensible character is particularly essential for improved gas retention during the process of baking which results in a good loaf volume [40].

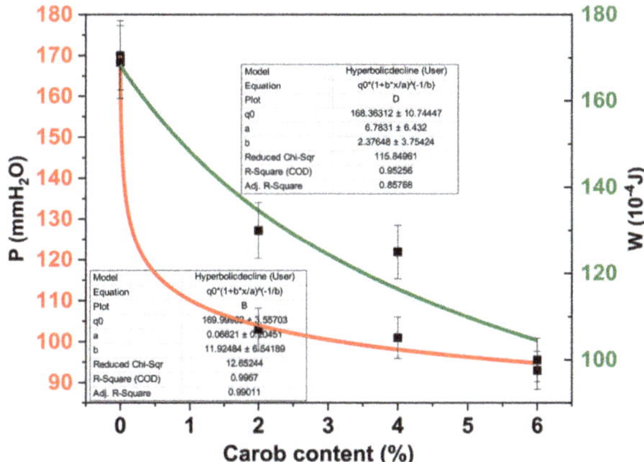



**Figure 2.** Variation of rheological parameters P and W with the amount of carob.

The tenacity of the dough decreases by up to 40% with increasing carob content (the P and W values, Figure 2 and Table 2), which is probably due to the higher fiber content of the additive compared to neat semolina flour. However, this decrease is normal for a non-gluten additive and does not significantly alter the processability of the dough. The increased extensibility of the carob-containing flour can be attributed to the plasticizing effect of the carbohydrates present in the product. The formulations with 2% and 4% carob flour may offer a good balance between processability and extensibility [42].

The beneficial effects of enrichment with different flours/powders after frozen storage conditions have been reported in the literature [43]. The physical characteristics of frozen dough and semi-baked frozen samples with the addition of commercial soluble fibers or whole oat flour were determined after baking and compared with the fresh samples. The results highlighted that in semi-baked frozen samples the crumb elasticity increased by 18% in comparison to the respective fresh ones. Additionally, samples containing whole oats presented an increased water adsorption capacity. Further studies are needed to assess the influence of carob powder on the quality of frozen dough.

*3.2. Physical and Chemical Properties*

Titratable acidity, protein content, moisture and TPC of obtained samples are reported in Table 3.

**Table 3.** Samples assessment after 2 h from the pasta production process (dry pasta).

| Parameter | 2 Hours | | | |
|---|---|---|---|---|
| | PM | P2CP | P4CP | P6CP |
| Titratable acidity (%) | 2.21 ± 0.015 [a] | 1.80 ± 0.010 [b] | 1.42 ± 0.015 [c] | 1.21 ± 0.015 [d] |
| Protein content (%) | 14.03 ± 0.058 [c] | 14.17 ± 0.153 [bc] | 14.30 ± 0.100 [ab] | 14.45 ± 0.050 [a] |
| Moisture (%) | 30.23 ± 0.257 [c] | 32.03 ± 0.057 [b] | 33.07 ± 0.057 [b] | 35.87 ± 0.152 [a] |
| TPC (µg GAE/g) | 239.07 ± 0.010 [d] | 311.53 ± 0.025 [c] | 406.80 ± 0.100 [b] | 461.10 ± 0.100 [a] |
| Fiber (%) * | 2.00 | 2.04 | 2.08 | 2.12 |

PM: control sample; P2CP: sample with 2% (*w/w*) carob powder; P4CP: sample with 4% (*w/w*) carob powder; P6CP: sample with 6% (*w/w*) carob powder. Values followed by different letters differ significantly by ANOVA test ($p < 0.05$), * the fiber content was calculated according to the fiber content of raw materials and the recipe of pasta fabrication.

Samples assessed after 2 h containing different proportions of carob flour showed higher titratable acidity contents compared to the control sample. It can be observed that, between samples, the differences were notable, with the highest value being recorded for the sample without carob powder (PM-2.21%) and the lowest value being recorded for the pasta with the highest concentration of carob powder P6CP—1.21%. Samples P4CP and P2CP showed lower values of 1.42% and 1.80%. The protein content showed slight differences between samples with the addition of carob flour and the blank sample. The most relevant development was recorded for sample P6CP—14.45% protein content due to the amount of carob flour added. Gopalakrishnan et al. [44] reported that the addition of sweet potato and fish powder in pasta raised the protein content to 12.84%.

The highest moisture content was recorded in the case of the sample with the highest content of carob powder, P6CP, (35.87%) and the lowest moisture content was recorded in the case of pasta without carob powder, PM, with 30.23%. Biernacka et al. [27] have stated that the pasta enriched with carob flour in various increasing proportions (1% to 5%) showed a certain correlation between the total phenolic content and antioxidant activity in pasta samples. The antioxidant profile of carob flour added in pasta samples revealed that the greatest value of total polyphenol content was noted for the sample containing the highest proportion of carob flour—6%—a fact that supports the statement that the addition of carob flour improves the antioxidant profile of samples. Values were considerably superior compared to the control sample. Similar to the present study, research developed by Biernacka et al. [27], Boroski et al. [12] and Zhu et al. [45] showed that the addition of carob powder in pasta in various proportions showed positive correlations concerning the TPC value. Makris et al. [46] reported that carob powder contains a higher ratio of antioxidants than red wines and can be considered a valuable source for pharmaceutical products due to its polyphenolic content. Issaoui et al. [47] reported that carob flour showed lower moisture values (13.40%, 13.50% and 13.700%) for carob pulp powder, (14.0%, 14.20%) respectively, for carob seed powder that was reflected in the final product's moisture content.

It was observed that the higher the concentration of carob is in a food product, the higher the phenolic concentration will be. For example, Aydin et al. [48] reported that the addition of 42% carob flour in spread samples was 615.28 mg GAE/100 g, considerably higher compared to the results obtained in the current study, in which the highest addition content was 6% of carob powder. Additionally, Sebecic et al. [49] mentioned that the total phenol content of biscuit samples obtained with the addition of 25% carob flour was 5.53 g/kg biscuit compared to the sample with wheat flour, with 1.10 g/kg biscuit, and 1.60 g/kg of wheat whole grain flour.

The present study reported a superior value of TPC for the sample containing carob flour in proportion of 6% (461.10 mg GAE/100 g) in contrast with the control sample, with 239.07 mg GAE/100 g. The increase in the total polyphenol content of the samples containing carob powder was closely related to the method for obtaining the carob powder, the amount of carob powder introduced into the products, the methodology used to obtain the products and the biochemical reactions developed in the product matrix [50–52].

The results shown in Table 4 show that fortifying pasta with carob increased the level of individual phenolic compounds.

All analyzed compounds are present in the P6CP sample, which contains the highest level of carob. Similar values were obtained when pasta was fortified with grape pomace [53]. The addition of 3–9% grape marc to the pasta recipe resulted in an increase of individual polyphenolic compounds depending on the percentage added. For example, Gaita et al. [29] reported the following values for individual polyphenols in pasta enriched with grape marc: gallic acid (27.95–88.22 µg/g), caffeic acid (0.66–58.11 µg/g), epicatechin (10.29–14.92 µg/g), coumaric acid (0. 32–0.85 µg/g), ferulic acid (6.33–15.75 µg/g), rutin (0.14–18.56 µg/g), rosmarinic acid (26.1–38.53 µg/g), resveratrol (31.15–34.38 µg/g), quercetin (2.89–7.3 µg/g), kaempferol (19.73–54.24 µg/g).

The addition of carob flour has been found to increase the content of polyphenols; however, the increase is not linear but exponential.

**Table 4.** Polyphenolic compounds identified in the dry pasta samples.

| Compound | Ion Mode | Polyphenolic Compounds (µg/g d.s.) | | | |
|---|---|---|---|---|---|
| | | PM | P2CP | P4CP | P6CP |
| Gallic acid | 169 | nd | nd | 6.97 ± 0.07 [a] | 31.44 ± 0.18 [b] |
| Protocatechuic acid | 153 | nd | nd | 19.93 ± 0.16 [a] | 51.14 ± 0.08 [b] |
| Caffeic acid | 179 | nd | nd | 0.86 ± 0.05 [a] | 8.56 ± 0.11 [b] |
| Epicatechin | 289 | nd | nd | 17.15 ± 0.13 [a] | 24.31 ± 0.17 [b] |
| p-Coumaric acid | 163 | nd | nd | 0.30 ± 0.03 [a] | 0.30 ± 0.09 [a] |
| Ferulic acid | 193 | nd | nd | nd | 0.23 ± 0.13 [a] |
| Rutin | 609 | nd | nd | 1.07 ± 0.06 [a] | 1.41 ± 0.16 [b] |
| Rosmarinic acid | 359 | nd | nd | nd | 224.24 ± 0.32 [a] |
| Resveratrol | 227 | nd | nd | nd | 0.85 ± 0.07 [a] |
| Quercetin | 301 | nd | 0.54 ± 0.02 [a] | 11.49 ± 0.11 [b] | 243.66 ± 0.25 [c] |
| Kaempferol | 285 | nd | nd | 63.76 ± 0.17 [a] | 231.51 ± 0.32 [b] |

nd—not detected; PM: control sample; P2CP: sample with 2% ($w/w$) carob powder; P4CP: sample with 4% ($w/w$) carob powder; P6CP: sample with 6% ($w/w$) carob powder. Values followed by different letters differ significantly by ANOVA test ($p < 0.05$).

This has been determined for quercetin, for which detectable values are available for all compositions. The increase may depend on the degree of mixing of the carob flour with the semolina dough and the distribution of these compounds in the flour.

Our findings are consistent with those of other authors such as Gaita et al. [29], who have stated that the improvement with natural phenolics can be influenced by a lot of factors, among which includes binding with food matrix components. In addition, Gaita et al. [29] and Sęczyk et al. [28] specified that the content of bioactive compounds may be influenced by the combination of proteins and phenolics content, such as hydrophobic interactions and hydrogen and ionic bonding. Frühbauerova et al. [54] conducted a study on the bioaccessibility of phenolics from carob pod powder prepared by cryogenic and vibratory grinding, and they showed that, from the 13 compounds involved in the UHPLC analysis, three were phenolic acids (vanillic, ferulic and cinnamic), three flavons (luteolin, apigenin and chrysoeriol), naringenin (flavanone) and quercitrin (glycoside form of flavonoid quercetin). The highest amount of quercitrin (44.54–64.68 µg/g), cinnamic acid (27.48–31.40 µg/g) and chrysoeriol (8.60–9.82 µg/g) have been found. The amount of the rest of the phenolic constituents ranged from 1.88 to 10.14 µg/g [55–57].

In our study, the highest number of compounds is also found in quercitrin, in P6CP (243.66 µg/g). The results shown in Table 4 show that fortifying pasta with carob increased the level of phenolic compounds.

### 3.3. Compression Resistance

The textural characteristics of products are crucial for fulfilling the acceptance of the consumers. During mastication, the brain processes the food's physical features and evaluates its texture. The sensation of food texture is important in influencing consumers' liking and preference for a food product [58].

Analyzing the values from the chart in Figure 3, it can be observed that the evolution of the values required for the crushing force of carob pasta is similar.

The differences were observed for the maximum values of crushing force, with the highest value being recorded for the sample with the lowest amount of carob powder added—P2CP with 2% of carob powder. The lowest value of the crushing force was recorded for the sample containing 6% carob powder—P6CP—compared to the control sample, where the crushing force was lower.

Regarding the pasta with 6% added carob powder, the elastoplastic curve remains constant for a longer period of time and the deformation of the carob pasta reaches 2.12 mm. After exceeding the flow area, the distinctive curves develop a new ascending tendency, the gradient being approximately identical.

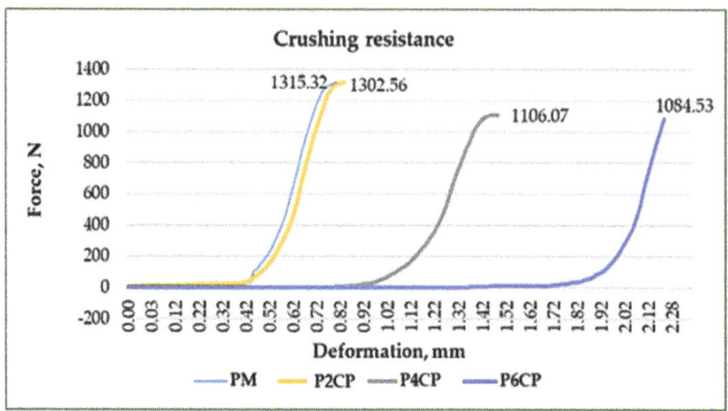



**Figure 3.** Crushing resistance of pasta samples with the addition of carob powder. PM: control sample; P2CP: sample with 2% ($w/w$) carob powder; P4CP: sample with 4% ($w/w$) carob powder; P6CP: sample with 6% ($w/w$) carob powder.

Some researchers [58] used seedless carob flour for obtaining pastry filling and observed that the major content of carob flour added to the samples' composition resulted in an increase in consistency and firmness, a fact that can be compared to the results of the crushing test from the current study, where the greatest deformation of samples was observed for the highest addition of carob powder—2.12 mm for samples with 6% carob powder addition. Additionally, a linear decrease in pasta cutting force was observed as the proportion of carob fiber increased. The same findings regarding the decreasing of crushing resistance from 1315.32 N in control to 1084.53 N in P6CP were observed in our study and can be attributed to the increase in the fiber content in the sample by 6% carob. The fiber content increased from 2.00 to 2.12% (Table 3) with the addition of carob powder in the pasta manufacturing recipe.

Research conducted by Biernacka et al. [27] showed that the cutting force required for carob flour pasta decreased while the content of carob flour in samples increased. In line with the current research, it can be stated that the high fiber content contributes to a weaker structure of the products due to the addition of fiber in the starch structure.

### 3.4. Sensory Analysis

Figure 4 illustrates the sensory characteristics in terms of overall acceptability, consistency, attractiveness, aftertaste, appearance, color, flavor, taste and texture of pasta samples with different percentages of carob flour.

It can be observed that the pasta with a medium carob powder, P4CP, with the addition of 4% of carob powder was the most preferred due to its attributes, with a mean value of 8.62 ± 3.8 out of 9 points. This indicates that the consumer prefers a small amount of carob powder in pasta. The addition of a higher amount of carob powder was rejected; samples with 6% carob flour were undervalued by panelists. Based on texture, taste, color and consistency, sample P4CP was the most appreciated, being ranked with the highest value out of a possible 9 points. This was very close to sample P6CP, which had flavor, color and texture scoring 8 ± 7.3 points. Samples with 6% carob powder addition received the lowest scores due to the affected attractiveness and appearance. The sensory analysis is an important parameter to evaluate the adaptation of several pulses' capacity in fresh pasta and if the addition of novel constituents in the structure of products influences the properties of final products and consumers' perception.

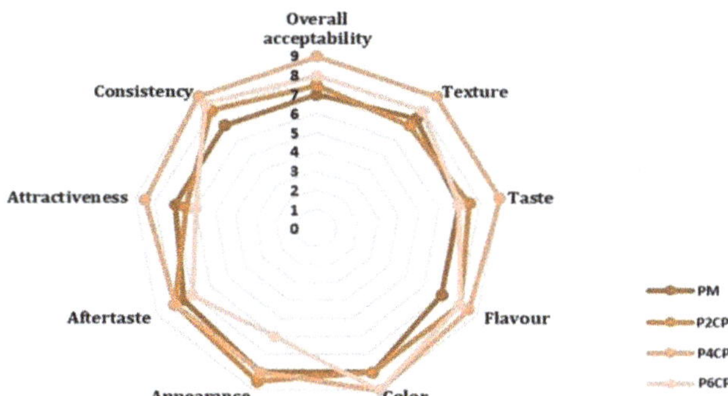


**Figure 4.** Sensory analysis score of pasta samples with the addition of carob powder. PM: control sample; P2CP: sample with 2% (*w*/*w*) carob powder; P4CP: sample with 4% (*w*/*w*) carob powder; P6CP: sample with 6% (*w*/*w*) carob powder.

It has been demonstrated that the addition of gluten-free flour to a pasta recipe alters the gluten network and reduces the overall structure of the pasta, resulting in a negative effect on sensory properties [53,55]. Furthermore, it may result in increased solid substance losses from pasta in cooking water. Darker color was observed when pasta was supplemented with 10% carob fruit which negatively influenced consumers' perceptions [41]. Consumers rejected products with excessively high concentrations due to their dark color and bitter taste [58]. According to Dulger Altiner, D. and Hallac, S., 2020, the highest values in terms of sensory analysis were determined in the 20% carob flour addition to pasta [41].

## 4. Conclusions

The addition of 2%, 4% and 6% (*w*/*w*) carob flour to pasta recipes resulted in an increase in total polyphenol content. A more significant increase in the individual polyphenolic content is observed from the concentration of 4% carob flour. At a concentration of 2%, compared to the control sample, the increase was very small. It was also found that all the individual polyphenolic compounds analyzed are present in the P6CP sample, which contains the highest concentration of carob. Up to 6% carob flour in durum wheat flour strongly influenced the color of pasta. However, due to the difficulties that could arise during pasta processing, it is not recommended to use a concentration of carob flour greater than 6% in the pasta recipe because the consistency of the pasta dough will no longer be achieved based on the recipe used. In order to use a larger amount of carob flour, the basic recipe must be modified and more flour added, as the pasta dough becomes softer as the concentration of carob flour increases. Carob flour affects the kneading, modeling and drying processes due to a reduction in the elasticity of the dough caused by a reduction in the gluten content. The sample with a medium carob flour addition (4%) was the most appreciated for its properties, with an average score of 8.62 out of 9 points. At a lower concentration (2%), the color was much more intense compared to the sample without carob flour, but the taste was not very permeating. At a concentration of 6% carob flour, the color of the pasta was very intense and the taste was very carob. The results of the physico-chemical analyses carried out on all the pasta types were within the limits set by the standards.

Regarding the recommended amount of carob flour to be used in the pasta technological process, correlating the results of the physico-chemical properties, dough rheological properties, compression resistance and sensory analysis, a concentration of 4% is the most recommended. It can be concluded that carob flour has enhanced health-promoting properties and could be a useful additive for the production of pasta from common wheat.

In order to improve the functionality of carob pasta, future studies will be considered to supplement the fiber intake, but also to improve the sensory properties by adding natural aromatic compounds. The synergism created by the active principles of the ingredients will also be another direction for future research.

**Author Contributions:** Conceptualization, M.I.L. and C.M.C.; methodology, V.P.; software, C.B.; validation, E.A., M.-A.P. and M.I.L.; formal analysis, M.I.L.; investigation, C.P. and A.M.; resources, C.P.; data curation, A.M. and C.C.; writing—original draft preparation, M.I.L. and A.B.; writing—review and editing, M.I.L., E.A. and M.-A.P.; visualization, V.P.; supervision, M.I.L., E.A., M.-A.P., C.M.C. and V.P. All authors have read and agreed to the published version of the manuscript.

**Funding:** This research received no external funding.

**Institutional Review Board Statement:** Not applicable.

**Informed Consent Statement:** Not applicable.

**Data Availability Statement:** Not applicable.

**Acknowledgments:** The chemical research was performed in the Interdisciplinary Research Platform belonging to the University of Life Sciences "King Mihai I" from Timisoara, where the analyses were made.

**Conflicts of Interest:** The authors declare no conflict of interest.

## References


1. Kaur, G.; Sharma, S.; Nagi, H.P.S.; Ranote, P.S. Enrichment of pasta with different plant proteins. *J. Food Sci. Technol.* **2013**, *50*, 1000–1005. [CrossRef]
2. Fuad, T.; Prabhasankar, P. Role of Ingredients in Pasta Product Quality: A Review on Recent Developments. *Crit. Rev. Food Sci. Nutr.* **2010**, *50*, 787–798. [CrossRef]
3. IPO. 2019. Available online: https://internationalpasta.org/annual-report/ (accessed on 5 March 2023).
4. Falciano, A.; Sorrentino, A.; Masi, P.; Di Pierro, P. Development of Functional Pizza Base Enriched with Jujube (Ziziphus jujuba) Powder. *Foods* **2022**, *11*, 1458. [CrossRef] [PubMed]
5. Statista Global Consumer Survey. Available online: https://www.statista.com/chart/19776/gcs-pasta-consumption/ (accessed on 6 October 2020).
6. Montalbano, A.; Tesoriere, L.; Diana, P.; Barraja, P.; Carbone, A.; Spano, V.; Parrino, B.; Attanzio, A.; Livrea, M.A.; Cascioferro, S.; et al. Quality characteristics and in vitro digestibility study of barley flour enriched ditalini pasta. *LWT* **2016**, *72*, 223–228. [CrossRef]
7. Abdel-Aal, E.-S.M.; Young, J.C.; Rabalski, I. Anthocyanin Composition in Black, Blue, Pink, Purple, and Red Cereal Grains. *J. Agric. Food Chem.* **2006**, *54*, 4696–4704. [CrossRef]
8. Spinelli, S.; Padalino, L.; Costa, C.; Del Nobile, M.A.; Conte, A. Food by-products to fortified pasta: A new approach for optimization. *J. Clean. Prod.* **2019**, *215*, 985–991. [CrossRef]
9. Fares, C.; Menga, V. Effects of toasting on the carbohydrate profile and antioxidant properties of chickpea (*Cicer arietinum* L.) flour added to durum wheat pasta. *Food Chem.* **2012**, *131*, 1140–1148. [CrossRef]
10. Lorusso, A.; Verni, M.; Montemurro, M.; Coda, R.; Gobbetti, M.; Rizzello, C.G. Use of fermented quinoa flour for pasta making and evaluation of the technological and nutritional features. *LWT* **2017**, *78*, 215–221. [CrossRef]
11. Rizzello, C.G.; Verni, M.; Koivula, H.; Montemurro, M.; Seppa, L.; Kemell, M.; Katina, K.; Coda, R.; Gobbetti, M. Influence of fermented faba bean flour on the nutritional, technological and sensory quality of fortified pasta. *Food Funct.* **2017**, *8*, 860–871. [CrossRef]
12. Boroski, M.; de Aguiar, A.C.; Boeing, J.S.; Rotta, E.M.; Wibby, C.L.; Bonafé, E.G.; de Souza, N.E.; Visentainer, J.V. Enhancement of pasta antioxidant activity with oregano and carrot leaf. *Food Chem.* **2011**, *125*, 696–700. [CrossRef]
13. Teterycz, D.; Sobota, A.; Zarzycki, P.; Latoch, A. Legume flour as a natural colouring component in pasta production. *J. Food Sci. Technol.* **2020**, *57*, 301–309. [CrossRef] [PubMed]
14. Testa, M.; Malandrino, O.; Santini, C.; Supino, S. Chapter 6—Nutraceutical and functional value of carob-based products The LBG Sicilia Srl Case Study. In *Case Studies on the Business of Nutraceuticals, Functional and Super Foods*; Woodhead Publishing Series in Consumer Science & Strategy Market; Woodhead Publishing: Sawston, UK, 2023; pp. 107–120, ISBN 9780128214084.
15. Ortega, N.R.; Marcia, A.; Romero, M.; Reguant, J.; Motilva, M.J. Matrix composition effect on the digestibility of carob flour phenols by an in-vitro digestion model. *Food Chem.* **2011**, *124*, 65–71. [CrossRef]
16. Azab, A. Carob (*Ceratonia siliqua*): Health, medicine and chemistry. *Eur. Chem. Bull.* **2017**, *610*, 456–469. [CrossRef]
17. Boublenza, I.; Lazouni, H.A.; Ghaffari, L.; Ruiz, K.; Fabiano-Tixier, A.S.; Chemat, F. Influence of Roasting on Sensory, Antioxidant, Aromas, and Physicochemical Properties of Carob Pod Powder (*Ceratonia siliqua* L.). *J. Food Qual.* **2017**, *2017*, 1–10. [CrossRef]
18. Papaefstathiou, E.; Agapiou, A.; Giannopoulos, S.; Kokkinofta, R. Nutritional characterization of carobs and traditional carob products. *Food Sci. Nutr.* **2018**, *6*, 2151–2161. [CrossRef]
19. Tsatsaragkou, K.; Yiannopoulos, S.; Kontogiorgi, A.; Poulli, E.; Krokida, M.; Mandala, L. Effect of Carob Flour Addition on the Rheological Properties of Gluten-Free Breads. *Food Bioproc. Tech.* **2014**, *7*, 868–876. [CrossRef]

20. Ayaz, F.A.; Torun, H.; Glew, R.H.; Bak, Z.D.; Chuang, L.T.; Presley, J.M.; Andrews, R. Nutrient content of carob pod (*Ceratonia siliqua* L.) flour prepared commercially and domestically. *Plant Foods Hum. Nutr.* **2009**, *64*, 286–292. [CrossRef]

21. Bengoechea, C.; Romero, A.; Villanueva, A.; Moreno, G.; Alaiz, M.; Millán, F.; Puppo, M.C. Composition and structure of carob (*Ceratonia siliqua* L.) germ proteins. *Food Chem.* **2008**, *107*, 675–683. [CrossRef]

22. Biner, B.; Gubbuk, H.; Karhan, M.; Aksu, M.; Pekmezci, M. Sugar profiles of the pods of cultivated and wild types of carob bean (*Ceratonia siliqua* L.) in Turkey. *Food Chem.* **2007**, *100*, 1453–1455. [CrossRef]

23. Durazzo, A.; Turfani, V.; Narducci, V.; Azzini, E.; Maiani, G.; Carcea, M. Nutritional characterisation and bioactive components of commercial carobs flours. *Food Chem.* **2014**, *153*, 109–113. [CrossRef]

24. Kumazawa, S.H.K.; Taniguchi, M.A.S.A.T.; Suzuki, Y.A.S.; Shimura, M.A.S.; Kwon, M.I.U.N.K.; Nakayama, T.S.N. Antioxidant activity of polyphenols in carob pods. *J. Agricult. Food Chem.* **2002**, *50*, 373–377. [CrossRef] [PubMed]

25. Germec, M.; Turhan, I.; Karhan, M.; Demirci, A. Ethanol production via repeated-batch fermentation from carob pod extract by using Saccharomyces cevisiae in biofilm reactor. *Fuel* **2015**, *161*, 304–311. [CrossRef]

26. Haddarah, A.; Ismail, A.; Bassal, A.; Hamieh, T.; Ioannou, I.; Ghoul, M. Morphological and chemical variability of Lebanese carob varieties. *Eur. Sci. J.* **2013**, *9*, 353–369.

27. Biernacka, B.; Dziki, D.; Gawlik-Dziki, U.; Roozy, R.; Siastala, M. Physical, sensorial, and antioxidant properties of common wheat pasta enriched with carob fiber. *LWT* **2017**, *77*, 186–192. [CrossRef]

28. Seczyk, L.; Swieca, M.; Gawlik-Dziki, U. Effect of carob (*Ceratonia siliqua* L.) flour on the antioxidant potential, nutritional quality, and sensory characteristics of fortified durum wheat pasta. *Food Chem.* **2016**, *194*, 637–642. [CrossRef] [PubMed]

29. Gaita, C.; Alexa, E.; Moigradean, D.; Conforti, F.; Poiana, M.A. Designing of high value-added pasta formulas by incorporation of grape pomace skins. *Rom. Biotechnol. Lett.* **2020**, *25*, 1607–1614. [CrossRef]

30. Gruber, W.; Sarkar, A. Chapter 8—Durum Wheat Milling. In *American Associate of Cereal Chemists International, Durum Wheat*, 2nd ed.; Sissons, M., Abecassis, J., Marchylo, B., Carcea, M., Eds.; AACC International Press: Washington, DC, USA, 2012; pp. 139–159.

31. Hellin, P.; Escarnot, E.; Mingeot, D.; Gofflot, S.; Sinnaeve, G.; Lateur, M.; Godin, B. Multiyear evaluation of the agronomical and technological properties of a panel of spelt varieties under different cropping environments. *J. Cereal Sci.* **2023**, *109*, 103615. [CrossRef]

32. Singleton, V.L.; Orthofer, R.; Lamuela-Raventos, R.M. Analysis of total phenols and other oxidation substrates and antioxidants by means of Folin-Ciocalteu reagent. *Meth. Enzymol.* **1999**, *299*, 152–178.

33. Abdel-Hameed, E.S.; Bazaid, S.A.; Salman, M.S. Characterization of the phytochemical constituents of Taif rose and its antioxidant and anticancer activities. *Biomed Res. Int.* **2013**, *2013*, 1–13. [CrossRef]

34. Rios, M.B.; Iriondo-DeHond, A.; Iriondo-DeHond, M.; Herrera, T.; Velasco, D.; Gomez-Alonso, S.; Callejo, M.J.; Dolores del Castillo, M. Effect of Coffee Cascara Dietary Fiber on the Physicochemical, Nutritional and Sensory Properties of a Gluten-Free Bread Formulation. *Molecules* **2020**, *25*, 1358. [CrossRef]

35. ISO 712:2009 Cereals and Cereal Products–Determination of Moisture Content–Reference Method. Available online: https://www.sis.se/api/document/preview/911751/ (accessed on 15 December 2021).

36. Heeger, A.; Kosinska-Cagnazzo, A.; Cantergiani, E.; Andlauer, W. Bioactives of coffee cherry pulp and its utilisation for production of Cascara beverage. *Food Chem.* **2017**, *221*, 969–975. [CrossRef] [PubMed]

37. Calculator de Calorii—Numararea Caloriilor—Klorii. Available online: https://klorii.ro/ (accessed on 8 March 2023).

38. Yazar, G.; Duvarci, O.; Tavman, S.; Kokini, J.L. Non-linear rheological behavior of gluten-free flour doughs and correlations of LAOS parameters with gluten-free bread properties. *J. Cereal Sci.* **2017**, *74*, 28–36. [CrossRef]

39. Dube, N.M.; Xu, F.; Zhao, R. The efficacy of sorghum flour addition on dough rheological properties and bread quality: A short review. *Grain Oil Sci. Technol.* **2022**, *4*, 164–171.

40. Sibanda, T.; Ncube, T.; Ngoromani, N. Rheological properties and bread making quality of white grain sorghum-wheat flour composites. *Int. Food Sci. Nutr. Eng.* **2015**, *5*, 176–182.

41. Arribas, C.; Cabellos, B.; Cuadrado, C.; Guillamón, E.; Pedrosa, M.M. Cooking Effect on the Bioactive Compounds, Texture, and Color Properties of Cold-Extruded Rice/Bean-Based Pasta Supplemented with Whole Carob Fruit. *Foods* **2020**, *9*, 415. [CrossRef]

42. Bchir, B.; Karoui, R.; Danthine, S.; Blecker, C.; Besbes, S.; Attia, H. Date, Apple, and Pear By-Products as Functional Ingredients in Pasta: Cooking Quality Attributes and Physicochemical, Rheological, and Sensorial Properties. *Foods* **2022**, *11*, 1393. [CrossRef] [PubMed]

43. Mandala, I.; Polaki, A.; Yanniotis, S. Influence of frozen storage on bread enriched with different ingredients. *J. Food Eng.* **2009**, *92*, 137–145. [CrossRef]

44. Gopalakrishnan, J.; Menon, R.; Padmaja, G.; Sajeev, M.S.; Moorthy, S.N. Nutritional and Functional Characteristics of Protein-Fortified Pasta from Sweet Potato. *Food Sci. Nutr.* **2011**, *2*, 944–955. [CrossRef]

45. Zhu, F.; Li, J. Physicochemical and sensory properties of fresh noodles fortified with ground linseed (Linum usitassimum). *LWT* **2019**, *101*, 847–853. [CrossRef]

46. Makris, D.P.; Kefalas, P. Carob pods (*Ceratonia siliqua* L.) As a source of polyphenolic antioxidants. *Food Technol. Biotechnol.* **2004**, *42*, 105–108.

47. Issaoui, M.; Flamini, G.; Delgado, A. Sustainability Opportunities for Mediterranean Food Products through New Formulations Based on Carob Flour (*Ceratonia siliqua* L.). *Sustainability* **2021**, *13*, 8026. [CrossRef]

48.  Aydin, S.; Ozdemir, Y. Development and Characterization of Carob Flour Based Functional Spread for Increasing Use as Nutritious Snack for Children. *J. Food Qual.* **2017**, *2017*, 1–7. [CrossRef]
49.  Sebecic, B.; Vedrina-Dragojevic, I.; Vitali, D.; Hecimovic, M.; Dragicevic, I. Raw Materials in Fibre Enriched Biscuits Production as Source of Total Phenols. *Agric. Conspec. Sci.* **2007**, *72*, 265–270.
50.  Bastida, S.; Sánchez-Muniz, F.J.; Olivero, R.; Pérez-Olleros, L.; Ruiz-Roso, B.; Jiménez-Colmenero, F. Antioxidant activity of Carob fruit extracts in cooked pork meat systems during chilled and frozen storage. *Food Chem.* **2009**, *116*, 748–754. [CrossRef]
51.  Bissar, S.; Ozcan, M.M. Determination of quality parameters and gluten free macaron production from carob fruit and sorghum. *Int. J. Gastron. Food Sci.* **2022**, *27*, 100460. [CrossRef]
52.  Brassesco, M.E.; Brandao, T.R.S.; Silva, C.L.M.; Pintado, M. Carob bean (*Ceratonia siliqua* L.): A new perspective for functional food. *Trends Food Sci. Technol.* **2021**, *114*, 310–322. [CrossRef]
53.  Tolve, R.; Pasini, G.; Vignale, F.; Favati, F.; Simonato, B. Effect of Grape Pomace Addition on the Technological, Sensory, and Nutritional Properties of Durum Wheat Pasta. *Foods* **2020**, *9*, 354. [CrossRef]
54.  Frühbauerova, M.; Cervenka, L.; Hajek, T.; Pouzar, M.; Palarcík, J. Bioaccessibility of phenolics from carob (*Ceratonia siliqua* L.) pod powder prepared by cryogenic and vibratory grinding. *Food Chemistry* **2022**, *377*, 131968. [CrossRef]
55.  Benković, M.; Bosiljkov, T.; Semić, A.; Ježek, D.; Srečec, S. Influence of Carob Flour and Carob Bean Gum on Rheological Properties of Cocoa and Carob Pastry Fillings. *Foods* **2019**, *8*, 66. [CrossRef]
56.  Rayas-Duarte, P.; Mock, C.M.; Satterlee, L.D. Quality of spaghetti containing buckwheat, amaranth, and lupin flours. *Cereal Chem.* **1996**, *73*, 381–387.
57.  Marinelli, V.; Padalino, L.; Nardiello, D.; Del Nobile, M.A.; Conte, A. New approach to enrich pasta with polyphenols from grape marc. *J. Chem.* **2015**, *2015*, 1–8. [CrossRef]
58.  Chen, J.; Rosenthal, A. Modifying food texture. In *Volume 1: Novel Ingredients and Processing Techniques*; Woodhead Publishing: Cambridge, UK, 2015; pp. 3–22.

*Article*

# Dehydrated Sauerkraut Juice in Bread and Meat Applications and Bioaccessibility of Total Phenol Compounds after In Vitro Gastrointestinal Digestion

**Liene Jansone *, Zanda Kruma, Kristine Majore and Solvita Kampuse**

Department of Food Technology, Faculty of Food Technology, Latvia University of Life Sciences and Technologies, Rigas iela 22, LV-3004 Jelgava, Latvia
* Correspondence: liene.jansone@llu.lv

**Abstract:** The aim of this study was to evaluate dehydrated sauerkraut juice (DSJ) in bread and meat applications and investigate bioaccessibility (BAC) of TPC in the analyzed products. In current research, sauerkraut juice, dehydrated sauerkraut juice, and bread and meat products prepared with dehydrated sauerkraut juice were analyzed. For all of the samples, total phenol content, antiradical activity by $ABTS^+$, bioaccessibility, and volatile compound profile were determined. Additionally, sensory evaluation was performed to evaluate the degree of liking bread and meat with dehydrated sauerkraut juice. The addition of DSJ increased TPC in bread and meat samples. The bioaccessibility was higher for the control samples compared to DSJ samples. It exceeded 1 and is considered as good. DSJ did not promote bioaccessibility. Benzaldehyde was the highest peak area for the Bread DSJ and Meat DSJ samples, giving a roasted peanut and almond aroma. There were no significant differences in degree of liking for structure, taste, and aroma between the control bread and the Bread DSJ, while Meat DSJ was more preferable in sensory evaluation. DSJ could be used in food applications, but further research is necessary.

**Keywords:** fermented cabbage; spray-drying; in vitro digestion; food

**Citation:** Jansone, L.; Kruma, Z.; Majore, K.; Kampuse, S. Dehydrated Sauerkraut Juice in Bread and Meat Applications and Bioaccessibility of Total Phenol Compounds after In Vitro Gastrointestinal Digestion. *Appl. Sci.* **2023**, *13*, 3358. https://doi.org/10.3390/app13053358

Academic Editors: Joanna Trafialek and Mariusz Szymczak

Received: 20 January 2023
Revised: 28 February 2023
Accepted: 2 March 2023
Published: 6 March 2023

## 1. Introduction

With a rising food crisis worldwide, it is crucial to exploit the maximum of the grown crop including a wise utilization of waste or by-products. Fermented cabbage juice is considered an industrial by-product of the technological process of sauerkraut. Sauerkraut juice (SJ) contains vast bioactive complexes such as phenolic compounds and glucosinolates [1] organic acids, sugars and biogenic amines [2,3], vitamins, especially vitamin C, and minerals [4,5]. Research into new technologies on how to exploit this valuable by-product was carried out. The dehydrated sauerkraut juice (DSJ) was obtained by spray-drying [6], and remains rich in minerals, potassium, and calcium being the most abundant, but also iron and magnesium, as well as vitamin C [7]. Physical and chemical composition of SJ and DSJ is described in authors' previous studies [4,6]. To the best of our knowledge, there is scarce research done on a similar product, so all of the presented results are based on authors' previous studies [6,7].

To provide a reliable knowledge for industrial production, investigations about DSJ application in various products is necessary.

In our previous study, DSJ was evaluated in olive oil and sour cream as a salt alternative and compared with the control samples with NaCl. Sauerkraut juice and dehydrated sauerkraut juice contains sugars such as glucose and fructose and also 10–12% NaCl, that is used to ensure the fermentation process. In the sensory evaluation of the samples, tastes such as sweet, sour, salty, garlic, yogurt, and mayonnaise were mentioned in the DSJ samples and overall liking [7]. Therefore, further research was conducted into staple foods such as bread and meat. There are numerous studies on the smart utilization of by-products simultaneously enhancing the functional properties of food products. The beneficial effect of various plant-based additives in bread and meat applications have been investigated before, such as lyophilized kale powder [8], Moringa peregrina seed husk in wheat bread [9], melon peel juice powder [10], enriched wheat bread with quinoa leaves powder [11], and with mushrooms [12], quinoa flour, nettle leaves [13], green coffee, berries, lyophilized pomegranate peel in minced beef [14], and many others.

As the cabbage and sauerkraut has a very specific aroma, it also remains in the obtained dehydrated sauerkraut juice (DSJ) and sequentially in the food applications. Volatile compounds deliver not only aroma or odor, but also take part in health promoting, for example with antibacterial, antioxidant, and antifungal attributes that metabolize from precursors such as fatty acids, carotenoids, leucine, etc. [15]. For example, thioglucosidase, which is present in Brassicaceae vegetables, acts as anticancer substance.

Considering the health attributes of fermented products, in vitro tests were carried out to determine the bioaccessibility of sauerkraut juice, DSJ, and the products made with the addition of DSJ. To investigate the DSJ in vitro, common foods such as wheat bread and minced meat have been chosen. Wheat bread, with a well-known and neutral taste, is one of the staple foods in most parts of the world [16]. Enrichment of such foods with functional properties would gain the most benefit because of the vast consumption [17].

However, plant-based bioactive compounds, in their prime state, can be easily degraded under environmental influence or can be inconstant under severe medium of the stomach. Spray-drying with a wall material encapsulates the core material, thus preventing its bioactive compounds from premature degradation, sustaining and handling the release, etc. [14,18]. For example, the release of polyphenols of encapsulated fruit juices, had a positive effect in simulated GIT [18].

Determining the bioaccessibility (BAC) of the active compounds can be achieved by simulating the gastrointestinal tract (GIT) to assess their release from a substance and availability for absorption [19,20]. BAC is a comparison of compound content before and after GIT [21]. GIT or in vitro methods are swift, cost and labor effective, and exclude ethical restrictions [19] while ensuring in vivo—a living organism condition–including oral, gastric, and intestinal phases, considering enzymes, pH, time, and bile salts as some of the main factors [20].

Research proves that DSJ is a nutritionally valuable product, therefore a study into its application in food products and in vitro tests to assess its health benefits as a potential functional ingredient would be useful.

The aim of this study was to evaluate dehydrated sauerkraut juice (DSJ) in bread and meat applications and investigate bioaccessibility (BAC) of TPC in the analyzed products.

## 2. Materials and Methods

### 2.1. Sauerkraut Juice and Dehydrated Sauerkraut Juice

Sauerkraut juice, for the analyses and spray-drying, was obtained from the production plant Ltd. "Dimdiņi" (Lizums, Latvia), which ferments cabbage by the traditional recipe. Cabbage is shredded and mixed with NaCl, grated carrots and caraway, pressed, and left to ferment for 14 days. Dehydrated sauerkraut juice was obtained in a vertical Mini Spray-dryer Buchi 290 (Buchi, Flawil, Switzerland), with a starch solution as a carrying agent [6]. For the starch solution, starch (Pure, soluble starch, 162.10 g mol$^{-1}$, CHEMPUR) and deionized water (1:20, accordingly) was heated for 30 min at 90 °C and cooled overnight. It was then mixed with the sauerkraut juice, with the core-to-wall ratio 1:1.5, and the mixture was constantly stirred on a magnetic stirrer while spray drying. The dehydrated sauerkraut juice was kept in two Ziplock bags to be used for further experiments and analyses.

For the sauerkraut juice (SJ) and dehydrated sauerkraut juice (DSJ), total phenol content (TPC–using Folin–Ciocalteu reagent, described below), antiradical activity by ABTS$^+$ assay, vitamin C by iodometric titration, and salt (NaCl) were determined by Mohr's method, all described in previous studies [4,6].

*2.2. Product Preparation*

### 2.2.1. Bread

The bread samples were made according to the following formulation: for the control sample 1% of NaCl, 2% sugar, 3% yeast, and 60% water ads to the necessary amount (100%) of wheat flour. The ingredients for the dough were placed in a spiral-type dough mixer KM400 (Kenwood Havant, Hampshire, UK) and kneaded for 7 min, let to rest for 10 min, formed in a baguette-like loaf, and proofed for 45 min in a Sveba Dahlen proofing cabinet (Sveba Dahlen AB, Fristadt, Sweden) at $35 \pm 5$ °C and 80% humidity. The loaves were then baked in a preheated rotary oven (Sveba Dahlen, Fristad, Sweden) at $200 \pm 3$ °C for 17 min, and then cooled down to room temperature for further analysis. For the sample with DSJ–1%, NaCl and 2% of sugar were substituted with 9% DSJ. The DSJ used in the sample, was calculated to obtain the necessary amount of NaCl and sugar, according to the formulation of the control sample. Due to our previous studies, salt content in DSJ used in this experiment is 11 g 100 g$^{-1}$. For the bread formulation, salt equivalence was calculated as 2 g of salt in the control sample, multiplied by 100 g, and divided by 11 g of salt in 100 g DSJ, according to formulation:

$$\text{Salt equivalent} = \frac{2 \times 100}{11}$$

For the Bread DSJ sample, 1000 g flour, 180 g DSJ, 60 g yeast, and 1100 g water were used.

### 2.2.2. Meat Samples–Minced Pork Sausage

Minced meat samples were prepared as follows: for the control sample blank minced pork was used, obtained from the local shop; for the DSJ90 g of DSJ was added to a 1000 g minced meat. The prepared meat was formed into long sausage-like loafs. Baked in the preheated rotary oven (Sveba Dahlen, Fristad, Sweden) at $200 \pm 3$ °C for 15 min.

### 2.2.3. Samples and Abbreviations

For further use in the text, abbreviations for the samples were made and are described in Table 1.

**Table 1.** The list of samples and abbreviations.

| Sample | Abbreviation |
| --- | --- |
| Sauerkraut juice | SJ |
| Dehydrated sauerkraut juice | DSJ |
| Bread control sample | Bread C |
| Bread with dehydrated sauerkraut juice | Bread DSJ |
| Meat control sample | Meat C |
| Meat with dehydrated sauerkraut juice | Meat DSJ |

*2.3. Analytical Methods*

### 2.3.1. Total Phenol Content and Antiradical Activity

For the determination of total phenol content (TPC), a Folin–Ciocalteu reagent was used [22], and TPC was calculated as gallic acid equivalent (GAE) mg 100 g$^{-1}$. To extract TPC, 10 mL of sauerkraut juice, 0.5 g of DSJ, 5 g of GIT outcome from juice and DSJ, and 3 g of bread and meat GIT outcome/content were used. Samples were prepared by magnetically stirring with 20 mL ethanol: water (80:20) for 2 h, then filtered. Next, 0.5 g of DSJ was stirred with 20 mL of deionized water. To determine TPC absorption in the obtained filtrates, a spectrophotometer JENWAY 6300 (Baroworld Scientific Ltd., Staffordshire, UK) was used. The ABTS$^{+}$ antiradical activity was assessed by a method described by Rokayya [23] with slight modifications. Then, 0.05 mL sample extract was added to 5 mL ABTS$^+$ solution and absorbance readings were measured at 734 nm after ten minutes of initial mixing.

### 2.3.2. Determination of Volatile Compounds

A method described by Galoburda [24] was used to determine volatile compounds in the raw material (DSJ and sauerkraut juice) and the experimental samples–bread and meat. First, 0.5 g of DSJ and 5 g of sauerkraut juice, 5 g of control and experimental bread and meat samples were placed in a 20 mL glass vial. They were heated and stirred for 10 min at $35 \pm 2$ °C to equilibrate headspace and 30 min with CarboxenTM/Polydimethylsiloxane (CAR/PDMS) fiber (Supelco Inc., Bellefonte, PA, USA) to extract volatiles. Solid-phase microextraction (SPME) techinique was used. The obtained compound data was identified using mass spectral library Nist98.

### 2.3.3. Sensory Evaluation

For the sensory evaluation of bread and meat samples, 52 consumers, based on the complexity/simplicity of the tasting samples [25] and who were willing to try the products, were invited to evaluate the overall liking, structure, taste, and aroma of the bread and meat samples. A 5-point hedonic scale for consumer preferences was used (1–dislike very much and 5–like extremely). Of the customers, 38% were 25 years old or older, and others were 18 to 25 years old. Seventeen percent of all the consumers were male. The tasting samples were prepared/cooked as explained above. Samples were cut into bite-size pieces (to fit easily in a mouth with one bite), and drinking water to cleanse the palate was available at consumers' disposal.

### 2.4. Static In Vitro Digestion Method

For digestibility of sauerkraut juice, DSJ, and experimental bread and meat samples, a standardized static in vitro digestion method by Minekus [20] has been used. It was performed in a model environment of the gastrointestinal tract (GIT)—a bioreactor Multifors 2 (Infors-HT, Bottmingen, Basel, Switzerland) in which digestibility and transit of nutrients are simulated. The process is controlled by the computer program Iris 6 Pallalel bioprocess Control Software (Infors-HT, Bottmingen, Basel, Switzerland). First, 30 g of the sauerkraut juice, bread and meat samples, and 1 g of DSJ were placed in the bioreactor with a pH and temperature control, and simulated saliva fluid (SSF) is then added and kept for 2 min at 37 °C. The transition to the stomach is simulated by introducing the simulated gastric fluid (SGF), which consists of a concentrated electrolyte solution, the enzyme pepsin, $CaCl_2$, and deionized $H_2O$. Gastric acid secretion was simulated by adding 1 M HCl and adjusting pH to $3.0 \pm 0.2$. Digestibility in the stomach is simulated for 2 h. Next, the stomach content is neutralized to pH $7.0 \pm 0.2$ by adding 1 M $NaHCO_3$ and simulated intestinal fluid (SIF), which consists of a concentrated electrolyte solution, enzymes (trypsin, chymotrypsin, α-amylase, lipase), bile salts, $CaCl_2$, and deionized $H_2O$, thus simulating the transit to the duodenum. Digestibility in the small intestine was simulated for 2 h. After the digestion process, the content is frozen to stop the enzymatic activity.

The bioavailability index (BAC) was calculated using the following equation [26]:

$$BAC = \frac{CGE}{CBE}$$

where CGE—the TPC and antiradical activity after gastrointestinal digestion; and CBE—the TPC and antiradical activity in the samples before digestion.

### 2.5. Statistical Analyses

The results are shown as the mean value $\pm$ standard deviation. Significant differences are considered as significant at $p \leq 0.05$ among the acquired samples and were determined by a $t$-test in the sensory tests. Analyses of variance (ANOVA) and Tukey's test is used to evaluate the effect of tested factors and to determine differences among the samples.

## 3. Results

### 3.1. Description of Sauerkraut Juice and Dehydrated Sauerkraut Juice

#### 3.1.1. Total Phenols, Antiradical Activity and Bioaccessibility of the SJ and DSJ Samples

Total phenol content (TPC) is the highest in the sauerkraut juice before processing, as shown in Table 2. After the encapsulation process, in the dehydrated sauerkraut juice (DSJ), the TPC decreases by half, which could be explained by the high core-to-wall ratio, which helps to protect bioactive compounds. As scientists previously have confirmed that modified starch entraps the core material in the starch granule [27]. The antiradical activity is also affected by the optimum choice of spray-drying parameters that can retain the activity [6].

**Table 2.** Total phenol content and antiradical activity by ABTS$^+$ in sauerkraut juice (SJ) and dehydrated sauerkraut juice (DSJ).

| Parameters | SJ | DSJ |
|---|---|---|
| TPC, mg 100 g GAE, dw * | 713.7 ± 43.2 a ** | 359.5 ± 7.7 b |
| ABTS, mg TE 100$^{-1}$, dw | 15.50 ± 1.84 a | 28.62 ± 2.03 b |

* mg GAE–gallic acid equivalent, dry weight (dw). ** Values with different letters are significantly different ($p \leq 0.05$).

Bioaccessibility (BAC) is defined as a share of bioactive compounds that is released from the food matrix and become available for absorption after ingestion [28]. Bioaccessibility is calculated as equality of the analyses after the GIT against the analyses before GIT. The bioaccessibility is considered high if the BAC index is higher than 1 [17].

For both analyzed products–SJ and DSJ—the bioaccessibility for ABTS$^+$ exceeds 1.2 and is considered as high, as shown in Figure 1.

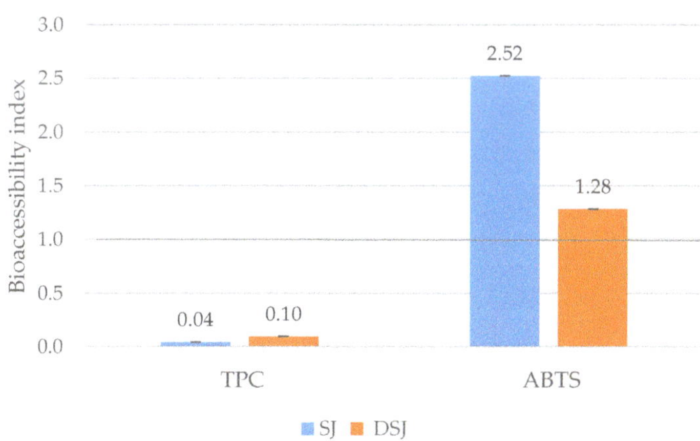


**Figure 1.** Bioaccessibility index (BAC) of the sauerkraut juice (SJ) and dehydrated sauerkraut juice (DSJ) based on TPC and ABTS$^+$ scavenging activity after in vitro digestion.

The bioaccessibility based on TPC is very low for SJ and DSJ and is below 0.2. This can be explained by the compound interaction with gastric juices and enzymes as well as the wall material or the combination of several, which plays an important role in the release of the phenolic compounds in the simulated gastrointestinal tract [18].

#### 3.1.2. Aromatic Volatiles in Sauerkraut Juice and Dehydrated Sauerkraut Juice

Profiles for aromatic volatiles was determined for sauerkraut juice and dehydrated sauerkraut juice and are presented in Figure 2.

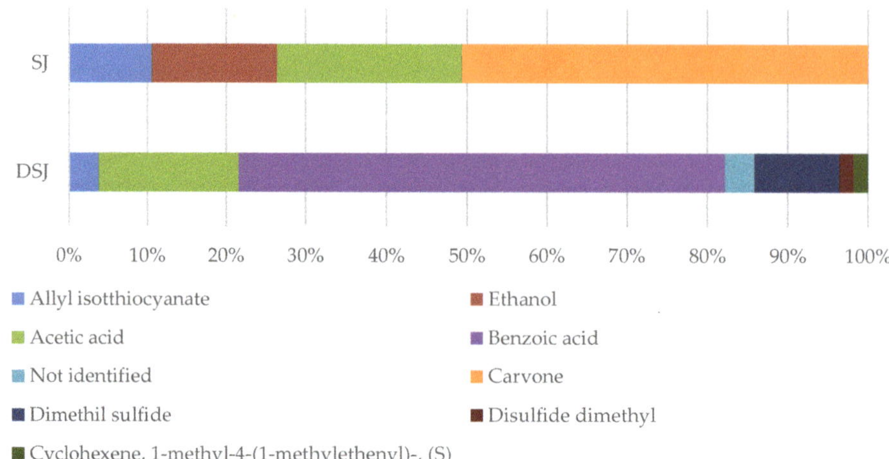


**Figure 2.** The percentage of volatile compound peak areas in sauerkraut and dehydrated sauerkraut juice.

There were two common compounds in sauerkraut juice before and after spray-drying: acetic acid with a sour taste, and allyl isothiocyanate with a pungent taste of mustard, horseradish, and wasabi (mean values and standard deviations in Table S1, Supplementary Materials). The metabolized products of glucosinolates–isotthiocyanates are the prime sources of the characteristic flavor of the Brassica vegetables [3,28]. The highest peak area in sauerkraut juice was carvone with a caraway and spearmint-like odor, while for dehydrated sauerkraut juice it was benzoic acid with a faint, pleasant odor. Other compounds are characteristic to each of the raw material.

### 3.2. Bread Samples

#### 3.2.1. Total Phenols, Antiradical Activity and Bioaccessibility of the Bread Samples

The aim of this experiment was to evaluate, if the addition of DSJ affects the bread quality, and can increase the bioactive compound content and the bioaccessibility in the wheat bread. The results, shown in Table 3, represent a significant influence on the total phenol content by the addition of DSJ to the wheat bread. The TPC content and antiradical activity by $ABTS^+$ in the Bread DSJ sample is higher by 66% and 56% accordingly. It has previously been reported that plant materials and food production by-products added to flour products increase the TPC and antiradical activity [29,30]. In our study, supplementing wheat bread with DSJ increases the TPC content in the sample, thus the food matrix provided suitable conditions for the release (or interaction) of the compounds.

**Table 3.** Total phenol content and antiradical activity by $ABTS^+$ in the bread samples.

| Parameters | Bread C | Bread DSJ |
|---|---|---|
| TPC, mg 100 g GAE, dw * | 54.36 ± 1.33 a ** | 82.56 ± 0.98 b |
| ABTS, mg TE 100⁻, dw | 4.614 ± 0.241 a | 8.232 ± 0.563 b |

* mg GAE–gallic acid equivalent, dry weight (dw). ** Values with different letters are significantly different ($p \leq 0.05$).

Although the TPC was higher in the Bread DSJ, the bioaccessibility index BAC, shown in Figure 3. is higher in the Bread C–1.10 while in the DSJ sample it is 0.65. This suggests that wheat bread is rich in bioavailable phenolic compounds, however addition of DSJ may interact with other compounds to form indigestible compounds and do not protect the TPC from the severe environment of the stomach [31].

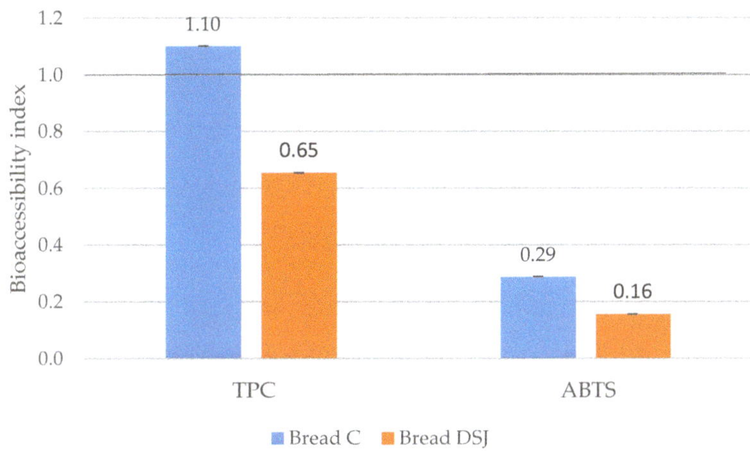

**Figure 3.** Bioaccessibility index (BAC) of the bread samples based on TPC and ABTS[+] scavenging activity after in vitro digestion.

That is contrary to the bread samples enriched with green coffee [17], where all of the samples were highly bioaccessible in vitro, but the control sample was higher than two. The interactions among the food matrix, phenolic compounds, and GIT enzymes is still under investigation [28].

### 3.2.2. Aromatic Volatiles in the Bread Samples

There were nine volatile compounds detected in the bread samples, and eight of them exceeded 5% and are shown in Figure 4. Hexanal did not exceed 1%. The highest peak area was for benzaldehyde, giving volatile oil-of-almond odor, and it was higher with the DSJ addition, being 29.15% and 33, 60%, accordingly (mean values and standard deviations in Table S2, Supplementary Materials). A very distinct nuance—a caraway-like odor–in the bread sample was detected with the DSJ addition. There is 1% of caraway added in the production process of sauerkraut in Ltd. "Dimdiņi" (Lizums, Latvia), and this volatile compound is so strong to remain through the spray-drying process and the bread baking. A freshly baked wheat bread, with no caraway or DSJ added, gives it a rose-like aroma.

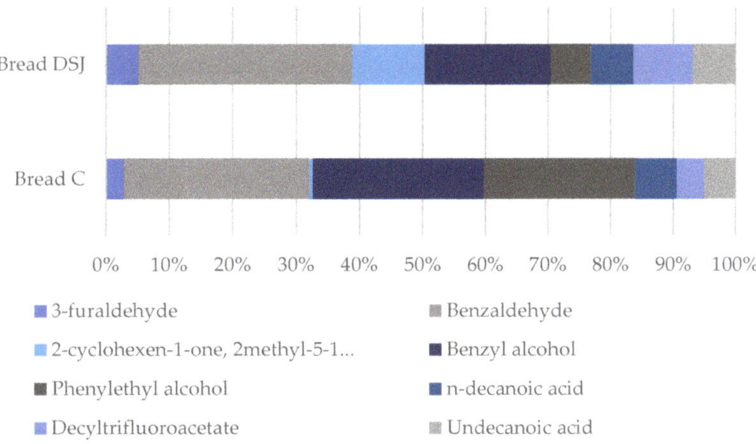

**Figure 4.** The percentage of volatile compound peak areas in the bread samples.

A sensory evaluation was carried out for bread with dehydrated sauerkraut juice. Fifty-two consumers were invited to rate overall liking, structure, taste, and aroma on a 5-point hedonic scale.

There are no significant differences in overall liking of Bread C and Bread DSJ; both samples are equally liked, and no significant differences are observed in structure, taste and aroma, as shown in Figure 5.

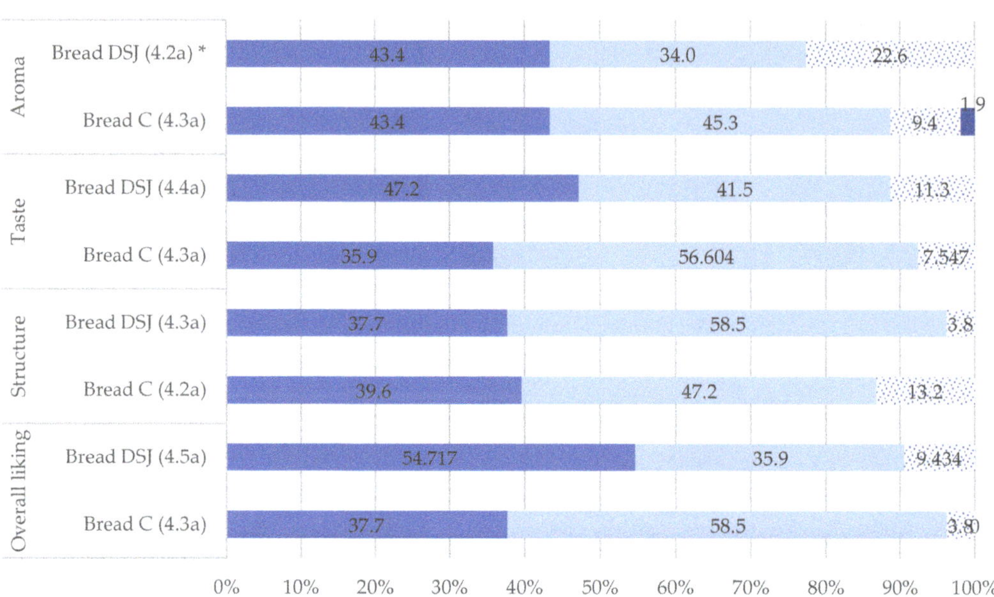


**Figure 5.** Hedonic evaluation of the bread samples. * Value in brackets present mean value of hedonic evaluation and the same letter along the values shows that the difference between the means between two samples is not statistically significant.

Other authors discussed that mean in some cases is not representative for hedonic evaluation due to differences in frequency of evaluations [32]. Therefore, Figure 5 presents the frequency of each evaluation in percentages and a few trends could be observed. It could be concluded that for overall liking and taste of Bread DSJ, more consumers selected evaluation "like extremely", 54.7% and 47.2%, respectively, and it is higher compared to Bread C. More people extremely like bread with DSJ. Additionally, it could be seen that aroma of Bread DSJ got 22.6% of evaluation–neither like nor dislike—showing that the aroma of this sample is not preferable for consumers in comparison to Bread C.

*3.3. Meat Samples*

3.3.1. Total Phenols, Antiradical Activity and Bioaccessibility of the Meat Samples

Mincing or grinding of meat influences total phenol content in meat products and its chemical composition consists of lipids, proteins, and polysaccharides that may interact with phenols and change their extractability from the samples [33]. In our study, there is a significant influence ($p < 0.01$) of DSJ addition in the meat samples on the TPC. The TPC content and antiradical activity by ABTS$^+$ in the Meat DSJ sample is higher by 64% and 51%, accordingly, as shown in Table 4.

**Table 4.** Total phenol content and antiradical activity by ABTS$^+$ in the meat samples.

| Parameters | Meat C | Meat DSJ |
| --- | --- | --- |
| TPC, mg 100 g GAE, dw * | 39.37 ± 1.62 a ** | 61.21 ± 1.03 b |
| ABTS, mg TE 100$^-$, dw | 2.924 ± 0.121 a | 5.721 ± 0.171 b |

* mg GAE–gallic acid equivalent, dry weight (dw). ** Values with different letters are significantly different ($p \leq 0.05$).

The bioaccessibility index for TPC of the meat samples is significantly higher than 1, as shown in Figure 6., and thus the compounds are available for absorption. The bioaccessibility of the control sample is significantly higher than the DSJ sample. The combination of proteins and phenolic compounds can affect the bioaccessibility of TPC, and is influenced by the specific compound interactions [34,35].

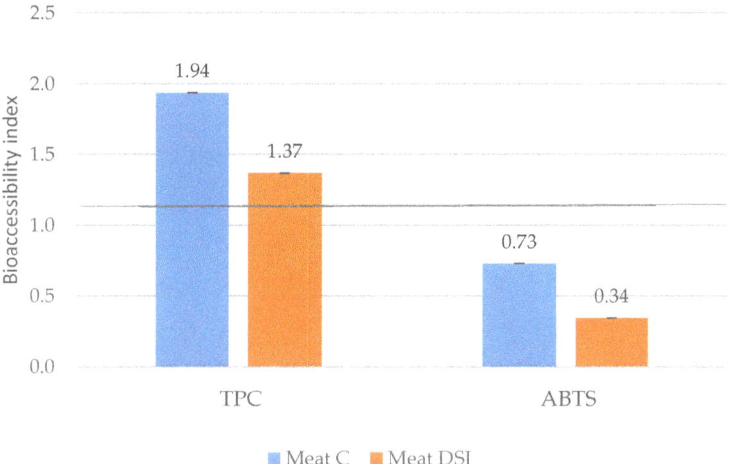


**Figure 6.** Bioaccessibility index (BAC) of the meat samples based on TPC and ABTS$^+$ compound activity after in vitro digestion.

As Nagar [36] has studied, dissolved oxygen levels in the intestinal phase and bile contribute to the decrease in bioaccessibility; the absence of oxygen increases the bioaccessibility of polyphenols.

Additionally, the bioaccessibility for ABTS$^+$ is higher in the control sample–0.73, while in the DSJ sample, it is 0.34, and thus the BAC of the antiradical activity is considered as low.

### 3.3.2. Aromatic Volatiles in the Meat Samples

There were eight volatile compounds detected in meat samples. Seven of them exceeded 5% and are shown in Figure 7. The highest peaks for the Meat C sample were hexanal, mostly formed by oxidation of linoleic acid [37], and 3-furaldehyde, giving a fruity, grass-, and almond-like odor, also containing the volatile oil-of-almond. The highest peaks in Meat DSJ sample were benzaldehyde, characterized by a roasted peanut aroma [37], and benzyl alcohol, giving a faint aromatic and volatile oil-of-almond aroma, as well as furfural with an almond-like odor, and n-decanoic acid with a rancid, unpleasant odor, mostly formed in the process of lipid hydrolysis and oxidation [37] (mean values and standard deviations in Table S2, Supplementary Materials). There is no caraway odor found in the meat sample.

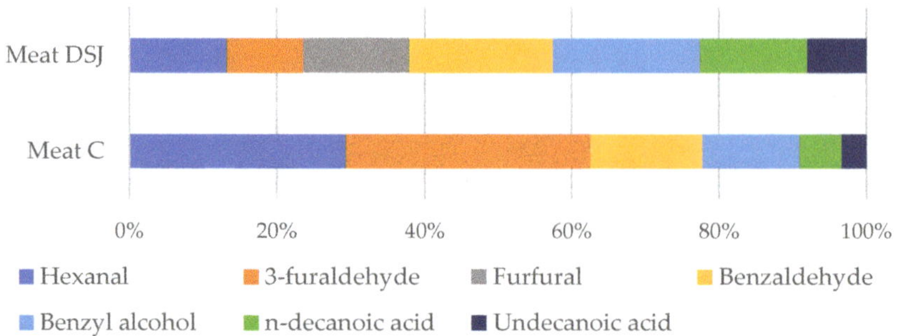



**Figure 7.** The percentage of volatile compound peak areas in the meat samples.

A sensory evaluation was carried out in the meat samples just as described above in the bread samples applying the Hedonic scale. Due to DSJ specific aroma and taste properties, it was useful to distinguish the effect and differences between the control and DSJ sample. In the bread samples, statistics showed no significant differences between samples, but for meat samples, a different trend was observed, as shown in Figure 8.

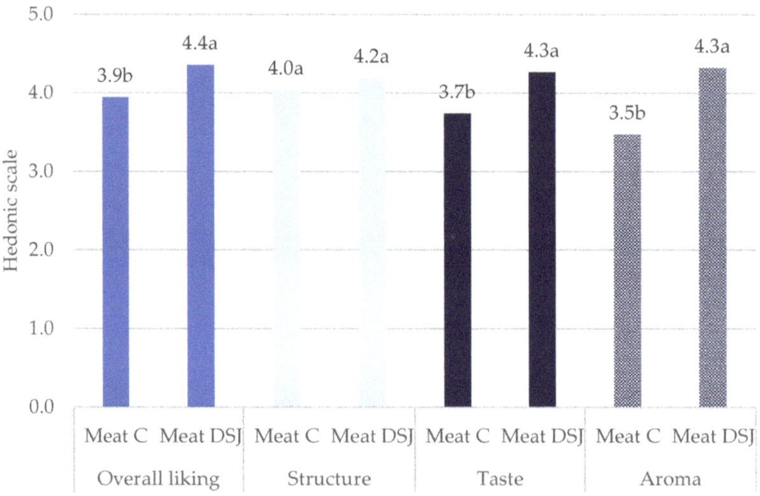



**Figure 8.** Hedonic evaluation of the meat samples. The same letter along the values shows that the difference between the means of two samples is not statistically significant.

No significant differences were in the structure ($p > 0.05$) of the samples whereas significant differences in the overall liking, taste, and aroma of the samples ($p < 0.05$) were determined and in all cases, higher evaluation was for a sample with added DSJ. It could be explained by traditional cooking and serving of meat with sauerkraut, and these tastes paired for Latvian consumers that the flavor was considered acceptable and highly evaluated. Therefore, meat products could be prospective for application of DSJ in food.

## 4. Discussion

Sauerkraut juice (SJ) and dehydrated sauerkraut juice (DSJ) has an ample amount of bioactive compounds, and contains phenolic compounds, vitamins, and minerals.

However, this study shows that the bioaccessibility of TPC is very low. There are numerous studies on phenolic compound bioaccessibility, mostly because the compounds are considered to have low-bioavailability [36], thus, the potential health benefits are under discussion [36,38]. The bioaccessibility of TPC is complex [28,36,39], since it is influenced by many factors and depends on a diversity of plant polyphenols, the interaction of compounds, and demeanor in the digestive system [34,40] where the gut microbiota's hydrolytic activity can increase activity and bioavailability of polyphenols [41].

Sauerkraut juice and its products have a very distinct flavor and aroma that is composed of various volatile compounds, such as aldehydes, alcohols, sulfur compounds, esters, ketons, terpenes, furans, etc. [42], and can be metabolized from bioactive compounds, delivering health promoting attributes [15]. The main volatile compound of sauerkraut juice was carvone with a spearmint like odor and ethanol, both of them are characteristic to sauerkraut. The main volatile compound for DSJ was benzoic acid with a faint, pleasant odor. There were two common volatile compounds in SJ and DSJ–acetic acid with a sour odor, and allyl isothiocyanate with a pungent taste of mustard, horseradish, and wasabi. Allyl isothiocyanate, with antioxidant and antimicrobial properties, is one of the major volatile compounds of a specific cabbage flavor [43], however, the compound was not identified in the bread and meat samples tested in this study.

Dehydrated sauerkraut juice (DSJ) was tested in foods such as wheat bread and minced pork, and the effect of total phenol content (TPC) and antiradical activity by $ABTS^{+}$ in in vitro gastrointestinal digestion, and the compound bioaccessibility were evaluated.

The addition of DSJ significantly enriched wheat bread with TPC and antiradical activity, as it is shown in this study and is in agreement with previous observations of other authors, that adding vegetables with high phenol content to wheat bread can increase its total phenol content [44].

The bioaccessibility of TPC in the wheat-based food matrix is different than in meat-based products. Wieca [17] have confirmed that wheat flour contains bound phenolics that are easily released during simulated digestion. However, bioaccessibility is affected by the choice of wall material [45]. Dehydrated sauerkraut juice was acquired via spray-drying, and starch solution was used as a wall material. Encapsulated ascorbic acid in enzyme-hydrolyzed starch [27] and rutin in debranched lentil starch coating material is released in the intestinal digestion phase [46]. The choice and combination of wall material is crucial, and the desired release of the bioactive compounds and the interaction of the food matrix is applied to [35,45]. The addition of DSJ did not promote TPC bioaccessibility of bread and meat samples.

The volatile compounds of bread samples are characteristic of the product, the main compound being benzaldehyde with rose- and almond-like odor, ref. [47] yet the addition of supplements affects the profile. In our study, a caraway-like odor was detected in the Bread DSJ. Caraway is used in the production process of sauerkraut.

There were no significant differences in the sensory evaluation of structure, taste, aroma, and overall liking of the Bread samples C and DSJ. However, enriching wheat bread with lyophilized kale had a decrease in acceptability of taste and aroma, but it was influenced by individual preferences [8]. Additionally, adding broccoli leaf powder to gluten free sponge cakes increased their antioxidant capacity, but sensory quality was affected [48]. With this, it can be concluded, that the addition of DSJ to wheat bread does not affect sensory characteristics, but increases TPC and antiradical activity.

DSJ also significantly increased the TPC content and antiradical activity of the meat samples. Studies show, that incorporating plant-based additives in meat products, significantly increase their TPC [49–51], however, sensory profile is disputable. The bioaccessibility of TPC in the meat samples exceeded 1 and is considered as good.

Studies of meat protein and phenolics action in vitro are scarce to the best of our knowledge, but Rashidinejad [52] has investigated that milk proteins bind the phenols and can affect their release due to the interaction between them [35]. The TPC is affected differently in GIT, depending on the food matrix, its physicochemical characteristics, and its composition [39,51]. When testing poultry feed with the addition of lucerne or chicory, the total phenol content decreased in the gastric phase but increased in the intestinal phase [51]. Additionally, higher content of bioactive compounds in the food matrix does not always ensure the same results in and after the GIT [31], as well as the accessibility of the compounds is influenced by numerous conditions such as the complexity of phenolic compounds in food matrix, the metabolic pathway, etc. [28,39].

Furthermore, Cantele [40] has concluded that the absorption of phenols from the solid food matrices is more challenging, as opposed to liquid matrices, because they first need to undergo mechanical, chemical, and enzymatic processes, whereas phenols from the liquid matrices are available straight away.

As Flores [53] investigated in their study about microencapsulated blueberry anthocyanins during in vitro with two types of wall material, whey protein and gum arabic, most of the whey protein microcapsules phenolic compounds are degraded in the intestinal digestion, whereas gum arabic, being a complex heteropolysaccharide, remains minimally digested. as Additionally, the stability of total phenol content is higher in the gastric phase, yet it decreases in the small intestine, because of bile [36] (especially phenolic acids), or the compounds transform into other compounds during digestion [28,38,54].

Benzaldehyde and 3-furaldehyde with almond-like odors are dominant in the meat samples, yet the addition of DSJ amplifies the rancid, fatty odor with n-decanoic acid. Volatile compounds in meat are affected by cooking time and temperature and different reactions during the process (such as Maillard reaction) and a series of nutrient degradation [37].

In the sensory evaluation, there were significant differences in overall liking, taste, and aroma of the meat samples. The preference was for the Meat DSJ sample, and can be explained by the additional flavor and salty taste. However, adding dried cabbage to minced mutton patties caused a decline in the overall liking, flavor, and texture [49].

In conclusion, DSJ could potentially be used as a supplement in food applications, to enrich their functional properties, if it is acceptable for taste and aroma sensory attributes. Research on sauerkraut juice and its production products is scarce and further research is recommended.

**Supplementary Materials:** The following supporting information can be downloaded at: https://www.mdpi.com/article/10.3390/app13053358/s1, Table S1: Relative area of volatile compounds in SJ and DSJ; Table S2: Relative area of volatile compounds in bread and meat samples.

**Author Contributions:** Conceptualization, L.J. and Z.K.; methodology, Z.K. and K.M.; formal analysis, L.J. and Z.K.; investigation, L.J.; resources, S.K.; data curation, Z.K.; writing—original draft preparation, L.J.; writing—review and editing, Z.K. and S.K.; supervision, Z.K. All authors have read and agreed to the published version of the manuscript.

**Funding:** This research was funded by ESF project grant number 8.2.2.0/20/I/001/TOPIC ES32.

**Institutional Review Board Statement:** Not applicable.

**Informed Consent Statement:** Informed consent was obtained from all subjects involved in the study.

**Data Availability Statement:** Not applicable.

**Acknowledgments:** This research was financially supported by the ESF project Nr. 8.2.2.0/20/I/001/TOPIC ES32—"Latvia University of Life Sciences and Technologies transition to the new doctoral funding model" and the European Innovation Partnership for Agricultural Productivity and Sustainability Working Group Cooperation project 18-00-A01612-000020. Z.K.; S.K. and K.M. acknowledges the support.

**Conflicts of Interest:** The authors declare no conflict of interest.

## References

1. Peñas, E.; Martinez-Villaluenga, C.; Frias, J. Sauerkraut: Production, Composition, and Health Benefits. *Fermented Foods Health Dis. Prev.* **2017**, 557–576. [CrossRef]
2. Martínez, S.; Armesto, J.; Gómez-Limia, L.; Carballo, J. Impact of Processing and Storage on the Nutritional and Sensory Properties and Bioactive Components of Brassica Spp. A Review. *Food Chem.* **2020**, *313*, 126065. [CrossRef]
3. Satora, P.; Skotniczny, M.; Strnad, S.; Piechowicz, W. Chemical Composition and Sensory Quality of Sauerkraut Produced from Different Cabbage Varieties. *LWT* **2021**, *136*, 110325. [CrossRef]

4. Jansone, L.; Kampuse, S. Comparison of chemical composition of fresh and fermented cabbage juice. In Proceedings of the 13th Baltic Conference on Food Science and Technology "Food. Nutrition. Well-Being", Jelgava, Latvia, 2–3 May 2019; pp. 160–164. [CrossRef]
5. Martinez-Villaluenga, C.; Peñas, E.; Frias, J.; Ciska, E.; Honke, J.; Piskula, M.K.; Kozlowska, H.; Vidal-Valverde, C. Influence of Fermentation Conditions on Glucosinolates, Ascorbigen, and Ascorbic Acid Content in White Cabbage (*Brassica oleracea* Var. *capitata* Cv. Taler) Cultivated in Different Seasons. *J. Food Sci.* **2009**, *74*, C62–C67. [CrossRef]
6. Jansone, L.; Kampuse, S.; Kruma, Z.; Lidums, I. Quality Parameters of Horizontally Spray-Dried Fermented Cabbage Juice. *Proc. Latv. Acad. Sci. Sect. B. Nat. Exact Appl. Sci.* **2022**, *76*, 96–102. [CrossRef]
7. Jansone, L.; Kruma, Z.; Straumite, E. Evaluation of Chemical and Sensory Characteristics of Sauerkraut Juice Powder and Its Application in Food. *Foods* **2022**, *12*, 19. [CrossRef]
8. Korus, A.; Witczak, M.; Korus, J.; Juszczak, L. Dough Rheological Properties and Characteristics of Wheat Bread with the Addition of Lyophilized Kale (*Brassica oleracea* L. Var. *sabellica*) Powder. *Appl. Sci.* **2023**, *13*, 29. [CrossRef]
9. Sardabi, F.; Azizi, M.H.; Gavlighi, H.A.; Rashidinejad, A. The Effect of Moringa Peregrina Seed Husk on the in Vitro Starch Digestibility, Microstructure, and Quality of White Wheat Bread. *LWT* **2021**, *136*, 110332. [CrossRef]
10. Gómez-García, R.; Vilas-Boas, A.A.; Machado, M.; Campos, D.A.; Aguilar, C.N.; Madureira, A.R.; Pintado, M. Impact of Simulated in Vitro Gastrointestinal Digestion on Bioactive Compounds, Bioactivity and Cytotoxicity of Melon (*Cucumis melo* L. *inodorus*) Peel Juice Powder. *Food Biosci.* **2022**, *47*, 101726. [CrossRef]
11. Gawlik-Dziki, U.; Dziki, D.; Świeca, M.; Sęczyk, Ł.; Rózyło, R.; Szymanowska, U. Bread Enriched with Chenopodium Quinoa Leaves Powder—The Procedures for Assessing the Fortification Efficiency. *LWT* **2015**, *62*, 1226–1234. [CrossRef]
12. Liu, Y.; Zhang, H.; Brennan, M.; Brennan, C.; Qin, Y.; Cheng, G.; Liu, Y. Physical, Chemical, Sensorial Properties and in Vitro Digestibility of Wheat Bread Enriched with Yunnan Commercial and Wild Edible Mushrooms. *LWT* **2022**, *169*, 113923. [CrossRef]
13. Đurović, S.; Vujanović, M.; Radojković, M.; Filipović, J.; Filipović, V.; Gašić, U.; Tešić, Ž.; Mašković, P.; Zeković, Z. The Functional Food Production: Application of Stinging Nettle Leaves and Its Extracts in the Baking of a Bread. *Food Chem.* **2020**, *312*, 126091. [CrossRef] [PubMed]
14. Zhang, D.; Ivane, N.M.A.; Haruna, S.A.; Zekrumah, M.; Elysé, F.K.R.; Tahir, H.E.; Wang, G.; Wang, C.; Zou, X. Recent Trends in the Micro-Encapsulation of Plant-Derived Compounds and Their Specific Application in Meat as Antioxidants and Antimicrobials. *Meat Sci.* **2022**, *191*, 108842. [CrossRef]
15. Goff, S.A.; Klee, H.J. Plant Volatile Compounds: Sensory Cues for Health and Nutritional Value? *Science* **2006**, *311*, 815–819. [CrossRef] [PubMed]
16. Gawlik-Dziki, U.; Dziki, D.; Baraniak, B.; Lin, R. The Effect of Simulated Digestion in Vitro on Bioactivity of Wheat Bread with Tartary Buckwheat Flavones Addition. *LWT-Food Sci. Technol.* **2009**, *42*, 137–143. [CrossRef]
17. Wieca, M.S.; Gawlik-Dziki, U.; Dziki, D.; Baraniak, B. Wheat Bread Enriched with Green Coffee—In Vitro Bioaccessibility and Bioavailability of Phenolics and Antioxidant Activity. *Food Chem.* **2016**, *221*, 1451–1457. [CrossRef]
18. Jafari, S.; Jafari, S.M.; Ebrahimi, M.; Kijpatanasilp, I.; Assatarakul, K. A Decade Overview and Prospect of Spray Drying Encapsulation of Bioactives from Fruit Products: Characterization, Food Application and in Vitro Gastrointestinal Digestion. *Food Hydrocoll.* **2023**, *134*, 108068. [CrossRef]
19. Mackie, A.; Mulet-Cabero, A.I.; Torcello-Gomez, A. Simulating Human Digestion: Developing Our Knowledge to Create Healthier and More Sustainable Foods. *Food Funct.* **2020**, *11*, 9397–9431. [CrossRef]
20. Minekus, M.; Alminger, M.; Alvito, P.; Ballance, S.; Bohn, T.; Bourlieu, C.; Carrière, F.; Boutrou, R.; Corredig, M.; Dupont, D.; et al. A Standardised Static in Vitro Digestion Method Suitable for Food–An International Consensus. *Food Funct.* **2014**, *5*, 1113–1124. [CrossRef]
21. Jakubczyk, A.; Kiersnowska, K.; Ömeroglu, B.; Gawlik-Dziki, U.; Tutaj, K.; Rybczyńska-Tkaczyk, K.; Szydłowska-Tutaj, M.; Złotek, U.; Baraniak, B. The Influence of *Hypericum perforatum* L. Addition to Wheat Cookies on Their Antioxidant, Anti-Metabolic Syndrome, and Antimicrobial Properties. *Foods* **2021**, *10*, 1379. [CrossRef]
22. Singleton, V.L.; Orthofer, R.; Lamuela-Raventós, R.M. Analysis of Total Phenols and Other Oxidation Substrates and Antioxidants by Means of Folin-Ciocalteu Reagent. *Methods Enzymol.* **1999**, *299*, 152–178. [CrossRef]
23. Rokayya, S.; Li, C.J.; Zhao, Y.; Li, Y.; Sun, C.H. Cabbage (*Brassica oleracea* L. Var. *capitata*) Phytochemicals with Antioxidant and Anti-Inflammatory Potential. *Asian Pac. J. Cancer Prev.* **2013**, *14*, 6657–6662. [CrossRef]
24. Galoburda, R.; Straumite, E.; Sabovics, M.; Kruma, Z. Dynamics of Volatile Compounds in Triticale Bread with Sourdough: From Flour to Bread. *Foods* **2020**, *9*, 1837. [CrossRef] [PubMed]
25. Mammasse, N.; Schlich, P. Adequate Number of Consumers in a Liking Test. Insights from Resampling in Seven Studies. *Food Qual. Prefer.* **2014**, *31*, 124–128. [CrossRef]
26. Quatrin, A.; Rampelotto, C.; Pauletto, R.; Maurer, L.H.; Nichelle, S.M.; Klein, B.; Rodrigues, R.F.; Maróstica, M.R., Jr.; Fonseca, B.d.S.; de Menezes, C.R.; et al. Bioaccessibility and Catabolism of Phenolic Compounds from Jaboticaba (*Myrciaria trunciflora*) Fruit Peel during in Vitro Gastrointestinal Digestion and Colonic Fermentation. *J. Funct. Foods* **2020**, *65*, 103714. [CrossRef]
27. Leyva-López, R.; Palma-Rodríguez, H.M.; López-Torres, A.; Capataz-Tafur, J.; Bello-Pérez, L.A.; Vargas-Torres, A. Use of Enzymatically Modified Starch in the Microencapsulation of Ascorbic Acid: Microcapsule Characterization, Release Behavior and in Vitro Digestion. *Food Hydrocoll.* **2019**, *96*, 259–266. [CrossRef]

28. Tchabo, W.; Kaptso, G.K.; Ngolong Ngea, G.L.; Wang, K.; Bao, G.; Ma, Y.; Wang, X.; Mbofung, C.M. In Vitro Assessment of the Effect of Microencapsulation Techniques on the Stability, Bioaccessibility and Bioavailability of Mulberry Leaf Bioactive Compounds. *Food Biosci.* **2022**, *47*, 101461. [CrossRef]

29. Wieczorek, M.N.; Drabińska, N. Flavour Generation during Lactic Acid Fermentation of Brassica Vegetables—Literature Review. *Appl. Sci.* **2022**, *12*, 5598. [CrossRef]

30. Ahmed, Z.S.; Abozed, S.S. Functional and Antioxidant Properties of Novel Snack Crackers Incorporated with Hibiscus Sabdariffa By-Product. *J. Adv. Res.* **2015**, *6*, 79–87. [CrossRef]

31. Tomsone, L.; Galoburda, R.; Kruma, Z.; Majore, K. Physicochemical Properties of Biscuits Enriched with Horseradish (*Armoracia rusticana* L.) Products and Bioaccessibility of Phenolics after Simulated Human Digestion. *Pol. J. Food Nutr. Sci.* **2020**, *70*, 419–428. [CrossRef]

32. Wichchukit, S.; O'Mahony, M. The 9-Point Hedonic Scale and Hedonic Ranking in Food Science: Some Reappraisals and Alternatives. *J. Sci. Food Agric.* **2015**, *95*, 2167–2178. [CrossRef]

33. Cheng, J.-R.; Liu, X.-M.; Zhang, W.; Chen, Z.-Y.; Wang, X.-P. Stability of Phenolic Compounds and Antioxidant Capacity of Concentrated Mulberry Juice-Enriched Dried-Minced Pork Slices during Preparation and Storage. *Food Control.* **2018**, *89*, 187–195. [CrossRef]

34. Miedzianka, J.; Lachowicz-Wiśniewska, S.; Nemś, A.; Kowalczewski, P.Ł.; Kita, A. Comparative Evaluation of the Antioxidative and Antimicrobial Nutritive Properties and Potential Bioaccessibility of Plant Seeds and Algae Rich in Protein and Polyphenolic Compounds. *Appl. Sci.* **2022**, *12*, 8136. [CrossRef]

35. Draijer, R.; van Dorsten, F.A.; Zebregs, Y.E.; Hollebrands, B.; Peters, S.; Duchateau, G.S.; Grün, C.H. Impact of Proteins on the Uptake, Distribution, and Excretion of Phenolics in the Human Body. *Nutrients* **2016**, *8*, 814. [CrossRef]

36. Iqbal, Y.; Ponnampalam, E.N.; Le, H.H.; Artaiz, O.; Muir, S.K.; Jacobs, J.L.; Cottrell, J.J.; Dunshea, F.R. In Vitro Bioaccessibility of Polyphenolic Compounds: The Effect of Dissolved Oxygen and Bile. *Food Chem.* **2023**, *404*, 134490. [CrossRef]

37. Jiang, S.; Xue, D.; Zhang, Z.; Shan, K.; Ke, W.; Zhang, M.; Zhao, D.; Nian, Y.; Xu, X.; Zhou, G.; et al. Effect of Sous-Vide Cooking on the Quality and Digestion Characteristics of Braised Pork. *Food Chem.* **2022**, *375*, 131683. [CrossRef] [PubMed]

38. Gao, B.; Wang, J.; Wang, Y.; Xu, Z.; Li, B.; Meng, X.; Sun, X.; Zhu, J. Influence of Fermentation by Lactic Acid Bacteria and in Vitro Digestion on the Biotransformations of Blueberry Juice Phenolics. *Food Control.* **2022**, *133*, 956–7135. [CrossRef]

39. Herranz, B.; Fernández-Jalao, I.; Dolores Álvarez, M.; Quiles, A.; Sánchez-Moreno, C.; Hernando, I.; de Ancos, B. Phenolic Compounds, Microstructure and Viscosity of Onion and Apple Products Subjected to in Vitro Gastrointestinal Digestion. *Innov. Food Sci. Emerg. Technol.* **2019**, *51*, 114–125. [CrossRef]

40. Cantele, C.; Rojo-Poveda, O.; Bertolino, M.; Ghirardello, D.; Cardenia, V.; Barbosa-Pereira, L.; Zeppa, G. In Vitro Bioaccessibility and Functional Properties of Phenolic Compounds from Enriched Beverages Based on Cocoa Bean Shell. *Foods* **2020**, *9*, 715. [CrossRef]

41. Tuohy, K.M.; Conterno, L.; Gasperotti, M.; Viola, R. Up-Regulating the Human Intestinal Microbiome Using Whole Plant Foods, Polyphenols, and/or Fiber. *J. Agric. Food Chem.* **2012**, *60*, 8776–8782. [CrossRef]

42. Rajkumar, G.; Shanmugam, S.; de Sousa Galvâo, M.; Dutra Sandes, R.D.; Leite Neta, M.T.S.; Narain, N.; Mujumdar, A.S. Comparative Evaluation of Physical Properties and Volatiles Profile of Cabbages Subjected to Hot Air and Freeze Drying. *LWT* **2017**, *80*, 501–509. [CrossRef]

43. Banerjee, A.; Penna, S.; Variyar, P.S. Allyl Isothiocyanate Enhances Shelf Life of Minimally Processed Shredded Cabbage. *Food Chem.* **2015**, *183*, 265–272. [CrossRef]

44. Klopsch, R.; Baldermann, S.; Hanschen, F.S.; Voss, A.; Rohn, S.; Schreiner, M.; Neugart, S. Brassica-Enriched Wheat Bread: Unraveling the Impact of Ontogeny and Breadmaking on Bioactive Secondary Plant Metabolites of Pak Choi and Kale. *Food Chem.* **2019**, *295*, 412–422. [CrossRef] [PubMed]

45. Glube, N.; von Moos, L.; Duchateau, G. Capsule Shell Material Impacts the in Vitro Disintegration and Dissolution Behaviour of a Green Tea Extract. *Results Pharma Sci.* **2013**, *3*, 1–6. [CrossRef]

46. Ren, N.; Ma, Z.; Li, X.; Hu, X. Preparation of Rutin-Loaded Microparticles by Debranched Lentil Starch-Based Wall Materials: Structure, Morphology and in Vitro Release Behavior. *Int. J. Biol. Macromol.* **2021**, *173*, 293–306. [CrossRef]

47. Rusinek, R.; Gawrysiak-Witulska, M.; Siger, A.; Oniszczuk, A.; Ptaszyńska, A.A.; Ptaszyńska, P.; Knaga, J.; Malaga-Toboła, U.; Gancarz, M.; Dymerski, M. Effect of Supplementation of Flour with Fruit Fiber on the Volatile Compound Profile in Bread. *Sensors* **2021**, *21*, 2812. [CrossRef]

48. Drabińska, N.; Ciska, E.; Szmatowicz, B.; Krupa-Kozak, U. Broccoli By-Products Improve the Nutraceutical Potential of Gluten-Free Mini Sponge Cakes. *Food Chem.* **2018**, *267*, 170–177. [CrossRef]

49. Malav, O.P.; Sharma, B.D.; Kumar, R.R.; Talukder, S.; Ahmed, S.R.; Irshad, A. Antioxidant Potential and Quality Characteristics of Functional Mutton Patties Incorporated with Cabbage Powder. *Nutr. Food Sci.* **2015**, *45*, 542–563. [CrossRef]

50. Moroney, N.C.; O'grady, M.N.; Ad Lordan, S.; Stanton, C.; Kerry, J.P. Marine Drugs Seaweed Polysaccharides (Laminarin and Fucoidan) as Functional Ingredients in Pork Meat: An Evaluation of Anti-Oxidative Potential, Thermal Stability and Bioaccessibility. *Mar. Drugs* **2015**, *13*, 2447–2464. [CrossRef] [PubMed]

51. Iqbal, Y.; Ponnampalam, E.N.; Le, H.H.; Artaiz, O.; Muir, S.K.; Jacobs, J.L.; Cottrell, J.J.; Dunshea, F.R. Assessment of Feed Value of Chicory and Lucerne for Poultry, Determination of Bioaccessibility of Their Polyphenols and Their Effects on Caecal Microbiota. *Fermentation* **2022**, *8*, 237. [CrossRef]

52. Rashidinejad, A.; Birch, E.J.; Sun-Waterhouse, D.; Everett, D.W. Addition of Milk to Tea Infusions: Helpful or Harmful? Evidence from in Vitro and in Vivo Studies on Antioxidant Properties. *Crit. Rev. Food Sci. Nutr.* **2017**, *57*, 3188–3196. [CrossRef] [PubMed]
53. Flores, F.P.; Singh, R.K.; Kerr, W.L.; Pegg, R.B.; Kong, F. Total Phenolics Content and Antioxidant Capacities of Microencapsulated Blueberry Anthocyanins during In Vitro Digestion. *Food Chem.* **2013**, *153*, 272–278. [CrossRef] [PubMed]
54. Lou, X.; Xiong, J.; Tian, H.; Yu, H.; Chen, C.; Huang, J.; Yuan, H.; Hanna, M.; Yuan, L.; Xu, H. Effect of High-Pressure Processing on the Bioaccessibility of Phenolic Compounds from Cloudy Hawthorn Berry (*Crataegus pinnatifida*) Juice. *J. Food Compos. Anal.* **2022**, *110*, 104540. [CrossRef]

*applied*
*sciences*

*Article*

# Baby Food Purees Obtained from Ten Different Apple Cultivars and Vegetable Mixtures: Product Development and Quality Control

Alexandra Mădălina Mateescu [1], Andruța Elena Mureșan [1,*], Andreea Pușcaș [1], Vlad Mureșan [1], Radu E. Sestras [2] and Sevastița Muste [1]

[1] Department of Food Engineering, Faculty of Food Science and Technology, University of Agricultural Science and Veterinary Medicine Cluj-Napoca, 3-5 Calea Mănăștur, 400372 Cluj-Napoca, Romania
[2] Faculty of Horticulture, University of Agricultural Sciences and Veterinary Medicine Cluj-Napoca, 3-5 Calea Mănăștur Street, 400372 Cluj-Napoca, Romania
* Correspondence: andruta.muresan@usamvcluj.ro

**Abstract:** As is well known, apples are the most complex fruit in terms of nutritional compounds, with a high content of fiber, vitamins (vitamins C, A, B3), and minerals. Both fruits and vegetables are important sources of nutrients for infants' nutrition and healthy development. The purpose of this study was to develop and analyze baby food purees obtained from apples and vegetables. Ten types of baby purees were obtained from the most-consumed varieties of apples from Romania, along with purees of carrots, pumpkin and celery. The resulting samples were analyzed in terms of moisture, ash content, titratable acidity, and vitamin C content. The total polyphenol content was assessed by the Folin–Ciocalteu method, while total antioxidant capacity was determined by the DPPH method; moreover, the color parameters and textural properties were also assessed. Following the results obtained, the purees can be introduced into the diet of infants and children, providing them with the necessary vitamins and minerals for optimal development. The analyses performed on both fresh and sterilized products highlighted the effects of heat treatment on the components of the product. The most important changes were observed in the vitamin C content, which was decreased by 50–70% in all ten purees. Total polyphenol content (TPC) increased in sterilized samples up to 70 mg GAE/100 g. Antioxidant capacity (AC) almost doubled its value in some samples after the thermal process application. Regarding the adhesiveness and deformation at hardness, which represented the main parameters for baby's food, the value increased in the sterilized product, making the product more suitable for infants. This survey provides a detailed description of the development of baby food purees, showing the conveniences of developing purees for children based on fruits and vegetables.

**Keywords:** apples; vegetables; baby food; food quality; children nutrition; bioactive compounds

**Citation:** Mateescu, A.M.; Mureșan, A.E.; Pușcaș, A.; Mureșan, V.; Sestras, R.E.; Muste, S. Baby Food Purees Obtained from Ten Different Apple Cultivars and Vegetable Mixtures: Product Development and Quality Control. *Appl. Sci.* **2022**, *12*, 12462. https://doi.org/10.3390/app122312462

Academic Editors: Joanna Trafialek and Mariusz Szymczak

Received: 26 October 2022
Accepted: 29 November 2022
Published: 6 December 2022

**Publisher's Note:** MDPI stays neutral with regard to jurisdictional claims in published maps and institutional affiliations.

## 1. Introduction

There is ample evidence that the quality and composition of children's available food products contribute to the present and future health status of young children. Infant food, designed for children between 6 months and 2 years, is quite limited, so baby foods based on fruits and vegetables, which are commercially available, are an important source of energy, basic nutrients, vitamins, fiber and minerals [1,2].

In developing countries, micronutrient deficiency in the first two years of life is a general public health problem for infants and children [3]. A good diet in childhood is essential for optimal growth and development [4]. This is also the ideal time to introduce to the child's diet a variety of foods and flavors to develop healthy eating habits and a variety of diets [5].

Food based on fruits and vegetables provides the necessary vitamins, minerals, fiber and other nutrients essential for the proper functioning of the body, strengthens the immune system, provides protection and reduces the risk of cancer, cardiovascular disease and oxidative risk [6].

Fruit and vegetable purees are a safe and fast option, suitable to provide the extra energy needed by babies [7]. These products have appeared on the market as ready-to-consume products, which allows the consumer to save the time required for obtaining them at home. In recent years, the introduction of fruit and vegetable purees in children's diets has been increasing [8].

The rapid growth and normal and healthy development of children require a high intake of nutrients and energy [9,10]. Infants' nutrient reserves are limited, so low nutrient intake can impair neuronal development [11].

Therefore, due to all the above-mentioned advantages of infants consuming fruits and vegetables, and also because parents prefer to give children ready-to-eat products [12], the aim of the present study was to develop purees from ten varieties of autumn apples and a mixture of vegetables (carrots, pumpkin and celery), which are ideal for the nutrition of children with ages between four months and two years, with four months being the earliest age when the fruits should be introduced in the diet, according to [12]. The combination of these components, namely apples, which are rich in fiber, vitamin C and phenolic compounds, and the mixture of vegetables, which are rich in antioxidants, vitamins and minerals, led to a product that could ensure the harmonious development of children [13].

## 2. Materials and Methods

### 2.1. Raw Materials

In order to obtain these baby food products, the following ingredients were used: apples (*Malus* × *domestica* Borkh.), carrots (*Daucus carota* subsp. *sativus*), pumpkin (*Cucurbita* subsp. *Cucurbita moschata* Baby Pam) and celeriac (*Apium graveolens* var. *rapaceum*). Apples, which represent the main component, were obtained from the Horticulture Research and Development Institute, Mărăcineni, Pitești Mărului Street no. 402, Mărăcineni 117450, Romania (Figure 1), and the carrots, pumpkin and celeriac were obtained from the Horticulture Research and Development Institute, Cluj-Napoca, 3-5 Manastur 119 Street, Cluj-Napoca, 400372, Romania.

**Type of apples/Foto**

**Figure 1.** Varieties of apples studied.

### 2.2. Fresh Puree Manufacturing

As the current study determines the influence of apple variety in the composition of the baby food puree, it was included in the highest proportion in the formulation, as seen in Table 1. From the harvest time to utilization, apples were placed in vacuum bags of 1 kg and kept in the refrigerator at a temperature of 4–6 °C Vegetables were washed and dried, except for the pumpkin, then stored at 10–12 °C at a relative humidity of about 50–60% in the institute's storage facility.

**Table 1.** Details regarding raw materials and proportions used to obtain purees for children.

| Ingredients | Proportions (%) | Average Mass (g) | Harvest Period |
|---|---|---|---|
| **Apple** | | | |
| Champion | 70 | 182.14 ± 34.38 [a,b,c] | August–September |
| Elstar | 70 | 101.04 ± 15.46 [c,d,e] | August–September |
| Baujade | 70 | 100.52 ± 10.97 [c,d,e] | May |
| Grimmes Golden | 70 | 199.94 ± 29.81 [a] | September |
| Granny Smith | 70 | 190 ± 27.81 [a] | September–October |
| Pătul | 70 | 75.16 ± 17.00 [d,e,f] | September–October |
| Reinette Harbert | 70 | 127.70 ± 33.81 [b,c,d,e] | August–September |
| Juliana | 70 | 142.14 ± 24.38 [a,b,c] | September–October |
| Domnesc | 70 | 184 ± 21.91 [a,b,c] | September–October |
| Parmen Clone | 70 | 117.60 ± 31.23 [b,c,d,e] | September–October |
| **Carrot** | 15 | 74.13 ± 28.17 | August–October |
| **Pumpkin** | 10 | 1866.37 ± 374.89 | August–October |
| **Celeriac** | 5 | 126.21 ± 26.43 | August–October |

Identical lowercase superscripts within columns indicate no significant difference ($p > 0.05$).

All ingredients were washed, peeled, mashed, and subjected to evaporation-concentration operation in a vacuum chamber at 650–680 mmHg, which reduced the boiling temperature to 50 °C (Figure 2). This operation can reduce the amount of product to be stored and can ensure the shelf life of the product because its water activity is decreased below 0.7. To prevent the development of spoilage microorganisms, the evaporation-concentration operation was realized in a vacuum chamber, which can also prevent the loss of volatile compounds (Dai et al., 2020) [14].

**Figure 2.** Flow chart of the processing steps of puree samples.

The vegetables and apples were mashed in a Vegetable Masher Machine HL-617-1 (Zhengzhou Hongle Machinery Equipment Co., Ltd., Zhengzhou, China).

The purees obtained were dosed manually in glass containers of 125 g with the following dimensions: 5.1 cm diameter and 7.8 cm height. Twist-off lids were used to close the containers. Sterilization of the purees was performed in a vertical autoclave at 119 °C for 15 min [15–17].

### 2.3. Determination Methods of Physico-Chemical Characteristics

### 2.3.1. Moisture Determination by Oven-Dry Method

Moisture determination (H) was conducted according to the AOAC Official Method 934.06 [18]. The determination consists of drying a quantity of vegetal material at the temperature of $103 \pm 2$ °C until a constant mass is achieved.

Two parallel determinations were performed from the same sample preparation for analysis [19].

### 2.3.2. Ash Content Determination

The ash content method was also conducted according to the AOAC Official Method 934.06 [18]. The ash content was determined by calcination at 550–600 °C until a white, coal-free ash was obtained.

### 2.3.3. Titratable Acidity Determination

Titratable acidity was performed according to ISO 750:1998. The determination consists of the neutralization of acidity with 0.1 n sodium hydroxide in the presence of phenolphthalein as an indicator.

### 2.3.4. Vitamin C Dosing—Iodometric Method

This method was performed according to Zhao et al., 2020 [20]. Chemical methods of dosing vitamin C are based on its reducing properties. Ascorbic acid is converted by oxidation to dehydroascorbic acid. After vitamin C has been completely oxidized, the iodine generated will form with the starch, presenting an intense blue-colored inclusion complex; the color must persist for 30 s.

### 2.3.5. Total Polyphenol Content Determination (TPC)

This determination was performed using the Folin–Ciocalteu method. and the absorbance of the samples was measured at 750 nm (UV-VIS Schimadzu, Kyoto, Japonia) [21].

Extraction of total polyphenols was performed for all ten experimental puree variants. From each sample, 1 g was taken, homogenized with methanol and kept at refrigeration temperature for 24 h. The extracts obtained were filtered and concentrated at 35 °C under reduced pressure (Rotavapor Heidolph, Heidolph Instruments GmbH & Co., Schwabach, Germany). The concentrate was recovered in 9 mL of methanol. A quantity of 25 µL of the pre-prepared sample was mixed with 1.8 mL of distilled water and 120 µL of Folin–Ciocalteu reagent in a glass vial and homogenized. After 5 min, 340 µL of 7.5% $Na_2 CO_3$ solution in distilled water was added in order to create basic conditions (pH~10) for the redox reaction between the phenolic compounds and the Folin–Ciocalteu reagent. The samples were incubated at room temperature. The control sample was methanol.

The total polyphenol content of the raw materials used and the finished product was expressed in gallic acid equivalents, mg of GAE/100 g.

### 2.3.6. Antioxidant Capacity Determination (AC)

The antioxidant capacity was determined by evaluating the free radical scavenging effect on 1,1-diphenyl-2-picrylhydrazyl (DPPH) following Dibacto et al., 2021 [22].

An amount of 150 µL of the methanolic extract of the analyzed samples was mixed with 2850 µL of DPPH solution. The mixture was properly homogenized and kept in the dark for 60 min. The absorbance of the samples was measured at 515 nm (UV-VIS

Schimadzu) against a blank of methanol. The results were expressed as a percentage of the absorbance of the standard DPPH solution. Results are expressed in µm TE/g fresh mass (FM) [23].

### 2.3.7. Color Determination

The color of the puree was determined using an NR200 portable colorimeter (3NH, Shenzhen, China). The puree color was scanned 4 times for each sample, and then their average was calculated. The color value was expressed as L*, a* and b*, where L* (brightness) represents how dark the sample is, ranging from 0 (black) to 100 (white), a* represents the color variation from green ($-$) to red (+), and b* represents the color variation from blue ($-$) to yellow (+), according to the International Commission on Lighting. Measurements were performed using a D65 illuminant with an opening of 8 mm. The colorimeter was subjected to an automatic black-and-white calibration [24,25].

### 2.3.8. Texture Determination

Texture is important for fruit purees because it evaluates the specific sensorial attributes appreciated by the consumers. The texture analysis was performed using a CT3 Brookfield Texture Analyzer (Brookfield, Middleboro, MA, USA) equipped with a TA39 probe (2 mm diameter rod, stainless steel 5 g, 20 mm length, flat end). A compression test with a 5 mm target value was selected, the probe test speed was set to 0.5 mm/s, the trigger load was 4 g, and 10 points/second were registered. The deformation curves were analyzed, and 4 parameters were studied: cohesiveness (n.a), firmness (N), deformation (mm), and adhesiveness (mJ), according to ISO 11036:2020 [26]. The computed textural parameters are described in Table 2.

**Table 2.** Parameter definition and calculation for texture analysis.

| Parameters | Units | Definition |
| --- | --- | --- |
| Firmness | N | The force at the maximum peak during the compression test |
| Deformation | mm | The displacement at the maximum force during the puncture compression test |
| Adhesiveness | mJ | Area under the Load vs. Distance curve, measured from where cycle 1 first reaches zero load to where it ends |
| Cohesiveness | n.a. | Indicates the strength of internal bonds of the product, being computed as Hardness Work Done Cycle 2/Hardness Work Done Cycle 1 |

### 2.3.9. Statistical Analyses

Registration data for the puree quality and index analysis elements were treated as the mean and standard deviation (SD) and coefficient of variation for the traits. An analysis of variance (ANOVA) was applied to the analyzed characteristics. The test yielded a significant F statistic. Tukey's test ($\alpha < 0.05$) was used as a post hoc test for the analysis of differences. Data were subjected to multivariate statistical analysis, namely principal component analysis (PCA), using Unscrambler X v.10.5.1 software.

## 3. Results and Discussion

### 3.1. Moisture Determination by Oven-Dry Method

The high moisture content is attributed to the high binding water capacity of the apple varieties' starch, which is associated with weak molecular forces between the starch granules [27]. On the other hand, the relatively low moisture content in purees can be due to the low binding capacity of starch in apples resulting in the excessive loss of moisture during ripening and storage [28–30].

Following the determination, it was found that the highest moisture of the untreated puree was 87.27% for the Baujade apple variety, which decreased by 84.59%, and the lowest moisture value was 80.72% for the Juliana variety, which decreased by 78.14% (Table 3).

After sterilization, the puree moisture decreased in all varieties, but the largest decrease was observed in the Parmen Clone apple variety, from 82.05% to 76.54%.

Other researchers obtained similar values for baby foods purees, where moisture decreased between 1–5% in different types of apple purees [31–33].

### 3.2. Ash Content Determination

Total ash content determination in purees provides a measure of the total amount of minerals since minerals cannot be destroyed by heating. The mineral content varies and depends on many factors, such as the ripening grade of the apples and vegetables used, the variety of plant species, the soil where the plantation is located, as well as fertilization programs and the climate [34].

Regarding the ash content of fresh purees, we obtained very different values depending on the variety of apples used. Thus, the varieties with the highest ash content are Elstar and Juliana (18.76% and 15.39%, respectively). The lowest values were registered for the Champion, Granny Smith, Pătul and Parmen Clone varieties (2.91–4.09%) (Table 3).

Following the heat treatment applied, the ash content decreased in most of the varieties, except for the apple varieties Granny Smith and Parmen Clone, for which the ash content was higher after sterilization.

Other researchers reached similar values for some baby foods purees and also reported a 1–3% increase in the ash content after thermal processing [2,35,36].

### 3.3. Titratable Acidity Determination

Titratable acidity is an important parameter in determining the puree quality, which also denotes the degree of ripening of the apples and vegetables used. Previous studies have shown that the trend in producing fruit purees is that organic acids increase during the heat treatment process and during storage [37].

The acidity determined for the fresh puree was similar for all ten varieties of purees, with values ranging between 0.30% and 0.45%. Following the application of heat treatment, the acidity values decreased, reaching values of a maximum of 0.40% (Table 3).

Our findings were consistent with those reported by Usal and Sahan in 2020 [38], who studied 24 fruit-and-vegetable-based baby food purees, for which values between 0.23–0.54% were determined.

**Table 3.** Moisture, ash content and titratable acidity determination.

| Puree Sample | Moisture Determination% | | Ash Content% | | Acidity Determination (mg/100 g) | |
|---|---|---|---|---|---|---|
| | Unsterilized Product | Sterilized Product | Unsterilized Product | Sterilized Product | Unsterilized Product | Sterilized Product |
| 1 | 84.0 ± 0.2 | 81.1 ± 0.1 | 2.9 ± 0.1 | 3.3 ± 0.1 | 0.34 ± 0.01 | 0.29 ± 0.01 |
| 2 | 84.4 ± 0.1 | 82.3 ± 0.1 | 1.8 ± 0.1 | 2.7 ± 0.2 | 0.30 ± 0.02 | 0.26 ± 0.01 |
| 3 | 87.2 ± 0.2 | 84.5 ± 0.2 | 1.9 ± 0.2 | 3.0 ± 0.1 | 0.42 ± 0.01 | 0.27 ± 0.02 |
| 4 | 83.4 ± 0.4 | 82.6 ± 0.3 | 1.6 ± 0.3 | 1.1 ± 0.1 | 0.26 ± 0.01 | 0.19 ± 0.01 |
| 5 | 86.7 ± 0.2 | 83.7 ± 0.2 | 2.4 ± 0.1 | 2.7 ± 0.3 | 0.45 ± 0.02 | 0.4 ± 0.02 |
| 6 | 82.6 ± 0.1 | 81.1 ± 0.1 | 2.6 ± 0.3 | 2.9 ± 0.2 | 0.38 ± 0.01 | 0.37 ± 0.01 |
| 7 | 86.2 ± 0.1 | 84.7 ± 0.1 | 3.7 ± 0.2 | 2.6 ± 0.3 | 0.4 ± 0.02 | 0.21 ± 0.02 |
| 8 | 80.7 ± 0.3 | 78.1 ± 0.1 | 2.3 ± 0.1 | 3.9 ± 0.1 | 0.48 ± 0.01 | 0.32 ± 0.01 |
| 9 | 84.5 ± 0.3 | 81.0 ± 0.2 | 1.6 ± 0.2 | 3.4 ± 0.2 | 0.29 ± 0.02 | 0.24 ± 0.02 |
| 10 | 82.0 ± 0.2 | 76.5 ± 0.2 | 4.7 ± 0.1 | 4.0 ± 0.3 | 0.41 ± 0.01 | 0.34 ± 0.01 |

Puree sample column represents the apple variety used in order to obtain the puree: 1—Champion, 2—Elstar, 3—Baujade 4—Grimmes Golden, 5—Granny Smith, 6—Pătul, 7—Reinette Harbert, 8—Juliana, 9—Domnesc, 10—Parmen Clone.

### 3.4. Vitamin C Dosing—Iodometric Method

Regarding the vitamin C content of the fresh product, we obtained the highest values for the Domnesc and Granny Smith puree apple varieties, with 4.21 mg% and 3.5 mg%, respectively. The varieties of the baby food purees that had the lowest amounts of vitamin C were those containing Champion and Parmen Clone apple varieties (2.11 mg% and 2.57 mg%, respectively). Following sterilization, the vitamin C content suffered significant losses, between 50% and 70%, in all ten purees (Tables 4 and 5) since vitamin C is a thermolabile compound, which is very susceptible to thermal, chemical and enzymatic oxidation during processing. The vitamin C and oxidative enzymes might come into contact when the food matrix is disrupted by the thermal treatment. Therefore, the oxidative reactions of vitamin C in thermally treated puree could explain the lowest bioaccessibility and vitamin C content [39].

Other researchers obtained similar values for baby foods purees developed from fruit and vegetables, while baby foods from apples demonstrated 3.5% mg/100 g vitamin C, and the highest 20% mg/100 g was determined in purees with other components besides apples, such as banana, oat, plum and broccoli [40]. Additionally, another study showed similar values for vitamin C in baby puree with apple, apple juice, orange juice, banana, mango and pear, namely between 1.9–2.4 mg/100 g [41].

### 3.5. Total Polyphenol Content Determination (TPC)

Regarding the content of total polyphenols in fresh puree, results range between 59.91 mg GAE/100 g for the puree with the Pătul apple variety to 86.31 mg GAE/100 g for the purre with the Domnesc apple variety (Tables 4 and 5).

TPC almost doubled its values after thermal processing in some samples, such as the puree containing the Elstar apple variety, which increased its TPC concentration from 64.36 to 126.65 mg GAE/100 g. The range of TPC after the thermal treatment in other purees was between 76.87 and 139.67 mg GAE/100 g.

This increase is due to heat treatment, which breaks the cell walls and enhances the mobilization of polyphenols from the food matrix, which makes them more available [42].

Similar results have been shown by multiple studies that presented a significant increase in TLC in food products after heat treatment was applied [41,43,44] shows similar values for TPC in baby purees with apple, banana, apricot, orange juice, lemon and peach (values between 85 and 234.2 mg GAE/100 g).

### 3.6. Antioxidant Capacity Determination (AC)

The AC increased in line with the increase in the TCL based on the same consideration that many antioxidant compounds are mainly present as covalently bound forms with insoluble polymers [42].

The values obtained from the determination of the antioxidant capacity of the untreated puree varied depending on the apple variety used. The Juliana puree apple variety had the highest antioxidant capacity of 0.91% inhibited DPPH, compared to the Reinette Harbert puree apple variety, which had a much lower value of only 0.51% inhibited DPPH.

Antioxidant capacity was almost doubled in some cases after thermal process application, with results after sterilization ranging between 0.83% (Elstar apple variety) and 0.93% (Juliana apple variety) (Tables 4 and 5).

Heat treatment disrupts the cell wall and releases antioxidant compounds, leading to an increase in antioxidant capacity [45,46].

Researchers Carbonell-Capella [41] determined higher values for AC in baby puree with apple, banana, apricot, orange juice, lemon and peach, with values between 3.6 and 41.6%. Researcher Čížková [47] showed, for commercial fruit baby food, an AC between 1.87 and 4.75% in products based on apples, strawberries, bilberries, plums, pears and peaches. These increased values are due to the presence of berries and plums in the formulation, which have higher contents of natural antioxidants [48].

**Table 4.** Vit. C, TPC and AC determination in unsterilized product.

| Puree Sample | Vitamin C (mg/100 g) | Polyphenols Content (mg GAE/100 g) | Antioxidant Capacity ($\mu$m TE/g FM) |
|---|---|---|---|
| 1 | 2.11 ± 0.1 | 73.67 ± 1.67 [bc] | 0.62 ± 0.47 [f] |
| 2 | 2.91 ± 0.3 | 64.36 ± 0.2 [de] | 0.64 ± 0.31 [e] |
| 3 | 3.27 ± 0.1 | 65.63 ± 0.1 [de] | 0.88 ± 0.70 [b] |
| 4 | 2.58 ± 0.1 | 61.9 ± 4.88 [e] | 0.67 ± 0.31 [d] |
| 5 | 3.5 ± 0.2 | 69.588 ± 0.59 [cd] | 0.67 ± 0.47 [d] |
| 6 | 2.8 ± 0.1 | 59.91 ± 0.1 [e] | 0.81 ± 0.39 [c] |
| 7 | 3.27 ± 0.2 | 70.492 ± 0.62 [bcd] | 0.51 ± 0.15 [g] |
| 8 | 3.28 ± 0.3 | 77.32 ± 1.95 [b] | 0.91 ± 0.31 [a] |
| 9 | 4.21 ± 0.3 | 86.31 ± 0.59 [a] | 0.81 ± 0.78 [c] |
| 10 | 2.57 ± 0.2 | 64.12 ± 1.03 [de] | 0.81 ± 0.78 [c] |

[a–g]—mean values in columns indicated with various letters are significantly different. Puree sample column represents the apple variety used in order to obtain the puree: 1—Champion, 2—Elstar, 3—Baujade, 4—Grimmes Golden, 5—Granny Smith, 6—Pătul, 7—Reinette Harbert, 8—Juliana, 9—Domnesc, 10—Parmen Clone.

**Table 5.** Vit. C, TPC and AC determination in sterilized product.

| Puree Sample | Vitamin C (mg/100 g) | Polyphenols Content (mg GAE/100 g) | Antioxidant Capacity ($\mu$m TE/g FM) |
|---|---|---|---|
| 1 | 1.18 ± 0.1 | 125.63 ± 1.24 [b] | 0.87 ± 0.15 [b] |
| 2 | 1.88 ± 0.1 | 126.65 ± 1.21 [b] | 0.83 ± 0.47 [d] |
| 3 | 0.94 ± 0.01 | 139.67 ± 3.22 [a] | 0.86 ± 0.78 [bc] |
| 4 | 0.93 ± 0.1 | 125.44 ± 1.28 [b] | 0.92 ± 0.62 [a] |
| 5 | 0.7 ± 0.02 | 76.87 ± 3.08 [b] | 0.92 ± 0.94 [a] |
| 6 | 1.41 ± 0.1 | 80.57 ± 1.28 [d] | 0.91 ± 0.78 [a] |
| 7 | 1.41 ± 0.2 | 80.07 ± 1.86 [d] | 0.87 ± 0.55 [b] |
| 8 | 1.17 ± 0.1 | 81.66 ± 0.61 [d] | 0.93 ± 0.78 [a] |
| 9 | 1.87 ± 0.2 | 102.19 ± 0.62 [c] | 0.85 ± 0.71 [cd] |
| 10 | 1.87 ± 0.1 | 129.6 ± 1.06 [d] | 0.83 ± 0.16 [d] |

[a–d]—mean values in columns indicated with various letters are significantly different. Puree sample column represents the apple variety used in order to obtain the puree: 1—Champion, 2—Elstar, 3—Baujade, 4—Grimmes Golden, 5—Granny Smith, 6—Pătul, 7—Reinette Harbert, 8—Juliana, 9—Domnesc, 10—Parmen Clone.

### 3.7. Color Determination

Food color is a major factor that affects the quality and acceptance of food. It can also be used to predict the chemical and microbiological changes in food. In our case, the effect of sterilization on the color of the product was significant. In addition, the amount of yellow index and lightness decreased compared to the those of the unsterilized products. Enzymatic browning (catalyzed by PPO) and non-enzymatic browning (Maillard reactions) are also important factors regarding the color changing in the browning of the samples [49].

According to Särkkä-Tirkkonen et al. [50], sterilization, as a method used for baby puree thermal treatments, causes structural, taste and color defects. In the current study, we used a vertical autoclave for the sterilization process, and all values, L*, a* and b*, were decreased after the thermal treatment. Consumers prefer infant food with no artificial colors, revealing that over time, consumers became more aware of what they consume [4,51–53]. Therefore, the products with a lower color intensity did not affect the acceptability as long as the color did not represent a quality defect. For the fresh puree, which was not subjected to heat treatment, the highest brightness value was in the puree obtained from the Champion apple variety, 46.03, and the darkest in color was the Reinette Harbert variety, 32.72. The reddest variety of apples was the Elstar variety with a* value equal to 21.02, and the yellowest were the Granny Smith and Reinette Harbert varieties, with similar results of 5.18 (Table 6).

After the sterilization was applied, the color parameters underwent changes, as follows: the brightness of the finished product decreased by 2–3 units for each variety, and the values of the parameter a* decreased significantly, from the range 20.15–15.18 to 14.07–11.51, so the shades of reddish decreased. At the same time, the values of the parameter b* were much lower, from the range 21.71–14.42 to 20.98–14.26, and the shades of yellow decreased.

The thermal treatment caused a decrease in the L* value for the processed samples when compared to the control samples. The purees were characterized by a darkened color determined by various factors such as processing temperature and time, type of cultivar, vitamin C content, and some pigments found in the raw material. Therefore, the brightest (L* = 43.04) and reddest (a* = 14.07) puree was obtained from the Grimmes Golden apple variety, and the yellowest was obtained from the Domnesc variety, with a b* value of 20.98.

According to a study conducted on 1200 subjects, age is one of the factors that influence color preference. Generally, people prefer red when they are 1 to 10 years old [54].

For this reason, we tend to conclude that the purees from the Champion, and Elstar apple varieties could be more preferred by infants.

**Table 6.** Effect of thermal treatment on color.

| Puree Sample | Unsterilized Product Color | | | Sterilized Product Color | | |
|---|---|---|---|---|---|---|
| | L* | a* | b* | L* | a* | b* |
| 1 | 46.03 ± 0.74 [a] | 20.25 ± 1.9 [a] | 20.74 ± 1.81 [a] | 31.77 ± 1.19 [b] | 12.64 ± 0.29 [cde] | 16.15 ± 0.3 [de] |
| 2 | 43.56 ± 1.74 [a] | 21.2 ± 3.01 [a] | 19.72 ± 0.75 [a] | 38.31 ± 0.18 [ab] | 12.79 ± 0.09 [cd] | 15.64 ± 0.08 [ef] |
| 3 | 40.36 ± 0.38 [ab] | 15.21 ± 0.52 [cd] | 21.71 ± 1.44 [a] | 38.26 ± 0.33 [ab] | 12.4 ± 0.52 [cde] | 18.07 ± 0.34 [b] |
| 4 | 43.24 ± 0.56 [a] | 16.14 ± 0.71 [bcd] | 19.82 ± 0.52 [a] | 43.04 ± 3.88 [a] | 14.07 ± 0.15 [b] | 17.33 ± 0.29 [c] |
| 5 | 42.76 ± 0.61 [a] | 15.18 ± 0.59 [cd] | 20.16 ± 0.18 [a] | 40.97 ± 0.19 [ab] | 12.25 ± 0.11 [de] | 18.09 ± 0.3 [b] |
| 6 | 38.85 ± 0.59 [ab] | 15.33 ± 0.78 [cd] | 17.77 ± 0.75 [a] | 38.26 ± 2.11 [ab] | 11.51 ± 0.4 [f] | 14.26 ± 0.21 [g] |
| 7 | 32.72 ± 1.27 [b] | 15.18 ± 0.78 [cd] | 16.1 ± 0.83 [a] | 31.71 ± 0.16 [ab] | 12.99 ± 0.21 [c] | 16.47 ± 0.31 [d] |
| 8 | 36.88 ± 0.79 [ab] | 14.33 ± 0.73 [d] | 14.42 ± 0.42 [a] | 35.66 ± 0.12 [ab] | 12.01 ± 0.16 [ef] | 15.38 ± 0.26 [f] |
| 9 | 42.42 ± 2.13 [b] | 19.02 ± 1.27 [ab] | 21.41 ± 0.55 [a] | 42.35 ± 0.8 [a] | 14.9 ± 0.3 [a] | 20.98 ± 0.36 [a] |
| 10 | 36.58 ± 0.69 [ab] | 18.1 ± 1.73 [abc] | 17.3 ± 1.67 [a] | 38.06 ± 0.25 [ab] | 13.67 ± 0.09 [b] | 15.98 ± 0.41 [def] |

L* (brightness) represents how dark the sample is, ranging from 0 (black) to 100 (white); a* represents the color variation from green (−) to red (+); b* represents the color variation from blue (−) to yellow (+); [a–g]—mean values in columns indicated with various letters are significantly different. Puree sample column represents the apple variety used in order to obtain the puree: 1—Champion, 2—Elstar, 3—Baujade, 4—Grimmes Golden, 5—Granny Smith, 6—Pătul, 7—Reinette Harbert, 8—Juliana, 9—Domnesc, 10—Parmen Clone.

### 3.8. Texture Determination

After analyzing the textural parameters of the purees, we noticed a significant difference regarding the adhesiveness and firmness between unsterilized and sterilized products, while cohesiveness and deformation at hardness parameters suffered insignificant modifications.

Firmness was determined as the maximum penetration force registered for the first cycle. It was found that the firmness values of the purees obtained from the Champion and Juliana apple varieties before sterilization were the highest, having values of 2.46 N and 2.78 N, respectively. After sterilization, the puree containing the Champion apple variety registered a firmness of 3.5 N, while that containing the Juliana apple variety registered a value of 2.07 N. On the opposite side were the Domnesc varieties and Parmen Clone (Table 7). These results are due to the content of pectin in apples, which creates firm gels. Since some apple varieties contain lower levels of pectin compared to others, it induces a weaker texture of the puree samples.

Adhesiveness represents the work required to pull the probe away from the sample. Before sterilization, the highest adhesiveness was observed in the purees obtained from the Champion and Parmen Clone apple varieties, with values of 1.85 mJ and 2.3 mJ, while the Elstar apple variety had a much lower value of 0.35 mJ. After the sterilization process, the highest value remained for the same purees obtained from the Champion and Parmen Clone apple varieties, with 8.1 mJ and 5 mJ. This variation in adhesiveness is due to the variation in the moisture of the samples.

**Table 7.** Texture determination.

| Puree Sample | Unsterilized Product Texture | | | | Sterilized Product Texture | | | |
|---|---|---|---|---|---|---|---|---|
| | Cohesiveness (n.a) | Firmness (N) | Deformation at Hardness (mm) | Adhesiveness (mJ) | Cohesiveness (n.a.) | Firmness (N) | Deformation at Hardness (mm) | Adhesiveness (mJ) |
| 1 | 0.615 ± 0.049 [a] | 2.46 ± 0.39 [a] | 7.79 ± 0.03 [a] | 0.55 ± 0.07 [ab] | 0.32 ± 0.06 [a] | 3.49 ± 0.12 [a] | 7.97 ± 0.03 [a] | 8.1 ± 0.7 [a] |
| 2 | 0.619 ± 0.084 [a] | 2.08 ± 0.09 [a] | 7.985 ± 0.007 [a] | 0.35 ± 0.35 [b] | 0.355 ± 0.205 [a] | 1.49 ± 0.31 [bc] | 7.97 ± 0.03 [a] | 4.1 ± 0.9 [bc] |
| 3 | 0.425 ± 0.163 [a] | 1.055 ± 0.007 [a] | 7.68 ± 0.27 [a] | 1.85 ± 0.92 [ab] | 0.455 ± 0.106 [a] | 2.455 ± 0.332 [b] | 7.4 ± 0.8 [a] | 3.4 ± 0.7 [bc] |
| 4 | 0.585 ± 0.049 [a] | 1.88 ± 0.08 [a] | 7.945 ± 0.007 [a] | 0.55 ± 0.64 [ab] | 0.53 ± 0.27 [a] | 1.44 ± 0.12 [c] | 7.97 ± 0.03 [a] | 2.8 ± 0.4 [cd] |
| 5 | 0.55 ± 0.01 [a] | 1.99 ± 0.41 [a] | 7.97 ± 0.02 [a] | 0.7 ± 0.1 [ab] | 0.32 ± 0.01 [a] | 2.09 ± 0.29 [bc] | 7.97 ± 0.03 [a] | 3.55 ± 0.49 [bc] |
| 6 | 0.38 ± 0.05 [a] | 1.49 ± 0.15 [a] | 7.645 ± 0.431 [a] | 1.35 ± 0.49 [ab] | 0.475 ± 0.106 [a] | 1.635 ± 0.091 [bc] | 7.99 ± 0.01 [a] | 3.8 ± 0.3 [bc] |
| 7 | 0.525 ± 0.007 [a] | 1.5 ± 0.2 [a] | 7.99 ± 0.01 [a] | 0.45 ± 0.21 [ab] | 0.42 ± 0.01 [a] | 1.7 ± 0.2 [bc] | 7.995 ± 0.007 [a] | 1.95 ± 0.63 [cd] |
| 8 | 0.59 ± 0.05 [a] | 2.795 ± 1.393 [a] | 7.92 ± 0.01 [a] | 1.2 ± 0.57 [ab] | 0.41 ± 0.07 [a] | 2.07 ± 0.23 [bc] | 7.975 ± 0.035 [a] | 0.9 ± 0.1 [d] |
| 9 | 0.355 ± 0.106 [a] | 1.17 ± 0.21 [a] | 7.87 ± 0.11 [a] | 1.15 ± 0.07 [ab] | 0.405 ± 0.148 [a] | 1.375 ± 0.233 [c] | 7.862 ± 0.049 [a] | 2.85 ± 0.35 [bcd] |
| 10 | 0.411 ± 0.099 [a] | 2.2 ± 0.2 [a] | 7.795 ± 0.028 [a] | 2.3 ± 0.6 [a] | 0.585 ± 0.049 [a] | 1.74 ± 0.41 [bc] | 7.97 ± 0.03 [a] | 5.1 ± 0.1 [b] |

[a–d]—mean values in columns indicated with various letters are significantly different. Puree sample column represents the apple variety used in order to obtain the puree: 1—Champion, 2—Elstar, 3—Baujade, 4—Grimmes Golden, 5—Granny Smith, 6—Pătul, 7—Reinette Harbert, 8—Juliana, 9—Domnesc, 10—Parmen Clone.

Adhesiveness and firmness are the main parameters for baby foods [55,56]. Smooth baby foods and those with proper moisture have the highest acceptance rate by infants. This was also confirmed by researchers from Alberta Health Services and the Centre for Taste and Feeding Behavior, who conducted similar studies in 2017 and 2019, respectively [57,58]. The textural characteristics of the purees developed in the present study meet the textural properties required for baby food.

### 3.9. Chemometric Comparison and Classification

The PCA model was applied to vitamin C, TPC and AC data for unsterilized and sterilized purees to determine the most important variables that explain the relationships between the ten selected genotypes of apples used and to identify any group patterns (Figure 3). The results from the current study showed that using the chemometric technique PCA, the physico-chemical quality attributes can be more comprehensively understood. The results showed that with a relative distance of 10, all the genotypes were grouped into four main clusters.

Principal component analysis (PCA) uses the maximum variance of the data in order to transform the original axes (parameters) into new axes named principal components (PCs) or factors. The PCs are orthogonal, and the intercorrelations are removed. The graph representation of the translation coordinates provides the scores plot and reveals the similarity/dissimilarity between cases (the grouping of the samples). On the other hand, the graph representation of the cosines of the rotational angles provides the loadings plot, which reveals the importance of the original parameters to the classification.

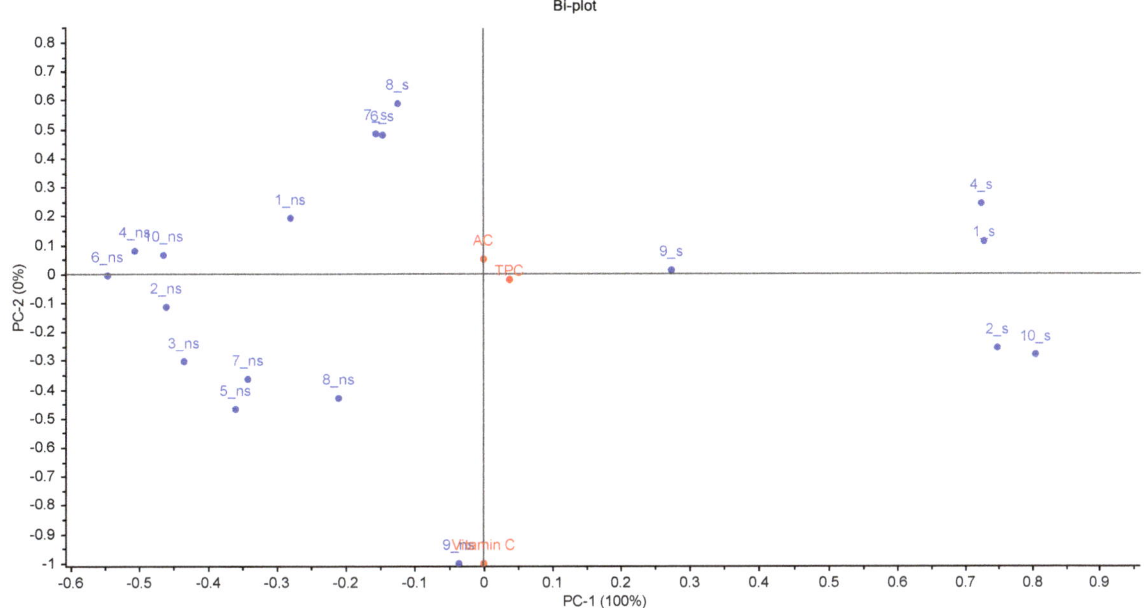



**Figure 3.** PCA Bi-plot for the sterilized (_s) and unsterilized (_ns) apple purees based on antioxidant capacity (AC), total phenolic compounds (TPC), and vitamin C. Numbers 1–10 represent the apple variety used in order to obtain the puree: 1—Champion, 2—Elstar, 3—Baujade, 4—Grimmes Golden, 5—Granny Smith, 6—Pătul, 7—Reinette Harbert, 8—Juliana, 9—Domnesc, 10—Parmen Clone.

## 4. Conclusions

Following the studies and determinations performed in this paper, we can conclude that a range of products for children obtained from apples, carrots, pumpkins and celery has been developed. In order to develop the product, ten experimental puree variants were developed, corresponding to the ten varieties of apples used. Although there are a number of similar products on the market, the current study also aimed to choose the right apple varieties for making a puree, which would ensure the proper functioning of the young body.

The final products were analyzed from a physico-chemical point of view, using methods of determination according to the standards in force, and a comparison was made between the apple varieties used to obtain the puree. Following the results obtained, the puree can be introduced into the diet of infants and children, providing them with the necessary bioactive compounds and minerals for optimal development.

The analyses performed on both the fresh and the sterilized products highlighted the effects of the heat treatment on the components of the product.

TPC content is directly related to the AC of baby food analyzed. This relationship indicates that the contribution of TPC to the free radical scavenging potential of baby food is significantly related. Having the results that show the high content of puree in polyphenol compounds and a high antioxidant capacity confirms the hypothesis that baby food, which is considered a convenient food, is a source of nutritional value, so it can be recommended for use and added effectively to children's diets.

Puree texture and color parameters were evaluated, these being two important factors in the perception of infants and children, being the ideal time to introduce the child to a variety of foods and flavors to develop healthy eating habits. Our product is very suitable in the parameters that correspond to baby food with the highest acceptance rate by infants, which proves the quality of the puree texture obtained in this research.

According to the present study, a suitable product for children based on Elstar apples, pumpkins, carrots and celery was produced. Baby puree obtained from the Elstar apple variety was, as the physico-chemical methods showed, the most suitable due to its high content of vitamin C, 3.27 mg/100 g, antioxidant capacity and polyphenol content in the sterilized product, 139.67 mg GAE/100 g, and 0.86 AC/100 g, respectively. Regarding the color and texture determinations, it has also been proved that the puree from this type of apple also fits in the parameters. We believe that this branch should be studied more, due to children's nutrition being an important factor in their development and fruits and vegetables being a major source of nutrients that play a key role in achieving this goal.

**Author Contributions:** Conceptualization, A.M.M. and A.E.M.; methodology, A.M.M. and A.E.M.; software, A.M.M., A.P., R.E.S. and A.E.M.; validation, A.M.M. and A.E.M.; formal analysis, A.M.M., R.E.S., V.M. and A.E.M.; investigation, A.M.M., A.P. and A.E.M.; resources, A.M.M., S.M., R.E.S. and A.E.M.; data curation, V.M.; writing—original draft preparation, A.M.M.; writing—review and editing, A.M.M. and A.E.M.; visualization, A.M.M. and A.E.M.; supervision, A.E.M. and S.M.; project administration, S.M. and R.E.S.; funding acquisition, A.E.M., R.E.S. and S.M. All authors have read and agreed to the published version of the manuscript.

**Funding:** This work was supported by a grant from the Ministry of Research, Innovation and Digitization, CNCS/CCCDI—UEFISCDI, project number PN-III-P1-1.1-PD-2019-1108, within PNCDI III.

**Institutional Review Board Statement:** Not applicable.

**Informed Consent Statement:** Not applicable.

**Conflicts of Interest:** The authors declare no conflict of interest. The funders had no role in the design of the study; in the collection, analyses, or interpretation of the data; in the writing of the manuscript, or in the decision to publish the results. The author A.M.M. (Alexandra Mădălina Mateescu) is an employee of MDPI; however, she is not working for the journal Applied Sciences at the time of submission and publication.

# References


1. Dhami, M.; Ogbo, F.; Diallo, T.; Olusanya, B.; Goson, P.; Agho, K.; on behalf of the Global Maternal and Child Health Research Collaboration (GloMACH). Infant and Young Child Feeding Practices among Adolescent Mothers and Associated Factors in India. *Nutrients* **2021**, *13*, 2376. [CrossRef] [PubMed]
2. Jiménez, M.D.; Lobo, M.O.; Sammán, N.C. Technological and Sensory Properties of Baby Purees Formulated with Andean Grains and Dried with Different Methods. *Proceedings* **2020**, *53*, 13. [CrossRef]
3. Hondru, G.; Laillou, A.; Wieringa, F.T.; Poirot, E.; Berger, J.; Christensen, D.L.; Roos, N. Age-Appropriate Feeding Practices in Cambodia and the Possible Influence on the Growth of the Children: A Longitudinal Study. *Nutrients* **2019**, *12*, 12. [CrossRef]
4. Elliott, C. Tracking Kids' Food: Comparing the Nutritional Value and Marketing Appeals of Child-Targeted Supermarket Products Over Time. *Nutrients* **2019**, *11*, 1850. [CrossRef] [PubMed]
5. Padarath, S.; Gerritsen, S.; Mackay, S. Nutritional Aspects of Commercially Available Complementary Foods in New Zealand Supermarkets. *Nutrients* **2020**, *12*, 2980. [CrossRef] [PubMed]
6. Mörkl, S.; Stell, L.; Buhai, D.; Schweinzer, M.; Wagner-Skacel, J.; Vajda, C.; Lackner, S.; Bengesser, S.; Lahousen, T.; Painold, A.; et al. 'An Apple a Day'?: Psychiatrists, Psychologists and Psychotherapists Report Poor Literacy for Nutritional Medicine: International Survey Spanning 52 Countries. *Nutrients* **2021**, *13*, 822. [CrossRef]
7. Johansson, U.; Öhlund, I.; Hernell, O.; Lönnerdal, B.; Lindberg, L.; Lind, T. Protein-Reduced Complementary Foods Based on Nordic Ingredients Combined with Systematic Introduction of Taste Portions Increase Intake of Fruits and Vegetables in 9 Month Old Infants: A Randomised Controlled Trial. *Nutrients* **2019**, *11*, 1255. [CrossRef]
8. Bernal, M.; Roman, S.; Klerks, M.; Haro-Vicente, J.; Sanchez-Siles, L. Are Homemade and Commercial Infant Foods Different? A Nutritional Profile and Food Variety Analysis in Spain. *Nutrients* **2021**, *13*, 777. [CrossRef]
9. Savarino, G.; Corsello, A.; Corsello, G. Macronutrient balance and micronutrient amounts through growth and development. *Ital. J. Pediatr.* **2021**, *47*, 1–14. [CrossRef]
10. Patel, J.K.; Rouster, A.S. Infant Nutrition Requirements and Options. In *StatPearls*; StatPearls Publishing: Treasure Island, FL, USA, 2021. Available online: https://www.ncbi.nlm.nih.gov/books/NBK560758/ (accessed on 7 May 2022).
11. Katiforis, I.; Fleming, E.; Haszard, J.; Hape-Cramond, T.; Taylor, R.; Heath, A.-L. Energy, Sugars, Iron, and Vitamin B12 Content of Commercial Infant Food Pouches and Other Commercial Infant Foods on the New Zealand Market. *Nutrients* **2021**, *13*, 657. [CrossRef]
12. Westland, S.; Crawley, H. Fruit and Vegetable Based Purées in Pouches for Infants and Young Children. Available online: https://www.firststepsnutrition.org (accessed on 27 October 2021).

13. Birch, L.; Savage, J.S.; Ventura, A. Influences on the Development of Children's Eating Behaviours: From Infancy to Adolescence. *Can. J. Diet. Pract. Res.* **2007**, *68*, s1–s56.
14. Dai, H.; Leung, C.E.; Corradini, M.; Xiao, H.; Kinchla, A.J. Increasing the nutritional value of strawberry puree by adding xylo-oligosaccharides. *Heliyon* **2020**, *6*, e03769. [CrossRef]
15. Al-Ghamdi, S.; Sonar, C.R.; Patel, J.; Albahr, Z.; Sablani, S.S. High pressure-assisted thermal sterilization of low-acid fruit and vegetable purees: Microbial safety, nutrient, quality, and packaging evaluation. *Food Control* **2020**, *114*, 107233. [CrossRef]
16. Gratz, M.; Sevenich, R.; Hoppe, T.; Schottroff, F.; Vlaskovic, N.; Belkova, B.; Chytilova, L.; Filatova, M.; Stupak, M.; Hajslova, J.; et al. Gentle Sterilization of Carrot-Based Purees by High-Pressure Thermal Sterilization and Ohmic Heating and Influence on Food Processing Contaminants and Quality Attributes. *Front. Nutr.* **2021**, *8*, 643837. [CrossRef]
17. Sevenich, R.; Kleinstueck, E.; Crews, C.; Anderson, W.; Pye, C.; Riddellova, K.; Hradecký, J.; Moravcova, E.; Reineke, K.; Knorr, D. High-Pressure Thermal Sterilization: Food Safety and Food Quality of Baby Food Puree. *J. Food Sci.* **2014**, *79*, M230–M237. [CrossRef]
18. AOAC Official Method 934.06 Moisture in Dried Fruits. Available online: https://dokumen.tips/documents/aoac-official-method-93406-moisture-in-dried-fruits.html (accessed on 27 October 2021).
19. Nirmaan, A.M.C.; Prasantha, B.D.R.; Peiris, B.L. Comparison of microwave drying and oven-drying techniques for moisture determination of three paddy (*Oryza sativa* L.) varieties. *Chem. Biol. Technol. Agric.* **2020**, *7*, 1–7. [CrossRef]
20. Zhao, W.-Z.; Cao, P.-P.; Zhu, Y.-Y.; Liu, S.; Gao, H.-W.; Huang, C.-Q. Rapid Detection of vitamin C content in fruits and vegetables using a digital camera and color reaction. *Quim. Nova* **2020**, *43*, 1421–1430. [CrossRef]
21. Carmona-Hernandez, J.C.; Taborda-Ocampo, G.; González-Correa, C.H. Folin-Ciocalteu Reaction Alternatives for Higher Polyphenol Quantitation in Colombian Passion Fruits. *Int. J. Food Sci.* **2021**, *2021*, 1–10. [CrossRef]
22. Dibacto, R.E.K.; Tchuente, B.R.T.; Nguedjo, M.W.; Tientcheu, Y.M.T.; Nyobe, E.C.; Edoun, F.L.E.; Kamini, M.F.G.; Dibanda, R.F.; Medoua, G.N. Total Polyphenol and Flavonoid Content and Antioxidant Capacity of Some Varieties of Persea americana Peels Consumed in Cameroon. *Sci. World J.* **2021**, *2021*, 1–11. [CrossRef]
23. Thaipong, K.; Boonprakob, U.; Crosby, K.; Cisneros-Zevallos, L.; Hawkins Byrne, D. Comparison of ABTS, DPPH, FRAP, and ORAC assays for estimating antioxidant activity from guava fruit extracts. *J. Food Compos. Anal.* **2006**, *19*, 669–675. [CrossRef]
24. Mureşan, A.E.; Sestras, A.F.; Militaru, M.; Păucean, A.; Tanislav, A.E.; Puşcaş, A.; Mateescu, M.; Mureşan, V.; Vlaic, R.A.M.; Sestras, R.E. Chemometric Comparison and Classification of 22 Apple Genotypes Based on Texture Analysis and Physico-Chemical Quality Attributes. *Horticulturae* **2022**, *8*, 64. [CrossRef]
25. Markovic, I.; Ilic, J.; Markovic, D.; Simonovic, V.; Kosanic, N. Color Measurement of Food Products Using CIE L*a*b* and RGB Color Space. *J. Hyg. Eng. Des.* **2013**, *4*, 50–53.
26. ISO 11036:2020(En), Sensory Analysis—Methodology—Texture Profile. Available online: https://www.iso.org/obp/ui/#iso:std: iso:11036:ed-2:v1:en (accessed on 27 October 2021).
27. Malavi, D.; Mbogo, D.; Moyo, M.; Mwaura, L.; Low, J.; Muzhingi, T. Effect of Orange-Fleshed Sweet Potato Purée and Wheat Flour Blends on β-Carotene, Selected Physicochemical and Microbiological Properties of Bread. *Foods* **2022**, *11*, 1051. [CrossRef] [PubMed]
28. Park, H.-S.; Chung, H.-S. Evaluation of the physicochemical properties of starch isolated from thinned young 'Fuji' apples compared to corn and potato starches. *Korean J. Food Preserv.* **2021**, *28*, 501–509. [CrossRef]
29. Doerflinger, F.C.; Miller, W.B.; Nock, J.F.; Watkins, C.B. Variations in zonal fruit starch concentrations of apples—A developmental phenomenon or an indication of ripening? *Hortic. Res.* **2015**, *2*, 15047. [CrossRef]
30. Carrín, M. Characterization of starch in apple juice and its degradation by amylases. *Food Chem.* **2004**, *87*, 173–178. [CrossRef]
31. Seidel, K.; Kahl, J.; Paoletti, F.; Birlouez, I.; Busscher, N.; Kretzschmar, U.; Särkkä-Tirkkonen, M.; Seljåsen, R.; Sinesio, F.; Torp, T.; et al. Quality assessment of baby food made of different pre-processed organic raw materials under industrial processing conditions. *J. Food Sci. Technol.* **2013**, *52*, 803–812. [CrossRef]
32. Asiimwe, A.; Kigozi, J.; Baidhe, E.; Muyonga, J. Optimization of refractance window drying conditions for passion fruit puree. *LWT* **2021**, *154*, 112742. [CrossRef]
33. Alzamora, S.M.; Chirife, J. The Water Activity of Canned Foods. *J. Food Sci.* **1983**, *48*, 1385–1387. [CrossRef]
34. Nor, N.Z.N.M.; Kormin, F.; Fuzi, S.F.Z.M.; Bin Abu Bakar, M.A.L. Comparison of Physicochemical, Antioxidant Properties and Sensory Acceptance of Puree from Tamarillo and Tomato. *J. Sci. Technol.* **2018**, *10*, 25–31. [CrossRef]
35. Summo, C.; De Angelis, D.; Rochette, I.; Mouquet-Rivier, C.; Pasqualone, A. Influence of the preparation process on the chemical composition and nutritional value of canned purée of kabuli and Apulian black chickpeas. *Heliyon* **2019**, *5*, e01361. [CrossRef] [PubMed]
36. Ettinger, L.; Keller, H.H.; Duizer, L.M. Characterizing Commercial Pureed Foods: Sensory, Nutritional, and Textural Analysis. *J. Nutr. Gerontol. Geriatr.* **2014**, *33*, 179–197. [CrossRef] [PubMed]
37. Cheng, X.-F.; Zhang, M.; Adhikari, B. Changes in Quality Attributes of Strawberry Purees Processed by Power Ultrasound or Thermal Treatments. *Food Sci. Technol. Res.* **2014**, *20*, 1033–1041. [CrossRef]
38. Usal, M.; Sahan, Y. In vitro evaluation of the bioaccessibility of antioxidative properties in commercially baby foods. *J. Food Sci. Technol.* **2020**, *57*, 3493–3501. [CrossRef]
39. Giannakourou, M.C.; Taoukis, P.S. Effect of Alternative Preservation Steps and Storage on Vitamin C Stability in Fruit and Vegetable Products: Critical Review and Kinetic Modelling Approaches. *Foods* **2021**, *10*, 2630. [CrossRef]

40. de Sousa, T.L.; da Silva, J.P.; Lodete, A.R.; Lima, D.S.; Mesquita, A.A.; de Almeida, A.B.; Placido, G.R.; Egea, M.B. Vitamin C, phenolic compounds and antioxidant activity of Brazilian baby foods. *Nutr. Food Sci.* **2020**, *51*, 725–737. [CrossRef]
41. Carbonell-Capella, J.M.; Barba, F.J.; Esteve, M.J.; Frígola, A. Quality parameters, bioactive compounds and their correlation with antioxidant capacity of commercial fruit-based baby foods. *Food Sci. Technol. Int.* **2013**, *20*, 479–487. [CrossRef]
42. Arfaoui, L. Dietary Plant Polyphenols: Effects of Food Processing on Their Content and Bioavailability. *Molecules* **2021**, *26*, 2959. [CrossRef]
43. Mongkontanawat, N.; Ueda, Y.; Yasuda, S. Increased total polyphenol content, antioxidant capacity and γ-aminobutyric acid content of roasted germinated native Thai black rice and its microstructure. *Food Sci. Technol.* **2022**, *42*. [CrossRef]
44. Wołosiak, R.; Drużyńska, B.; Piecyk, M.; Majewska, E.; Worobiej, E. Effect of Sterilization Process and Storage on the Antioxidative Properties of Runner Bean. *Molecules* **2018**, *23*, 1409. [CrossRef]
45. Chumyam, A.; Whangchai, K.; Jungklang, J.; Faiyue, B.; Saengnil, K. Effects of heat treatments on antioxidant capacity and total phenolic content of four cultivars of purple skin eggplants. *ScienceAsia* **2013**, *39*, 246–251. [CrossRef]
46. Choi, Y.; Lee, S.M.; Chun, J.; Lee, H.B.; Lee, J. Influence of heat treatment on the antioxidant activities and polyphenolic compounds of shiitake (Lentinus edodes) mushroom. *Food Chem.* **2006**, *99*, 381–387. [CrossRef]
47. Čížková, H.; Ševčík, R.; Rajchl, A.; Voldrich, M. Nutritional Quality of Commercial Fruit Baby Food. *Czech J. Food Sci.* **2009**, *27*, S134–S137. [CrossRef]
48. Huang, W.-Y.; Zhang, H.-C.; Liu, W.-X.; Li, C.-Y. Survey of antioxidant capacity and phenolic composition of blueberry, blackberry, and strawberry in Nanjing. *J. Zhejiang Univ. Sci.* **2012**, *13*, 94–102. [CrossRef] [PubMed]
49. De Moura, S.C.S.R.; Tavares, P.E.D.R.; Germer, S.P.M.; Nisida, A.L.A.C.; Alves, A.B.; Kanaan, A.S. Degradation Kinetics of Anthocyanin of Traditional and Low-Sugar Blackberry Jam. *Food Bioprocess Technol.* **2011**, *5*, 2488–2496. [CrossRef]
50. Särkkä-Tirkkonen, M.; Väisänen, H.-M.; Beck, A.; Kretzschmar, U.; Seidel, K. *Overview on Different Sterilization Techniques for Baby Food*; University of Helsinki: Helsinki, Finland, 2010.
51. Ahmed, M.A.; Al-Khalifa, A.S.; Al-Nouri, D.M.; El-Din, M.F.S. Dietary intake of artificial food color additives containing food products by school-going children. *Saudi J. Biol. Sci.* **2020**, *28*, 27–34. [CrossRef]
52. Stevens, L.J.; Burgess, J.R.; Stochelski, M.A.; Kuczek, T. Amounts of Artificial Food Dyes and Added Sugars in Foods and Sweets Commonly Consumed by Children. *Clin. Pediatr.* **2014**, *54*, 309–321. [CrossRef]
53. Environmental Health Hazard Assessment, Office. Health Effects Assessment Potential Neurobehavioral Effects of Synthetic Food Dyes in Children. 2021. Available online: https://oehha.ca.gov/risk-assessment/report/health-effects-assessment-potential-neurobehavioral-effects-synthetic-food (accessed on 13 May 2022).
54. Psicodinâmica Das Cores Em Comunicação—EDIÇAO REVISTA E AMPLIADA I Nebraska, F.—Academia.Edu. Available online: https://www.academia.edu/40122222/Psicodin%E2mica_das_Cores_em_Comunica%E7%E3o_-_EDI%C7%C3O_REVISTA_E_AMPLIADA (accessed on 17 June 2022).
55. Cappellotto, M.; Olsen, A. Food Texture Acceptance, Sensory Sensitivity, and Food Neophobia in Children and Their Parents. *Foods* **2021**, *10*, 2327. [CrossRef]
56. Jasim, A.; Ramaswamy, H.S. Viscoelastic and thermal characteristics of vegetable puree-based baby foods. *J. Food Process Eng.* **2006**, *29*, 219–233. [CrossRef]
57. Demonteil, L.; Tournier, C.; Marduel, A.; Dusoulier, M.; Weenen, H.; Nicklaus, S. Longitudinal study on acceptance of food textures between 6 and 18 months. *Food Qual. Prefer.* **2019**, *71*, 54–65. [CrossRef]
58. Alberta, H.S. Food Textures for Children Developed by Registered Dietitians Nutrition Services 404107-NFS. Available online: https://peas.albertahealthservices.ca/Uploads/Food%20Textures%20for%20Children.pdf (accessed on 13 May 2022).

 applied sciences

*Article*

# Natural Gum from Flaxseed By-Product as a Potential Stabilizing and Thickening Agent for Acid Whey Fermented Beverages

Łukasz Łopusiewicz [1,*], Izabela Dmytrów [2,*], Anna Mituniewicz-Małek [2], Paweł Kwiatkowski [3], Edward Kowalczyk [4], Monika Sienkiewicz [5] and Emilia Drozłowska [1]

[1]  Center of Bioimmobilisation and Innovative Packaging Materials, Faculty of Food Sciences and Fisheries, West Pomeranian University of Technology Szczecin, Janickiego 35, 71-270 Szczecin, Poland
[2]  Department of Toxicology, Dairy Technology and Food Storage, Faculty of Food Science and Fisheries, West Pomeranian University of Technology in Szczecin, Papieża Pawła VI Street 3, 71-459 Szczecin, Poland
[3]  Department of Diagnostic Immunology, Pomeranian Medical University in Szczecin, Powstańców Wielkopolskich 72, 70-111 Szczecin, Poland
[4]  Department of Pharmacology and Toxicology, Medical University of Łódź, 90-752 Łódź, Poland
[5]  Department of Pharmaceutical Microbiology and Microbiological Diagnostic, Medical University of Lodz, Muszynskiego 1, 90-151 Łódź, Poland
*  Correspondence: lukasz.lopusiewicz@zut.edu.pl (Ł.Ł.); izabela.dmytrow@zut.edu.pl (I.D.); Tel.: +48-91-449-6135 (Ł.Ł.); +48-91-449-6500 (I.D.)

Citation: Łopusiewicz, Ł.; Dmytrów, I.; Mituniewicz-Małek, A.; Kwiatkowski, P.; Kowalczyk, E.; Sienkiewicz, M.; Drozłowska, E. Natural Gum from Flaxseed By-Product as a Potential Stabilizing and Thickening Agent for Acid Whey Fermented Beverages. *Appl. Sci.* **2022**, *12*, 10281. https://doi.org/10.3390/app122010281

Academic Editors: Joanna Trafialek and Mariusz Szymczak

Received: 16 September 2022
Accepted: 10 October 2022
Published: 12 October 2022



**Abstract:** The valorization of food industry by-products is still a major challenge. Here, we report the production of acid whey fermented beverages stabilized with flaxseed gum (derived from oil industry by-product). Four variants of drinks were prepared: (1) fermented whey (W), (2) fermented whey with milk powder added (5% $w/v$) (WMP), (3) fermented whey with flaxseed gum added (0.5% $w/v$) (WFG1) and (4) fermented whey with flaxseed gum added (1.0 % $w/v$) (WFG2). The beverages were kept in refrigerated conditions (5 ± 1 °C) for 28 days. Alterations in lactic acid bacteria population, pH, titratable acidity, water activity, syneresis, viscosity, acetaldehyde content, color, consumer acceptance, bio-active compounds and antioxidant activity were identified. The findings revealed that flaxseed gum addition significantly enhanced bacteria survivability and improved the viscosity of acid whey at a level comparable with milk powder, meeting consumer acceptance criteria. The beverages were characterized by normative physicochemical properties and showed high antioxidant activity and free amino acids level. The use of valuable by-products from the dairy and oil industries opens up a promising route for the production of innovative beverages, which is in accordance with the principles of circular economy and the idea of zero waste.

**Keywords:** by-products; flaxseed; whey; natural thickeners; fermentation; zero waste

## 1. Introduction

Whey is a by-product of the dairy industry, produced in vast quantities and obtained during the production of rennet cheeses, cottage and acid curd cheeses (tvorogs) as well as milk protein preparations [1–4]. It is estimated that the processing of 1 kg of cheese produces ca. 9 L of whey [1,2,5–7]. It is a yellow-greenish liquid, which contains about 45–55% milk solids, including lactose, whey proteins and mineral compounds [6,7]. The fat content of whey is low, and the casein content, which forms the cheese curd, is negligible. The total solid content in whey is approximately 6–7% [1–3,6]. Whey comprises a significant number of nutritional values and innumerable health benefits, as it is a source of valuable proteins, bioactive peptides, amino acids, vitamins (such as folic acid, cobalamin and riboflavin), as well as minerals [1,6,8]. This product has a high biochemical and chemical oxygen demand, so discharging whey with wastewater directly into the water bodies degrades natural ecosystems. It must be disposed of and processed. The processing of whey consists mainly of the use of membrane processes, such as reverse osmosis, nanofiltration and ultrafiltration, and drying processes, in order to obtain the concentrates and isolates of whey proteins, as well as powdered whey, which are used in many branches of the food industry. These products are obtained primarily from the so-called "sweet whey" obtained when rennet cheeses are produced. The management of acid whey (AW), which is a by-product in the production of acid curd cheese (tvorog), is much more difficult due to its low pH, lower lactose content, higher mineral content and a more sour and distinct taste compared to sweet whey [1,2,6,9]. AW presents less favorable processing properties compared with rennet whey, and it presents a sustainability challenge to the dairy industry [1,6,9,10]. Unprocessed whey is also not attractive to consumers because of its sensory properties [1,6]; therefore, it seems promising to use it as a raw material for the production of unfermented and fermented beverages [1,2,11]. Various whey-based beverages are being successfully developed around the world, which include fermented, unfermented, alcoholic, non-alcoholic, carbonated, plain and fruit-flavored products [10–13]. Whey fermentation leads to advantages, such as a partial hydrolysis of whey protein (which may cause allergies), increase in shelf life (due to lactic acid production) and the production of volatile compounds that improve the sensory features [2,8]. On the other hand, the production of whey-based fermented beverages becomes a challenge due to the high water content, and therefore, the low dry matter content, and consequently, the watery consistency of the final product [2,5,14,15]. The texture and mouthfeel of fermented whey beverages (FWB) tend to be weak and watery compared to fermented milk, since liquid whey contains a low percentage of total solids (ca. 6 g/100 g) [16]. Until now, the evaporation, centrifugation or ultrafiltration processes have been used to increase the dry matter of milk intended for the production of fermented milk. An alternative to the above-mentioned processes is adding, among others, milk powder, whey powder, whey protein concentrates, casein and caseinates. The addition of milk powder or whey, however, significantly increases the lactose content in the product, which is not well perceived by consumers who have problems with this sugar intolerance [17]. Additives of non-dairy origin were also used in the past, e.g., modified starch. The current trends in creating new products are directed toward "clean label". Therefore, the most commonly used structure-forming compounds are substances of natural origin. This has created a requirement for using exopolysaccharide-producing starter cultures or the addition of hydrocolloids [12]. Hydrocolloids are defined as hydrophilic molecules, which have a high molecular weight. Their role in food products is the formation of texture, water and oil binding, as well as stabilization [18].

Against this background, more attention has been paid to the by-products and co-products of flaxseed (*Linum usitatissimum* L.) [19]. Due to its high content of bioactive compounds, it has been categorized as a functional or bioactive food [20]. The mucilage obtained from flaxseed seeds contains L-galactose, D-xylose, L-arabinose, L-rhamnose and D-galacturonic acid, and it demonstrates a high ability to hold water and bind oil, which is important when stabilizing the food systems [13,21–25]. The polysaccharides that make up the mucilage are responsible for the formation of a multi-form structure and better resistance to environmental stresses. From a nutritional perspective, flaxseed mucilage is a fine source of dietary fiber and has many health benefits, such as the prevention of diabetes, obesity or colon cancer, decreasing blood cholesterol levels and improving insulin sensitivity. Flaxseed polysaccharides are called flaxseed gum (FG), as the main component of soluble dietary fiber in flaxseed (3–9 wt% of the flaxseed) [16,22]. FG is composed of 80% neutral and acidic polysaccharides and 9% protein. Neutral monosaccharides consist of xylose, glucose, arabinose and galactose, while acidic monosaccharides consist of rhamnose, galactose, fucose and galacturonic acid [16,20,25]. The protein in FG is predominantly conlinin [23]. FG is most commonly used as a thickener, stabilizer, gelling agent and emulsifier [16,22,23,25]. From a functional point of view, FG more closely resembles gum Arabic or guar gums than other commonly used gums and can be used to replace most non-gelling gums in food and non-food applications due to its "weak gel"-like properties and remarkable ability to retain water [16,20,25–29]. FG is a promising new dairy ingredient, which acts, for instance, as a natural stabilizer in stirred yogurts [24]. However, very little is understood regarding its effect when added to dairy by-products. Moreover, FG also has the advantage of relatively low cost compared with most commercial gums [28]. Hence, based on the technological characteristics of FG, it can be considered that it can potentially be used as a stabilizer and thickener for acid whey. It was reported that FG can be obtained from flaxseed oil cake (FOCE)—a cheap by-product from flaxseed oil production [25,30]. This is particularly meaningful, since a large amount of press cakes and residues are available, and their management meets the criteria of a "zero waste" economy [21].

To the best of our knowledge, there are no reports of studies on the use of FG as a natural stabilizer for fermented beverages based on acid whey. We hypothesized that the application of FG as a natural stabilizer would allow obtaining new products with high added value. Hence, the goal of the research was to produce fermented beverages from acid whey with various FG concentrations and evaluate their microbiological, physicochemical properties, as well as bioactivity and sensory acceptance during refrigerated storage ($5 \pm 1\ ^{\circ}\mathrm{C}$) for 28 days.

## 2. Materials and Methods

### 2.1. Materials and Reagents

Flaxseed oil cake (FOC) was procured from ACS Sp. z o.o. (Bydgoszcz, Poland). Sodium hydroxide, disodium phosphate, monosodium phosphate, 2,2-diphenyl-1-picrylhydrazyl (DPPH), 2,2′-azino-bis(3-ethylbenzothiazoline-6-sulfonic acid) (ABTS), potassium persulphate, methanol, ethanol, hydrochloric acid, phenolphthalein, sodium chloride, 3,5-dinitrosalicylic acid, sodium tartate tetrahydrate, glucose, iron chloride, glycine, ninhydrin, glacial acetic acid, sodium acetate and cadmium chloride were purchased from Merck (Darmstad, Germany). Sulphuric acid 96%, dimethylsulfoxide, acetone, sodium tungstate dihydrate and 3-methyl-2-benzothiazolinone hydrazinone hydrochloride hydrate were obtained from P.P.H. Stanlab Sp. J. (Lublin, Poland). All reagents were of analytical grade. MRS (de Man, Rogosa and Sharpe) agar was obtained from Merck (Darmstad, Germany).

### 2.2. Acid Whey Production

Acid whey (AW) was derived from the production of acid curd cheese (ACC). Full-fat pasteurized ($85\ ^{\circ}\mathrm{C}$, 15 s) and homogenized (15 MPa, $55\ ^{\circ}\mathrm{C}$) cow's milk (containing 3.2% fat, 3.1% protein and 4.5% lactose) was procured at a local market. ACC was produced with traditional technology [31], using a freeze-dried DVS starter to direct the inoculation of milk containing *Lactococcus lactis* ssp. *lactis* and *Lactococcus lactis* ssp. *cremoris* (Lactoferm MSO Cheese-Tek®, Biochem s.r.l., Monterotondo (Rome), Italy). The production of ACC in laboratory conditions started with heating the milk to $23\ ^{\circ}\mathrm{C}$ followed by the addition of 2.5% ($v/v$) of the activated starter. The inoculated milk was incubated ($23\ ^{\circ}\mathrm{C}$, 12 h) until the curd reached pH 4.5. The curd was gently heated to separate it from the walls of the cheese tub; then, it was cut into cuboidal shapes with dimensions of approx. $120 \times 120$ mm. The curd was then gently stirred and gradually heated ($1\ ^{\circ}\mathrm{C}/10$ min) to $40\ ^{\circ}\mathrm{C}$ in order to intensify the separation of whey. The curd mass was divided into disposable, polyethylene cheese cloths and allowed to drain. Subsequently, the obtained ACCs were pressed with a laboratory press for 45 min (1 kg per 1 kg of cheese) [31].

### 2.3. Flaxseed Gum Preparation

Flaxseed gum (FG) was obtained from the flaxseed oil cake extract (FOCE) produced from FOC, as described elsewhere [26]. In the first step, flaxseed protein was precipitated by 0.1 M HCl from FOCE at flaxseed protein isoelectric point (4.2); then, the crude flaxseed gum was extracted from FOCE with 96% ethanol at a ratio 1:1. The obtained pellet was washed with distilled water at a ratio 1:1 and centrifuged for 10 min at 5000 rpm. The supernatant was carefully decanted, and FG was dried for 12 h at $55\ ^{\circ}\mathrm{C}$, then pulverized.

*2.4. Fermented Whey Beverages Production*

The fermented whey beverages (FWB) were produced on the day of AW production. Whey was submitted to a thermal treatment at 72 °C for 10 min and cooled to 20 °C. Four variants of FWB were prepared: (i) fermented whey (W), (ii) fermented whey with milk powder added (5% $w/v$) (WMP), (iii) fermented whey with flaxseed gum added (0.5% $w/v$) (WFG1) and (iv) fermented whey with flaxseed gum added (1.0% $w/v$) (WFG2). The levels of FG addition were based on preliminary trials, and the intention was to obtain FWB with a consistency similar to a drinkable yogurt. To increase the dry matter content in the WMP variant, skim milk powder (MP) with 0.2% fat, 35% protein and 96% total solids was used (SM Gostyń, Gostyń, Poland). In the case of the WFG1 and WFG2 variants, the previously prepared FG was added to the whey in amounts of 0.5% and 1.0%, respectively. Whey (W), whey with milk powder (WMP) and whey with both concentrations of FG (WFG1 and WFG2) were heated to 40 °C and inoculated with 0.6 g/L of direct vat set starter cultures of *Streptococcus salivarius* ssp. *thermophilus*, *Lactobacillus delbrueckii* ssp. *bulgaricus* (Lactoferm YO 122 YogurtTek®, Biochem s.r.l., Italy). Then, the samples were poured into sterile, low-density polyethylene cups (50 mL capacity), tightly sealed and incubated at $42 \pm 1$ °C for 7 h. After incubation, the samples were cooled and stored at $5 \pm 1$ °C in the dark for 28 days.

*2.5. Fermented Whey Beverages Characterization*

2.5.1. Microbial Analyses

Microbial analyses were performed as described elsewhere [32]. Briefly, a series of dilutions of the samples were prepared with physiological saline (0.85%), and the lactic acid bacteria counts were enumerated in triplicate on MRS medium (Merck, Darmstad, Germany) after incubation at 37 °C under anaerobic conditions for 72 h. The viable bacteria counts were expressed as CFU/mL of the samples.

2.5.2. Determination of Titratable Acidity, pH, Water Activity and Acetaldehyde Content

The following physicochemical properties of the FWB were evaluated: titratable acidity (% $w/w$ lactic acid) was assessed by titration with 0.25 N NaOH using phenolphthalein as an indicator [33,34]; pH was determined using a pH meter (CP-411, Elmetron, Zabrze, Poland); water activity ($a_w$) was measured at a temperature of 23 °C with a HygroLab C1 hygrometer (Rotronic, Bassersdorf, Switzerland). Acetaldehyde content was determined using a diffusive method [2].

2.5.3. Viscosity, Syneresis and Particle Size Measurements

Viscosity measurements were carried out with a rheometer (AR G2, TA Instruments Ltd., Lukens Drive, New Castle, USA). The temperature was kept at 20 °C during the measurements. Flow experiments were performed with a stainless-steel cone plate of 62 mm diameter at 60° for 50 s$^{-1}$. Data from rheological measurements were collected using TA Rheology Advantage Data Analysis equipment software V 5.4.7.

Syneresis was examined for all variants during the storage according to Yu et al. [35], with some modifications. A 10 mL aliquot of the sample was poured into 15 mL Falcon tubes and centrifuged at 890 rpm for 10 min at 4 °C, followed by reading the amount of the clear supernatant. Syneresis (%) was expressed as the percentage volume of the supernatant over the initial volume of the sample.

Particle size distribution measurements of the samples were performed using a Mastersizer 2000 (Malvern Instrument Ltd., Worcestershire, UK) [15]. The samples were dispersed in distilled water (stirring speed—2000 rpm) until an obscuration rate of 10% was obtained. The optical properties of the sample were defined as follows: refractive index 1.500 and absorption 1.00.

### 2.5.4. Color Analysis

The color of FWB was assessed with the objective method using colorimeter WR 18 (FRU®, Shenzhen Wave Optoelectronics Technology Co., Ltd.) [36] based on the white standard plate (L* = 92.4; a* = −0.04; b* = +1/9) and CIE L*a*b*, illuminant D65, observer 10°, illumination mode d/8 and caliber 8 mm. The total color difference (ΔE), hue (h) and (C*) chroma of the color were calculated based on the following formulae:

$$\mathbf{C}^* = \sqrt{\mathbf{a}^2 + \mathbf{b}^2}$$

$$\mathbf{H}^0 = \tan^{-1}\left(\frac{\mathbf{b}}{\mathbf{a}}\right)$$

$$\Delta\mathbf{E} = \left[\left(\mathbf{L}_{\text{standard}} - \mathbf{L}_{\text{sample}}\right)^2 + \left(\mathbf{a}_{\text{standard}} - \mathbf{a}_{\text{sample}}\right)^2 + \left(\mathbf{b}_{\text{standard}} - \mathbf{b}_{\text{sample}}\right)^2\right]^{0.5}$$

### 2.5.5. Sensory Evaluation

Ten panelists (male and female) of different age groups, tested for taste sensitivity and experienced in sensory evaluation of dairy products, participated in the sensory evaluation. The panelists were asked to indicate how much they liked or disliked each variant of FWB on a 5-point hedonic scale (5 = like extremely; 1 = dislike extremely). The appearance, consistency, smell and taste were evaluated according to standards [37,38]. The samples used for the analysis were selected randomly. The evaluation was carried out in a room that was free of any foreign odors; each panelist had a separate test stand and distilled water to rinse the mouth. The sensory analysis was performed by the same group of panelists each time. The results for each descriptor were added together and were expressed as an arithmetic mean. The mean scores for each attribute were used to calculate the overall sensory quality.

### 2.5.6. Preparation of Supernatants

To obtain clear supernatants for analyses, the samples were prepared based on a methodology described elsewhere [32]. Samples were transferred to 1.5 mL Eppendorf tubes and centrifuged at 14,000× $g$ rpm for 10 min at 20 °C (Centrifuge 5418 Eppendorf, Warsaw, Poland). After centrifugation, the supernatants were filtered through 0.22 μm nylon membrane filters (Merck, Darmstad, Germany) and used for further determinations.

### 2.5.7. Determination of Reducing Sugars Content and Total Free Amino Acids Level

The reducing sugars content (RSC) was determined by the DNS (3,5-dinitrosalicylic acid) method, as described in a previous study [32]. A total of 10 g of DNS was dissolved in 200 mL of distilled water by continuous stirring; then, 16 g of NaOH (first dissolved in 150 mL of distilled $H_2O$) was slowly added. The mixture was incubated at 50 °C with stirring to obtain a clear solution. Then, 403 g of potassium sodium tartrate tetrahydrate was added in small portions. The mixture was filtered using a paper filter, and the volume was raised to 1000 mL with distilled water. A volume of 1 mL of the supernatant was mixed with 1 mL of 0.05 m acetate buffer (pH 4.8), and 3 mL of the DNS reagent was added, then vigorously shaken. The mixtures were incubated in boiled water for 5 min, then cooled at room temperature. The absorbance was measured at 540 nm (UV-Vis Thermo Scientific Evolution 220 spectrophotometer). Glucose in acetate buffer was used for the calibration curve.

Total free amino acids level (TFAAL) was analyzed based on a methodology described elsewhere [32]. A quantity of 1 mL of the supernatants was mixed with 2 mL of a Cd-ninhydrin reagent (0.8 g ninhydrin was dissolved in a mixture of 80 mL ethanol and 10 mL glacial acetic acid, followed by the addition of 1 g $CdCl_2$ dissolved in 1 mL of distilled water). The mixtures were shaken and heated at 84 °C for 5 min and cooled in ice water; then, the absorbance was measured at 507 nm. The results were expressed as milligram Gly per mL of the sample by reference to a standard curve, which was first prepared using glycine at various concentrations.

### 2.5.8. Determination of DPPH and ABTS Radicals Scavenging Activity

The scavenging activity against DPPH and ABTS radicals was determined as described elsewhere [32]. Briefly, the DPPH radical scavenging activity was determined by mixing 1 mL of the supernatants with 1 mL of 0.01 mM DPPH methanolic solution and incubation for 30 min. Subsequently, the absorbance was measured at 517 nm. A volume of 3 mL of $ABTS^{+\cdot}$ solution was mixed with 50 µL of the supernatants, and the absorbance was measured at 734 nm after 6 min of incubation.

### 2.6. Statistical Analysis

All analyses were performed at least in triplicate. All data were expressed as mean $\pm$ standard deviation (SD). The results were subjected to a statistical analysis carried out using Statistica software (StatSoft Polska, Kraków, Poland), using the Tukey test and a two-factor analysis of variance with repetition (ANOVA). All the tests were performed at a significance level of $p = 0.05$.

## 3. Results and Discussion

### 3.1. The Lactic Acid Bacteria (LAB) Survivability during Cold Storage

One of the most pivotal factors that affect the quality of fermented beverages is the number of live cells of lactic acid bacteria (LAB) [2]. LAB counts, as well as their survivability during refrigerated storage, are summarized in Table 1. Generally, whey has been reported to be a good medium for the growth of yogurt bacteria [2]. The minimum level of bacteria required for a functional product was reported to be at least $10^6$ CFU/mL, and it cannot be lower during storage, until expiration date [2]. After fermentation, the lowest bacterial concentration was noted for sample W ($7.65 \times 10^6 \pm 1.05$ CFU/mL), and the highest LAB counts were observed for samples with FG ($2.65 \times 10^8 \pm 0.07$ CFU/mL and $2.75 \times 10^8 \pm 1.63$ CFU/mL, for samples WFG1 and WFG2, respectively). There is a lack of reports regarding the effect of the addition of FG to whey on LAB growth. However, the survival rate of LAB in fermented milk with the addition of mucilage and whey-based beverages containing ingredients of plant origin was examined [39,40]. According to the results described in the literature, LAB were characterized by high survival rates in whey and in whey-based beverages [39,41]. For instance, Sady et al. [41] found a high survival rate of probiotic bacteria in drinks produced from whey combined with orange, apple and blackcurrant juice. The results of their research, as well as those of other authors, confirm that the addition of plant-based ingredients may create favorable conditions for the growth of LAB [3,39]. Additionally, Madhavi and Shah [42] observed high survivability of Lactobacilli counts in symbiotic whey drinks during refrigerated storage for 28 days. Hadi Nezhad et al. [43] showed that soluble dietary fiber from flaxseed functions as a good prebiotic, promoting LAB growth in a kefir model. Basiri et al. [44] reported that flaxseed mucilage enhances the starter bacterial counts of stirred yogurt. It can be assumed that, also in our studies, the significantly higher survival of LAB in samples with FG during refrigerated storage ($p < 0.05$) compared to W and WMP variants resulted from the presence of FG presumably acting as a prebiotic.

**Table 1.** The lactic acid bacteria (LAB) survivability during storage time.

| Sample | Time of Storage (Days) | | | | |
|---|---|---|---|---|---|
| | 1 | 7 | 14 | 21 | 28 |
| | LAB (CFU/mL) | | | | |
| W | $7.65 \times 10^6 \pm 1.05$ [Aa] | $5.05 \times 10^6 \pm 0.71$ [Aa] | $4.86 \times 10^6 \pm 0.79$ [Aa] | $4.50 \times 10^5 \pm 2.15$ [Ba] | $4.60 \times 10^5 \pm 0.85$ [Ba] |
| WMP | $6.40 \times 10^7 \pm 1.55$ [Ab] | $2.47 \times 10^7 \pm 0.20$ [BCb] | $5.30 \times 10^7 \pm 0.05$ [ABb] | $1.49 \times 10^7 \pm 0.01$ [Cb] | $3.32 \times 10^6 \pm 0.09$ [Cb] |
| WFG1 | $2.65 \times 10^8 \pm 0.07$ [Ac] | $7.25 \times 10^7 \pm 2.33$ [Bc] | $1.15 \times 10^8 \pm 0.07$ [Cc] | $7.50 \times 10^7 \pm 0.71$ [Bc] | $1.49 \times 10^7 \pm 0.32$ [Ec] |
| WFG2 | $2.75 \times 10^8 \pm 1.63$ [Av] | $1.18 \times 10^8 \pm 0.06$ [Bd] | $1.37 \times 10^8 \pm 0.14$ [Cc] | $4.67 \times 10^7 \pm 0.40$ [Dd] | $7.90 \times 10^7 \pm 0.99$ [Ed] |

W—fermented whey, WMP—fermented whey with milk powder added (5% *w/v*), WFG1—fermented whey with flaxseed gum added (0.5% *w/v*), WFG2—fermented whey with flaxseed gum added (1.0% *w/v*). Values are means $\pm$ standard deviation of triplicate determinations. Means with different lowercase letters in the same column are significantly different at $p < 0.05$. Means with different uppercase letters in the same row are significantly different at $p < 0.05$.

### 3.2. Titratable Acidity and pH

The changes in titratable acidity (TA) and pH are listed in Table 2. The addition of FG significantly increased the content of lactic acid and the active acidity (pH) of the samples ($p < 0.05$). Generally, the WMP sample had the lowest titratable acidity during the storage period, which may be attributed to the buffering capacity of milk powder [45]. The pH of the W sample did not change until the last week of storage, when an increase in pH was observed ($p < 0.05$). This is partially consistent with the results of the study conducted by Terpou et al. [46] who did not find any significant differences concerning the acidity of whey beverages during 30 days of storage at 4 °C. A significant increase in pH ($p < 0.05$) was observed in the first half of the research period and a stabilization in the second half of the cycle for variants with the addition of FG (WFG1 and WFG2). An overall increase in pH was observed for these samples. The highest acidity during storage was observed for the WFG2 sample ($p < 0.05$). Generally, according to the information in the literature, the cooling storage of WFB was associated with an increase, decrease or stabilization of the acidity of the product, depending on the type of gum used and the additives. Khoshdouni Farahani et al. [47] and Tabibloghmany and Ehsandoost [48] noted that the presence of gums in the beverage medium can increase pH and decrease acidity. Additionally, Aamer and El-Kholy [49] observed a gradual change in titratable acidity during a monthly storage of WFB. In a study on whey-based mango herbal beverage, Alane et al. [50] found that the pH of the beverage decreased during the storage period; the lowest pH value was recorded on the last day of the study and was 4.15. The acidity of the beverages changed during storage and was found to increase from an initial value of 0.24% citric acid to a final value of 0.32% after 30 days. These results of pH and acidity are in line with those reported by other authors [51]. The pH of the WFB containing fruit juices during storage, according to Sady et al. [41], was in the range of 3.89–4.17, while the content of citric acid was 0.42–0.61%. The value of both parameters was stable during storage. The high acidity of WFB compared to values reported in the present study was due to the high acidity of the whey as well as the fruit juices as flavoring agents. In their work, Pescuma et al. [8] tested the pH of functional fermented whey-based beverages produced with lactic acid bacteria (*L. delbrueckii* subsp. *bulgaricus*, *S. thermophilus* and *L. acidophilus*) and reported on the stabilization, decrease or increase in the pH of beverages during storage (28 days) depending on the experimental variant used. The fluctuation of pH values could be connected to the buffering capacity of whey protein and dairy products. It is noted that this effect could be increased by milk products' treatment, such as heating, high-pressure treatment or salt addition [52]. The same effect was observed in the present study because the samples' pH was no lower than $4.26 \pm 0.02$ and no higher than $4.55 \pm 0.03$. On the other hand, flaxseed gum also consists of a protein fraction with a possible buffering effect [53].

**Table 2.** pH and titratable acidity of the samples during storage time.

| Sample | Time of Storage (Days) | | | | |
|---|---|---|---|---|---|
| | 1 | 7 | 14 | 21 | 28 |
| | | | pH | | |
| W | $4.40 \pm 0.01$ [Ab] | $4.39 \pm 0.02$ [Aab] | $4.42 \pm 0.03$ [Aa] | $4.43 \pm 0.02$ [Aac] | $4.48 \pm 0.03$ [Ba] |
| WMP | $4.51 \pm 0.02$ [Aa] | $4.52 \pm 0.02$ [Ad] | $4.52 \pm 0.01$ [Ab] | $4.52 \pm 0.01$ [Ab] | $4.55 \pm 0.03$ [ABb] |
| WFG1 | $4.32 \pm 0.02$ [Ac] | $4.37 \pm 0.02$ [Bbc] | $4.40 \pm 0.02$ [Cac] | $4.41 \pm 0.02$ [Cacd] | $4.44 \pm 0.01$ [Cc] |
| WFG2 | $4.26 \pm 0.02$ [Ad] | $4.34 \pm 0.02$ [Bc] | $4.38 \pm 0.01$ [Cc] | $4.39 \pm 0.01$ [Ccd] | $4.41 \pm 0.07$ [Cc] |
| | | | TA (%) | | |
| W | $0.86 \pm 0.00$ [Aa] | $0.88 \pm 0.03$ [ABa] | $0.89 \pm 0.01$ [ABa] | $0.90 \pm 0.01$ [Ba] | $0.82 \pm 0.05$ [Ca] |
| WMP | $0.65 \pm 0.01$ [Ab] | $0.64 \pm 0.01$ [Ab] | $0.63 \pm 0.01$ [Ab] | $0.60 \pm 0.01$ [Bb] | $0.60 \pm 0.01$ [Bb] |
| WFG1 | $0.94 \pm 0.01$ [Ac] | $0.95 \pm 0.01$ [Abc] | $0.96 \pm 0.01$ [Abc] | $0.97 \pm 0.01$ [Ac] | $0.96 \pm 0.04$ [Abc] |
| WFG2 | $1.00 \pm 0.01$ [ABd] | $0.99 \pm 0.01$ [Ad] | $1.03 \pm 0.01$ [BCd] | $1.04 \pm 0.00$ [Cd] | $0.97 \pm 0.02$ [Dd] |

W—fermented whey, WMP—fermented whey with milk powder added (5% *w/v*), WFG1—fermented whey with flaxseed gum added (0.5% *w/v*), WFG2—fermented whey with flaxseed gum added (1.0% *w/v*), TA—titratable acidity. Values are means $\pm$ standard deviation of triplicate determinations. Means with different lowercase letters in the same column are significantly different at $p < 0.05$. Means with different uppercase letters in the same row are significantly different at $p < 0.05$.

### 3.3. Water Activity, Syneresis, Viscosity and Particle Sizes

The water activity, syneresis, viscosity and particle sizes of the samples are summarized in Table 3. Water activity ($a_w$), which is a thermodynamic measure of the chemical potential of water, affects the durability and quality of products, influencing the potential of microorganisms to develop [54]. Fermented whey (W) had the highest $a_w$ during storage, whereas the lowest values were observed for the WFG2 sample. A significant decrease in the $a_w$ of all samples during storage was found ($p < 0.05$), and the greatest decrease was observed for the W sample. The use of FG decreased the $a_w$ of FWB, but the amount of FG did not significantly differentiate the WFG1 and WFG2 samples ($p > 0.05$). The lowest syneresis (serum release from the gel matrix) during refrigerated storage was observed for the sample containing 1% addition of FG (WFG2) and the highest for fermented whey (W) ($p < 0.05$). The addition of 0.5% FG limited the syneresis of the samples to a level similar to the addition of milk powder (MP). Skryplonek and Jasińska [55,56] reported that FWB are characterized by a significant whey leakage, amounting to one-third of the sample volume, which results in a low assessment of the appearance of this type of product. According to Dehghan et al. [57], the main problem in the production of FWB is precisely their colloidal instability caused by protein aggregation, mainly β-lactoglobulin. Changes in colloidal stability negatively affect the viscosity and appearance of the product. The results of studies by other authors also confirm the positive effect of FG on the release of whey from fermented beverages. Hashemi et al. [58] found that the addition of hydrocolloids positively influences the sensory characteristics of FWB. Zendeboodi et al. [59] found an increase in the apparent viscosity of the beverages, and thus the lowest release of whey, and recommended the use of gum in the industrial production of FWB. This is consistent with the finding that all adsorbent and non-adsorbent hydrocolloids can stabilize beverages and prevent phase separation by increasing the consistency of the continuous phase and chemically interacting with proteins [59]. The reduction in syneresis observed in our studies after adding FG to the samples is consistent with the results of Basiri et al. [44] who reported reduced syneresis of mixed yogurt supplemented with FG. The decreased hydrodynamic properties, as well as syneresis, can be linked to the tendency of FG biopolymers to associate due to the intermolecular associations via hydrogen bonding, leading to pseudogel behavior and the formation of complexes with proteins such as conlinin [60].

In dairy beverages, viscosity is an important characteristic for consumer acceptance, and it depends on the solids content, the type and concentration of the additives, and the fermentation conditions, such as time, temperature and the LAB strains used. The addition of some natural ingredients allows the design of the viscosity of the products [4]. AW has a low total solids content when compared to milk [2,13,21]. The addition of FG could fix this problem and may improve the texture parameters because biopolymer complexes formed by protein–polysaccharide interaction can serve as food texture modifiers in food products [61]. Cui and Mazza [62] as well as Chen et al. [63] found that under acidic conditions, the FG gel strength decreases as the pH value decreases, and under basic conditions, the gel strength decreases as the pH value increases. In our study, the addition of FG, despite the acidic environment, clearly improved the viscosity of the tested samples. As can be seen in Table 3, the viscosity of fermented whey significantly increased when MP was added ($p < 0.05$). Moreover, a significant increase in viscosity (8-fold and over 14-fold for samples WFG1 and WFG2, respectively) was noted ($p < 0.05$). However, a viscosity decrease was observed for all samples during cold storage, which can be attributed to proteins and carbohydrates hydrolysis. It should be pointed out that on day 28, the viscosity of the WFG2 sample ($1501.24 \pm 0.06$ Pa·s) was approximately 45-fold higher than the viscosity of whey ($33.52 \pm 0.10$ Pa·s) ($p < 0.05$). The largest particle size during storage was noted for the WMP sample, followed by the W sample. As can be seen in Table 3, the addition of FG caused a significant decrease in particle size ($p < 0.05$). The enhanced viscosity of the samples can be linked to FG's "weak gel-like" properties and the ability to form networks [23,60]. Moreover, it was reported that FG plays the role of a thickening agent, which is able to increase the viscosity and decrease the particle size of emulsions [26]. Electrostatic interactions produce profound effects on biopolymer interactions between the charged species [60]. In the case of a protein–anionic polysaccharide biopolymer system, electrostatic interactions can be attractive or repulsive. When the pH of the solution is greater than the pI of the protein, repulsive interactions will dominate, since both biopolymers have negative charges. However, attractive electrostatic interactions occur when the pKa of the anionic polysaccharide is less than the pH of the solution, and the pH, in turn, is less than the pI of the protein [60]. As reported, whey protein's isoelectric point is approximately at pH = 5.2, and below this value, it will be positively charged, whereas polysaccharide (FG) is negatively charged [27,64]. In an acidic environment, FG possesses a negative charge [27]. Thus, presumably, FG and whey proteins interacted and formed complexes, resulting in enhanced textural properties, syneresis and decreased hydrodynamic properties.

**Table 3.** Water activity, syneresis, viscosity and particle sizes of the samples during storage time.

| Sample | Time of Storage (Days) | | | | |
|---|---|---|---|---|---|
| | 1 | 7 | 14 | 21 | 28 |
| | $a_w$ (-) | | | | |
| W | 0.98 ± 0.01 [Aa] | 0.98 ± 0.00 [ABa] | 0.98 ± 0.00 [BCa] | 0.97 ± 0.00 [Ca] | 0.97 ± 0.01 [CDa] |
| WMP | 0.97 ± 0.00 [Ab] | 0.97 ± 0.01 [Ab] | 0.97 ± 0.01 [Aa] | 0.97 ± 0.03 [Bb] | 0.96 ± 0.00 [Bb] |
| WFG1 | 0.96 ± 0.00 [Ac] | 0.96 ± 0.01 [ABc] | 0.95 ± 0.00 [ABb] | 0.95 ± 0.02 [BCc] | 0.95 ± 0.01 [Cc] |
| WFG2 | 0.96 ± 0.03 [Ac] | 0.95 ± 0.00 [BCd] | 0.95 ± 0.00 [BCb] | 0.95 ± 0.02 [Bc] | 0.95 ± 0.02 [ACc] |
| | Syneresis (%) | | | | |
| W | 94.83 ± 0.30 [Aa] | 94.67 ± 0.30 [Aa] | 95.47 ± 0.50 [Aa] | 94.77 ± 0.40 [Aa] | 94.30 ± 1.10 [Aa] |
| WMP | 74.50 ± 0.50 [Ab] | 74.50 ± 0.50 [Ab] | 74.67 ± 2.10 [Ab] | 72.67 ± 2.50 [BCb] | 73.99 ± 1.60 [ABb] |
| WFG1 | 74.67 ± 0.60 [Ab] | 74.70 ± 0.30 [Ab] | 75.57 ± 1.30 [Ab] | 74.83 ± 0.20 [Ac] | 74.67 ± 0.70 [Ab] |
| WFG2 | 67.00 ± 1.00 [Ac] | 65.67 ± 0.80 [Ac] | 65.03 ± 0.30 [Bb] | 64.00 ± 0.40 [Cbc] | 61.82 ± 1.50 [Dc] |
| | Viscosity (Pa·s) | | | | |
| W | 125.31 ± 0.05 [Aa] | 122.30 ± 0.03 [Aa] | 89.21 ± 0.10 [Ba] | 41.48 ± 0.03 [Ca] | 33.52 ± 0.10 [Da] |
| WMP | 604.30 ± 0.02 [Ab] | 749.52 ± 0.02 [Bb] | 425.10 ± 0.05 [Cb] | 159.32 ± 0.02 [Db] | 89.30 ± 0.12 [Eb] |
| WFG1 | 1038.22 ± 0.10 [Ac] | 1012.32 ± 0.01 [Bc] | 989.22 ± 0.02 [Cd] | 975.13 ± 0.02 [Dc] | 911.34 ± 0.05 [Ec] |
| WFG2 | 1800.21 ± 0.05 [Ad] | 1740.90 ± 0.02 [Bd] | 1534.00 ± 0.01 [Ce] | 1522.23 ± 0.09 [Dd] | 1501.24 ± 0.06 [Ed] |
| | $D_{4,3}$ (µm) | | | | |
| W | 279.48 ± 1.14 [Aa] | 130.52 ± 0.85 [Ba] | 124.63 ± 0.36 [Ca] | 132.04 ± 0.59 [Da] | 133.22 ± 0.35 [Ea] |
| WMP | 261. 44 ± 1.12 [Ab] | 259.17 ± 0.48 [Bb] | 262.24 ± 0.79 [Cb] | 252.49 ± 0.10 [Db] | 257.12 ± 0.77 [Eb] |
| WFG1 | 110.48 ± 0.46 [Ac] | 115.18 ± 0.58 [Bc] | 114.16 ± 0.41 [Cc] | 115.97 ± 0.19 [Dc] | 116.79 ± 0.39 [Ec] |
| WFG2 | 81.20 ± 0.36 [Ad] | 70.76 ± 0.39 [Bd] | 69.47 ± 0.35 [Cd] | 68.86 ± 0.37 [Dd] | 70.92 ± 0.36 [Bd] |
| | $D_{3,2}$ (µm) | | | | |
| W | 38.86 ± 1.18 [Aa] | 22.52 ± 0.92 [Ba] | 21.48 ± 0.47 [Ca] | 22.58 ± 0.68 [Ba] | 21.00 ± 0.47 [Da] |
| WMP | 79.04 ± 1.21 [Ab] | 81.52 ± 0.58 [Bb] | 79.63 ± 0.79 [Ab] | 71.93 ± 0.11 [Cb] | 75.81 ± 0.78 [Db] |
| WFG1 | 14.42 ± 0.66 [Ac] | 17.25 ± 0.71 [Bc] | 14.97 ± 0.61 [Cc] | 16.32 ± 0.13 [Dc] | 14.99 ± 0.61 [Cc] |
| WFG2 | 16.89 ± 0.53 [Ad] | 14.85 ± 0.57 [Bd] | 15.29 ± 0.53 [Cd] | 14.52 ± 0.56 [Bd] | 14.76 ± 0.56 [Bc] |

W—fermented whey, WMP—fermented whey with milk powder added (5% *w/v*), WFG1—fermented whey with flaxseed gum added (0.5% *w/v*), WFG2—fermented whey with flaxseed gum added (1.0% *w/v*). Values are means ± standard deviation of triplicate determinations. Means with different lowercase letters in the same column are significantly different at $p < 0.05$. Means with different uppercase letters in the same row are significantly different at $p < 0.05$.

### 3.4. Changes in Acetaldehyde Content and Sensory Evaluation Results

The acetaldehyde content and sensory scores are presented in Table 4. As can be seen, the highest acetaldehyde content at the beginning and at the end of the storage was determined in the WFG1 variant ($0.44 \pm 0.10$ mg/dm$^3$ and $1.65 \pm 0.04$ mg/dm$^3$, respectively), whereas the lowest content was noted for the fermented whey (W) ($0.07 \pm 0.00$ mg/dm$^3$ and $1.34 \pm 0.04$ mg/dm$^3$, respectively). A significant increase ($p < 0.05$) in acetaldehyde content was observed during storage in all the FWB variants. The highest increase was found for the WMP (1.470 mg/dm$^3$). The addition of FG generally resulted in an increase in acetaldehyde content. Acetaldehyde is the main compound shaping the aroma of fermented milk, synthesized from lactose or amino acid threonine [2]. As reported in the literature, the acetaldehyde content in fermented milk varies, and in yogurt, it can be as high as $10 \div 15$ mg/dm$^3$, and the threshold of its detectability by the human sense of smell is 0.415 mg/dm$^3$ [65]. The values determined are comparable with those reported for sour milk, kefir, quark or crème fraiche but lower than those reported for ayran or yogurt [65].

From the perspective of the consumer, beverages based on by-products should be visually as homogeneous as dairy products [14,15]. Consumers pay attention to their sensory features and stability during storage. The most important sensory characteristics of fermented milk that determine their choice are texture, taste and aroma [66,67]. The highest overall sensory FWB rating was assigned to the WFG2 sample during storage, whereas the lowest was noted for the W sample (Table 4). The overall sensory scores of the WFG1 and WFG2 samples are comparable to those reported for whey drinks with fruit concentrates [68]. The WFG2 variant received the highest score for consistency (4.80–5.00) and appearance (4.30–4.70) but the lowest score for taste due to straw and grassy aftertaste (3.20–3.30). Chen et al. [63] reported that a creamy mouthfeel was noticed in whey-based beverages when xanthan gum was added. The lowest scores for appearance (3.20–3.50) and consistency (2.00–2.20) were given to fermented whey (W). The consistency of the beverage containing 0.5% FG (WFG1) was similar to that of the beverage produced with the addition of milk powder (WMP). The maximum score for smell was given to the W and WMP samples. The addition of FG lowered the score for smell, as the panelists emphasized the perceptible vegetable note, unusual for dairy products. Basiri et al. [44] reported that an increase in acidity and firmness of yogurt prevents the release of aromatic compounds. Similarly, Gallardo-Escamilla et al. [14] showed that the yogurt aroma of samples containing hydrocolloids (such as carboxymethyl cellulose and propylene glycol alginate) was perceived as less intense. Hydrocolloids can be used to increase the viscosity of whey-based dairy beverages, although they may mask the typical beverage flavor when using a yogurt culture [58]. Thus, it is reasonable to conclude that FG influenced the release of acetaldehyde, despite the highest content in samples with FG added, resulting in a lower score for the smell. Presumably, this can also be linked to the formation of FG–whey protein complexes. The highest rated taste was noted for the fermented whey containing the addition of milk powder (WMP). Hydrocolloids can be applied to enhance the viscosity of whey-based dairy beverages, although they may mask the typical beverage flavor when a yogurt culture is involved [58]. The effect of gum on the sensory characteristics of fermented whey beverages, based on the information available in the literature, may depend on the type of gum. Terpou et al. [46], studying the functional whey drinks containing Chios mastic gum, noted that all tasters could detect the taste and aroma of mastic gum in the whey drinks; however, they sensed no difference in smoothness compared to the control drinks. In fact, all whey drinks with Chios mastic gum were characterized as cooling beverages with a pleasant distinct aroma. The industry should focus on increasing the flavor acceptance of FWB with gums, as the flavor is directly related to the overall acceptance and should also develop the attributes that are linked to the most accepted samples [46].

### 3.5. Color Changes

The results of color determination are summarized in Table 5. As can be noted, the addition of FG had a significant effect on the color parameters, i.e., L*, a*, b*, while it did not significantly affect the color saturation (C) and the hue (h). The addition of FG as well as MP caused a significant increase in the L* parameter (brightness) ($p < 0.05$). The cooling storage of FWB resulted in a decrease in brightness for WMP and WFG1 and no significant changes in this parameter for the fermented whey (W) and the sample containing 1% flaxseed gum (WFG2). The redness/greenness parameter (a*) of the W sample did not change during storage ($p > 0.05$). In the case of the WMP variant, fluctuations in the a* parameter during storage were generally statistically insignificant ($p < 0.05$). When analyzing the color of the WFG1 and WFG2 samples, an initial stabilization of this parameter was observed until the third week and second week of storage, respectively, and a subsequent increase in the value of a*, respectively, by 0.49 and 0.56 units. The W and WMP samples differed significantly in the a* parameter, and the addition of FG increased its value in the WFG1 and WFG2 samples ($p < 0.05$). The b* parameter (yellowness/blueness) during cold storage of FWB ranged from $2.32 \pm 0.11$ (sample W on day

1) to 5.76 ± 0.39 (sample WFG2 on day 14). The highest value of parameter b* was noted for the WMP sample and the lowest for sample W. Yellow-greenish coloration is characteristic for whey due to the presence of riboflavin [1,2,13]. All the samples differed significantly in the b* parameter until the second week of storage; in the following two, there was no significant difference in the b* parameter between the samples containing the addition of FG (WFG1 and WFG2). The addition of FG caused a significant color change toward yellow ($p < 0.05$). The increase in the L* parameter in samples with FG can be linked to lower particle size, thus higher light scattering [22,26]. On the first day, the color differences (ΔE) between the fermented whey (W) and the other variants of FWB, i.e., WMP, WFG1 and WFG2, were 16.59 ± 2.48, 22.17 ± 0.62 and 23.49 ± 0.17, respectively. The ΔE for the samples containing the addition of FG (WFG1 and WFG2) and WMP was higher than 1 (which is considered as perceptible for the human eye), indicating a clear color difference between the samples [28]. The highest color saturation (C) on the first day was found in the WMP sample (6.71 ± 0.39), whereas the lowest was noted for the W sample (3.73 ± 0.07). Significant fluctuations of the C value during storage and an overall decrease for the WMP and WFG2 samples, at the same level (by 0.11), were found. The color saturation in the W sample did not change significantly ($p > 0.05$) during the 3 weeks of storage, but on day 28, a significant increase was observed (0.52) ($p < 0.05$) when compared to the initial value. The overall increase in color saturation (by 0.08) was also found in the WFG1 sample ($p < 0.05$). The refrigerated storage of the experimental samples was associated with an increase in the h parameter for the W sample (by 0.12) and a decrease for the WMP, WFG1 and WFG2 samples by 0.17, 0.09 and 0.15, respectively.

**Table 4.** Changes in acetaldehyde content and sensory evaluation results.

| Sample | Time of Storage (Days) | | | | |
|---|---|---|---|---|---|
| | 1 | 7 | 14 | 21 | 28 |
| | Acetaldehyde (mg/dm$^3$) | | | | |
| W | 0.075 ± 0.00 [Ac] | 0.393 ± 0.03 [Bb] | 0.491 ± 0.06 [Cd] | 0.475 ± 0.06 [Cd] | 1.343 ± 0.04 [Dc] |
| WMP | 0.088 ± 0.01 [Ab] | 0.923 ± 0.03 [Ba] | 0.984 ± 0.01 [Bc] | 0.964 ± 0.03 [Bc] | 1.558 ± 0.08 [Cb] |
| WFG1 | 0.441 ± 0.10 [Aa] | 0.952 ± 0.03 [Ba] | 1.389 ± 0.10 [Ca] | 1.346 ± 0.04 [Cb] | 1.645 ± 0.04 [Da] |
| WFG2 | 0.364 ± 0.01 [Ab] | 0.449 ± 0.01 [Bb] | 1.184 ± 0.03 [Cb] | 1.613 ± 0.02 [Da] | 1.638 ± 0.04 [Da] |
| | Consistency (points) | | | | |
| W | 2.20 ± 0.30 [Aa] | 2.30 ± 0.30 [Aa] | 2.20 ± 0.30 [Aa] | 2.20 ± 0.30 [Aa] | 2.00 ± 0.00 [Ba] |
| WMP | 2.70 ± 0.30 [Aa] | 3.20 ± 0.30 [Bb] | 3.30 ± 0.30 [Bb] | 3.20 ± 0.30 [Bb] | 3.50 ± 0.50 [BCb] |
| WFG1 | 4.00 ± 0.60 [Ab] | 4.20 ± 0.60 [Abc] | 4.30 ± 0.30 [BCc] | 4.30 ± 0.30 [BCc] | 4.50 ± 0.00 [Cc] |
| WFG2 | 4.80 ± 0.30 [ABc] | 4.80 ± 0.30 [ABd] | 5.00 ± 0.00 [Bd] | 5.00 ± 0.00 [Bd] | 5.00 ± 0.00 [Bc] |
| | Appearance (points) | | | | |
| W | 3.20 ± 0.30 [ABa] | 3.00 ± 0.00 [Aa] | 3.20 ± 0.30 [ABa] | 3.20 ± 0.30 [ABa] | 3.50 ± 0.50 [Ba] |
| WMP | 3.70 ± 0.30 [Ab] | 3.80 ± 0.30 [Ab] | 4.00 ± 0.00 [Ab] | 4.00 ± 0.00 [Ab] | 3.80 ± 0.30 [Ab] |
| WFG1 | 4.00 ± 0.50 [Abc] | 4.20 ± 0.30 [Bbc] | 4.2 ± 0.30 [Bb] | 4.20 ± 0.30 [Bb] | 4.30 ± 0.30 [Bc] |
| WFG2 | 4.30 ± 0.30 [ABc] | 4.50 ± 0.00 [BCc] | 4.70 ± 0.30 [Cc] | 4.70 ± 0.30 [Cc] | 4.70 ± 0.30 [Cc] |
| | Taste (points) | | | | |
| W | 3.80 ± 0.30 [Aa] | 3.80 ± 0.30 [Aa] | 3.80 ± 0.30 [Aa] | 3.40 ± 0.50 [Ba] | 3.40 ± 0.50 [Ba] |
| WMP | 3.80 ± 0.30 [Aa] | 3.80 ± 0.30 [Aa] | 3.80 ± 0.30 [Aa] | 4.20 ± 0.30 [Bb] | 4.20 ±0.30 [Bb] |
| WFG1 | 3.80 ± 0.30 [Aa] | 3.80 ± 0.30 [Aa] | 3.80 ± 0.30 [Aa] | 3.80 ± 0.30 [Ac] | 3.80 ± 0.30 [Ac] |
| WFG2 | 3.20 ± 0.30 [Ab] | 3.30 ± 0.30 [Ab] | 3.20 ± 0.30 [Ab] | 3.30 ± 0.30 [Ad] | 3.30 ± 0.30 [Ad] |
| | Smell (points) | | | | |
| W | 4.80 ± 0.30 [Aa] | 5.00 ± 0.00 [Ba] | 5.00 ± 0.00 [Ba] | 5.00 ± 0.00 [Ba] | 5.00 ± 0.00 [Ba] |
| WMP | 5.00 ± 0.00 [Ab] | 5.00 ± 0.00 [Aa] | 5.00 ± 0.00 [Aa] | 5.00 ± 0.00 [Aa] | 5.00 ± 0.00 [Aa] |
| WFG1 | 4.30 ± 0.30 [Ac] | 4.20 ± 0.30 [Ab] | 4.00 ± 0.00 [Bb] | 3.80 ± 0.30 [Cb] | 3.80 ± 0.30 [Cb] |
| WFG2 | 4.20 ± 0.30 [Ac] | 4.20 ± 0.30 [Ab] | 3.80 ± 0.30 [Bc] | 3.70 ± 0.30 [Bb] | 3.70 ± 0.30 [Bb] |
| | Overall sensory quality (points) | | | | |
| W | 3.50 ± 1.12 [Aa] | 3.54 ± 1.15 [Aa] | 3.54 ± 1.19 [Aa] | 3.44 ± 1.17 [Aa] | 3.48 ± 1.23 [Aab] |
| WMP | 3.79 ± 0.96 [Aa] | 3.96 ± 0.76 [ABa] | 4.04 ± 0.70 [Ba] | 4.08 ± 0.75 [Ba] | 4.13 ± 0.64 [Ba] |
| WFG1 | 4.04 ± 0.21 [Aa] | 4.08 ± 0.17 [Aa] | 4.08 ± 0.22 [Aa] | 4.04 ± 0.25 [ABa] | 4.13 ± 0.34 [Bb] |
| WFG2 | 4.13 ± 0.70 [Aa] | 4.21 ± 0.64 [Aa] | 4.17 ± 0.83 [Aa] | 4.17 ± 0.79 [Aa] | 4.17 ± 0.79 [Aab] |

W—fermented whey, WMP—fermented whey with milk powder added (5% *w/v*), WFG1—fermented whey with flaxseed gum added (0.5% *w/v*), WFG2—fermented whey with flaxseed gum added (1.0% *w/v*). Values are means ± standard deviation of triplicate determinations. Means with different lowercase letters in the same column are significantly different at $p < 0.05$. Means with different uppercase letters in the same row are significantly different at $p < 0.05$.

**Table 5.** Color changes of the samples during storage time.

| Sample | Time of Storage (Days) | | | | |
|---|---|---|---|---|---|
| | 1 | 7 | 14 | 21 | 28 |
| | | | L* | | |
| W | 48.82 ± 2.80 [Aa] | 50.13 ± 0.80 [Aa] | 49.92 ± 0.40 [Aa] | 48.51 ± 0.40 [Aa] | 49.78 ± 1.00 [Aa] |
| WMP | 65.50 ± 0.40 [Ab] | 66.67 ± 3.00 [Ab] | 65.27 ± 0.50 [Bb] | 61.85 ± 2.90 [Cb] | 57.81 ± 1.50 [Db] |
| WFG1 | 69.42 ± 0.50 [Ac] | 67.79 ± 1.60 [Bb] | 67.40 ± 0.70 [Ac] | 66.50 ± 6.10 [Ac] | 60.20 ± 1.20 [Cb] |
| WFG2 | 70.72 ± 0.00 [Acd] | 71.35 ± 2.20 [Ac] | 69.50 ± 1.10 [Ac] | 69.78 ± 1.20 [Ac] | 70.52 ± 0.40 [Ac] |
| | | | a* | | |
| W | 2.92 ± 0.09 [Aa] | 2.90 ± 0.11 [Aa] | 3.07 ± 0.05 [Aa] | 2.83 ± 0.06 [Aa] | 2.89 ± 0.07 [Aa] |
| WMP | 3.57 ± 0.34 [Ab] | 3.83 ± 0.16 [BCbc] | 3.54 ± 0.08 [Ab] | 3.49 ± 0.44 [Ab] | 3.67 ± 0.06 [ABb] |
| WFG1 | 4.06 ± 0.03 [Abc] | 4.08 ± 0.04 [Abc] | 4.07 ± 0.08 [Abc] | 4.04 ± 0.00 [Bc] | 4.28 ± 0.18 [ACc] |
| WFG2 | 4.19 ± 0.05 [ABcd] | 4.17 ± 0.21 [ABcd] | 4.18 ± 0.10 [ABc] | 4.36 ± 0.16 [Bd] | 3.92 ± 0.04 [Cb] |
| | | | b* | | |
| W | 2.32 ± 0.11 [Aa] | 2.50 ± 0.13 [Aa] | 2.38 ± 0.06 [Aa] | 2.38 ± 0.05 [Aa] | 3.00 ± 0.03 [Ba] |
| WMP | 5.68 ± 0.36 [ABb] | 5.72 ± 0.17 [ABb] | 5.24 ± 0.08 [Cb] | 5.83 ± 0.10 [Db] | 4.40 ± 0.11 [Ab] |
| WFG1 | 5.27 ± 0.15 [Ac] | 5.10 ± 0.21 [Abc] | 5.76 ± 0.39 [Cc] | 5.06 ± 0.07 [Abc] | 4.53 ± 0.19 [Dbc] |
| WFG2 | 4.73 ± 0.11 [ABd] | 5.50 ± 0.08 [Dd] | 4.86 ± 0.17 [ACd] | 4.91 ± 0.03 [ACc] | 4.74 ± 0.40 [ABc] |
| | | | C* | | |
| W | 3.73 ± 0.07 [Aa] | 3.83 ± 0.06 [Aa] | 3.88 ± 0.05 [ABa] | 3.70 ± 0.07 [Aa] | 4.25 ± 0.03 [Ba] |
| WMP | 6.71 ± 0.39 [Ab] | 6.88 ± 0.18 [Ab] | 6.32 ± 0.03 [Bb] | 6.79 ± 0.30 [Ab] | 6.60 ± 0.12 [Cb] |
| WFG1 | 6.65 ± 0.13 [Ab] | 6.53 ± 0.14 [Ac] | 7.05 ± 0.36 [Bc] | 6.47 ± 0.05 [Ac] | 6.73 ± 0.26 [Cb] |
| WFG2 | 6.32 ± 0.10 [ABc] | 6.90 ± 0.19 [Cd] | 6.41 ± 0.08 [ABb] | 6.57 ± 0.09 [Ac] | 6.21 ± 0.33 [Bc] |
| | | | h° | | |
| W | 0.67 ± 0.03 [ABa] | 0.71 ± 0.04 [Ca] | 0.66 ± 0.01 [Aa] | 0.70 ± 0.01 [BCa] | 0.79 ± 0.02 [Da] |
| WMP | 1.01 ± 0.04 [ABb] | 0.98 ± 0.02 [ACb] | 0.98 ± 0.02 [ACb] | 1.03 ± 0.05 [Bb] | 0.84 ± 0.01 [Dc] |
| WFG1 | 0.92 ± 0.01 [Ac] | 0.90 ± 0.02 [Abc] | 0.96 ± 0.03 [Cb] | 0.90 ± 0.01 [ABc] | 0.83 ± 0.00 [Db] |
| WFG2 | 0.85 ± 0.01 [Ad] | 0.92 ± 0.02 [Bc] | 0.86 ± 0.03 [Ac] | 0.84 ± 0.02 [Ad] | 0.70 ± 0.04 [Cc] |
| | | | ΔE | | |
| W | used as standard | 2.19 ± 0.76 [Aa] | 2.48 ± 0.65 [Aa] | 2.35 ± 1.42 [Ba] | 2.92 ± 0.65 [Aa] |
| WMP | 16.59 ± 2.48 [Aa] | 16.47 ± 3.15 [Ab] | 15.26 ± 0.62 [Ab] | 13.23 ± 2.46 [Ab] | 7.98 ± 0.47 [Bb] |
| WFG1 | 22.17 ± 0.62 [Aa] | 17.19 ± 1.33 [Bb] | 17.41 ± 1.07 [Bb] | 17.95 ± 5.71 [Abc] | 10.44 ± 0.57 [Cc] |
| WFG2 | 23.49 ± 0.17 [Aa] | 20.75 ± 0.85 [Bc] | 19.54 ± 1.41 [Cc] | 21.27 ± 1.65 [Bc] | 20.72 ± 1.15 [Bd] |

W—fermented whey, WMP—fermented whey with milk powder added (5% *w/v*), WFG1—fermented whey with flaxseed gum added (0.5% *w/v*), WFG2—fermented whey with flaxseed gum added (1.0% *w/v*). Values are means ± standard deviation of triplicate determinations. Means with different lowercase letters in the same column are significantly different at *p* < 0.05. Means with different uppercase letters in the same row are significantly different at *p* < 0.05.

The phenomena observed in our research are confirmed in the available literature. Keshtkaran et al. [69] found that gum concentration correlated with color parameters. Similarly, Behbahani and Abbasi [70] claimed that the yellowness–blueness index (b*) of the drink samples stabilized with Iranian native hydrocolloids (Persian gum and tragacanth gum) was significantly higher than that of the sample without hydrocolloids. This may be due to the color characteristics of hydrocolloids, their concentrations and possible interactions with salts and sugars [28].

### 3.6. Reducing Sugars Content, Total Free Amino Acids Level and Free Radicals Scavenging Activity

The changes of reducing sugars content (RSC), total free amino acids level (TFAAL) and free radicals scavenging activity during storage are presented in Table 6. As can be seen, the addition of MP caused a significant increase in RSC and TFAAL due to the presence of lactose and proteins, respectively. Interestingly, the addition of FG also caused a significant increase in TFAAL, as well as RSC (*p* < 0.05). Some fluctuations were observed during storage; however, on day 28, the highest TFAAL was noted for the WFG1 sample (3.83 ± 0.80 mg/mL) and was over four-fold higher than that observed for the W sample (0.92 ± 0.00 mg/mL), whereas the RSC of the WFG1 and WFG2 samples was comparable to that of the sample with MP addition. However, a significant decrease in TFAAL was observed from day 14 in the WFG2 sample, which can presumably be linked to the consumption of the released amino acids by LAB [8]. The results for the RSC are comparable to those reported by Alane et al. [50]. Although there is no clear evidence, this result was presumably linked to the prebiotic effect of FG, enhancing microbial activity, causing hydrolysis of polysaccharide (FG) as well as proteins [3,13]. Whey protein has a high biological value, mainly due to its high content of branched-chain essential amino acids (isoleucine, leucine and valine) [1,13,21]. Moreover, the generation of high TFAAL can also be linked to the production of bioactive peptides as intermediates [4]. The decrease in TFAAL in the W variant could be an effect of the lower content of nutrients for the bacteria. According to Neis et al., amino acids could also be used as an energy source for fermentation [71]. On the other hand, by introducing WMP, which is a source of protein, into the system, as well as FG, which also contains a protein fraction, it is thus possible to potentially increase the amino acids level [43]. However, some reports indicated that FG has a prebiotic potential; thus, presumably, it could have a stimulating effect on the growth of LAB as well as on their metabolic activity, including both amino acid production and proteolytic activity. However, this requires further in-depth analysis. The presence of bioactive peptides in beverages contributes not only to their physicochemical stability but also to health-promoting properties, i.e., cardiovascular, antihypertensive, antithrombotic, antioxidant and antimicrobial [4,8,13]. Generally, the addition of FG caused an increase in antioxidant activity ($p < 0.05$), which was more clearly seen in the case of the ABTS radical. This observation is consistent with the findings of Kadyan et al. [3]. Ansari et al. [72] also reported an increased antioxidant activity of whey drink fermented with *L. acidophilus* LA5 [72]. Whey proteins show an antioxidant potential by chelating transition metal ions through lactoferrin or scavenging free radicals through sulfhydryl-containing amino acids. Fermentation by starter cultures increases the release of reductants, such as cysteine, present in the peptide chains of whey protein, thus neutralizing the free radicals present in the system [3,73]. The enhanced antioxidant activity in FG samples is presumably multifactorial and can be attributed to the metabolic antioxidant capacity of the bacteria (production of antioxidant metabolites, such as glutathione, butyrate, etc., or chelation of metal ions) as well as the production of amino acids [3,13,73].

**Table 6.** Free radicals scavenging activity, reducing sugars content (RSC) and total free amino acids level (TFAAL) of the samples during storage time.

| Sample | Time of Storage (Days) | | | | | |
|---|---|---|---|---|---|---|
| | 0 | 1 | 7 | 14 | 21 | 28 |
| | DPPH (%) | | | | | |
| W | 96.17 ± 0.11 [ABab] | 96.22 ± 0.11 [Ba] | 95.99 ± 0.35 [Aa] | 95.67 ± 0.11 [Ca] | 95.71 ± 0.10 [Ca] | 95.53 ± 0.21 [Ca] |
| WMP | 95.90 ± 0.17 [ABEa] | 95.86 ± 0.05 [ACEb] | 96.25 ± 0.23 [Ca] | 96.10 ± 0.14 [Bb] | 95.63 ± 0.26 [Da] | 95.81 ± 0.11 [DEab] |
| WFG1 | 96.32 ± 0.08 [Ab] | 96.29 ± 0.04 [ACa] | 96.01 ± 0.14 [Ba] | 96.10 ± 0.16A [BCb] | 96.31 ± 0.06 [ACb] | 95.59 ± 0.17 [Da] |
| WFG2 | 96.16 ± 0.12 [ABab] | 96.91 ± 0.19 [Cc] | 95.54 ± 0.54 [Db] | 96.71 ± 0.07 [Cc] | 96.32 ± 0.15 [Ab] | 95.95 ± 0.17 [Bb] |
| | ABTS (%) | | | | | |
| W | 69.84 ± 1.70 [Aa] | 62.78 ± 2.10 [BCa] | 64.34 ± 2.30 [CDa] | 61.72 ± 1.20 [Ba] | 67.02 ± 1.90 [Ea] | 65.40 ± 2.20 [DEa] |
| WMP | 73.02 ± 1.80 [Aabc] | 71.61 ± 1.40 [Ab] | 72.95 ± 1.10 [ABb] | 72.39 ± 3.90 [Ab] | 72.67 ± 2.30 [ABb] | 74.79 ± 2.30 [Bb] |
| WFG1 | 76.77 ± 2.30 [Ab] | 75.56 ± 1.20 [ABb] | 72.39 ± 0.10 [Db] | 70.06 ± 1.40 [Cb] | 72.25 ± 2.50 [CDb] | 73.94 ± 2.10 [BDbc] |
| WFG2 | 77.54 ± 0.40 [Ac] | 72.46 ± 2.20 [Bb] | 75.07 ± 1.50 [Cb] | 71.89 ± 2.40 [CDb] | 73.52 ± 3.50 [BCb] | 72.46 ± 1.20 [Cbc] |
| | RSC (mg/mL) | | | | | |
| W | 141.46 ± 1.52 [Aa] | 133.46 ± 2.50 [Ba] | 150.35 ± 3.32 [Ca] | 162.15 ± 3.05 [Da] | 133.27 ± 1.03 [Ba] | 106.38 ± 2.18 [Ea] |
| WMP | 188.73 ± 3.86 [Ab] | 203.54 ± 2.39 [Bb] | 222.85 ± 5.87 [Cb] | 253.23 ± 5.66 [Db] | 244.85 ± 2.72 [Db] | 217.46 ± 8.81 [Bb] |
| WFG1 | 248.32 ± 1.23 [Ac] | 232.69 ± 12.29 [Bc] | 224.45 ± 2.21 [Cc] | 222.96 ± 0.82 [Cb] | 234.35 ± 3.97 [Bc] | 216.08 ± 4.79 [Dc] |
| WFG2 | 271.04 ± 5.60 [Ad] | 247.69 ± 5.33 [Bd] | 274.81 ± 4.51 [Ad] | 224.31 ± 0.65 [Cc] | 247.23 ± 2.39 [Bc] | 211.19 ± 2.23 [Dd] |
| | TFAAL (mg/mL) | | | | | |
| W | 1.51 ± 0.12 [ABa] | 1.60 ± 0.18 [ABa] | 1.72 ± 0.11 [Aa] | 1.63 ± 0.15 [Aa] | 1.32 ± 0.14 [Ba] | 0.92 ± 0.00 [Ca] |
| WMP | 2.22 ± 0.15 [Ab] | 2.48 ± 0.27 [ABb] | 2.66 ± 0.19 [Bb] | 2.68 ± 0.23 [Bb] | 2.23 ± 0.20 [Ab] | 2.31 ± 0.19 [Ab] |
| WFG1 | 2.59 ± 0.22 [ABbc] | 2.31 ± 0.24 [ACb] | 2.82 ± 0.27 [BDb] | 3.10 ± 0.30 [Db] | 2.21 ± 0.18 [Cb] | 3.83 ± 0.80 [Ec] |
| **WFG2** | 3.03 ± 0.31 [Ac] | 3.26 ± 0.46 [Ac] | 3.92 ± 0.32 [Bc] | 1.77 ± 0.11 [Ca] | 1.81 ± 0.18 [Cab] | 1.59 ± 0.15 [Cd] |

W—fermented whey, WMP—fermented whey with milk powder added (5% *w/v*), WFG1—fermented whey with flaxseed gum added (0.5% *w/v*), WFG2—fermented whey with flaxseed gum added (1.0% *w/v*). Values are means ± standard deviation of triplicate determinations. Means with different lowercase letters in the same column are significantly different at $p < 0.05$. Means with different uppercase letters in the same row are significantly different at $p < 0.05$.

## 4. Conclusions

The findings of the study indicated that acid whey can be successfully supplemented with flaxseed gum in a novel functional dairy beverage. With the aim to evaluate the properties of the novel beverage, several parameters, such as microbial survivability, physicochemical properties, color changes, viscosity, sensory profile and antioxidant potential, were analyzed. The addition of flaxseed gum as a natural stabilizer and thickening agent improved the viscosity of acid whey to a level comparable with milk powder, meeting the consumer acceptance criteria. Moreover, the added value of using FG as a whey additive was the high antioxidant activity and free amino acids level of the products. It should be emphasized that the consumption of the drinks may exert a potentially beneficial action on human organisms due to the highly beneficial microflora (lactic acid bacteria) viability, antiradical activity and amino acids level. However, further tests should be carried out focusing on proteolytic changes and the production of bioactive peptides, amino acids, as well as an in vivo analysis of potential health benefits for the functioning of the human body. Research should also focus on increasing the taste acceptance of FWB with gums, as taste is directly related to the overall acceptance of beverages to consumers. The use of valuable by-products from the dairy and oil industries opens up a promising direction for the production of innovative food products, which is in line with the principles of circular economy and the idea of zero waste.

**Author Contributions:** Ł.Ł.—Conceptualization, data curation, formal analysis, investigation, methodology, supervision, resources, visualization, writing—original draft; I.D.—conceptualization, investigation, methodology, resources, formal analysis, writing—original draft; E.D.—data curation, formal analysis, investigation; A.M.-M.—conceptualization, investigation, methodology, formal analysis; P.K.—investigation; M.S., E.K.—formal analysis. All authors have read and agreed to the published version of the manuscript.

**Funding:** This research received no external funding.

**Institutional Review Board Statement:** Not applicable.

**Informed Consent Statement:** Not applicable.

**Data Availability Statement:** The data presented in this study are available on request from the corresponding authors.

**Conflicts of Interest:** The authors declare no conflict of interest.

## References

1. Ryan, M.P.; Walsh, G. The biotechnological potential of whey. *Rev. Environ. Sci. Bio/Technol.* **2016**, *15*, 479–498. [CrossRef]
2. Skryplonek, K.; Dmytrów, I.; Mituniewicz-Małek, A. Probiotic fermented beverages based on acid whey. *J. Dairy Sci.* **2019**, *102*, 7773–7780. [CrossRef] [PubMed]
3. Kadyan, S.; Rashmi, H.M.; Pradhan, D.; Kumari, A.; Chaudhari, A.; Deshwal, G.K. Effect of lactic acid bacteria and yeast fermentation on antimicrobial, antioxidative and metabolomic profile of naturally carbonated probiotic whey drink. *LWT* **2021**, *142*, 111059. [CrossRef]
4. León-López, A.; Pérez-Marroquín, X.A.; Campos-Lozada, G.; Campos-Montiel, R.G.; Aguirre-Álvarez, G. Characterization of whey-based fermented beverages supplemented with hydrolyzed collagen: Antioxidant activity and bioavailability. *Foods* **2020**, *9*, 1106. [CrossRef] [PubMed]
5. Łopusiewicz, Ł.; Drozlowska, E.; Trocer, P.; Kostek, M.; Śliwiński, M.; Henriques, M.H.F.; Bartkowiak, A.; Sobolewski, P. Whey protein concentrate/isolate biofunctional films modified with melanin from watermelon (*Citrullus lanatus*) seeds. *Materials* **2020**, *13*, 3876. [CrossRef] [PubMed]
6. Macwan, S.R.; Dabhi, B.K.; Parmar, S.C.; Aparnathi, K.D. Whey and its utilization. *Int. J. Curr. Microbiol. Appl. Sci.* **2016**, *5*, 134–155. [CrossRef]
7. Yadav, J.S.S.; Yan, S.; Pilli, S.; Kumar, L.; Tyagi, R.D.; Surampalli, R.Y. Cheese whey: A potential resource to transform into bioprotein, functional/nutritional proteins and bioactive peptides. *Biotechnol. Adv.* **2015**, *33*, 756–774. [CrossRef]
8. Pescuma, M.; Hébert, E.M.; Mozzi, F.; de Valdez, G.F. Functional fermented whey-based beverage using lactic acid bacteria. *Int. J. Food Microbiol.* **2010**, *141*, 73–81. [CrossRef]
9. Djurić, M.; Carić, M.; Milanović, S.; Tekić, M.; Panić, M. Development of whey-based beverages. *Eur. Food Res. Technol.* **2004**, *219*, 321–328. [CrossRef]

10. Luo, S.R.; Demarsh, T.; Stelick, A.; Alcaine, S. Characterization of the Fermentation and Sensory Profiles of Novel Yeast-Fermented Acid Whey Beverages. *Foods* **2021**, *10*, 1204. [CrossRef]
11. Joshi, J.; Gururani, P.; Vishnoi, S.; Srivastava, A. Whey Based Beverages: A Review. *Octa J. Biosci.* **2020**, *8*, 30–37.
12. Borges, A.R.; Pires, A.F.; Marnotes, N.G.; Gomes, D.G.; Henriques, M.F.; Pereira, C.D. Dairy by-Products Concentrated by Ultrafiltration Used as Ingredients in the Production of Reduced Fat Washed Curd Cheese. *Foods* **2020**, *9*, 1020. [CrossRef]
13. Pires, A.F.; Marnotes, N.G.; Rubio, O.D.; Garcia, A.C.; Pereira, C.D. Dairy By-Products: A Review on the Valorization of Whey and Second Cheese Whey. *Foods* **2021**, *10*, 1067. [CrossRef]
14. Gallardo-Escamilla, F.J.; Kelly, A.L.; Delahunty, C.M. Mouthfeel and flavour of fermented whey with added hydrocolloids. *Int. Dairy J.* **2007**, *17*, 308–315. [CrossRef]
15. Farah, J.S.; Araujo, C.B.; Melo, L. Analysis of yoghurts', whey-based beverages' and fermented milks' labels and differences on their sensory profiles and acceptance. *Int. Dairy J.* **2017**, *68*, 17–22. [CrossRef]
16. Buriti, F.C.A.; Freitas, S.C.; Egito, A.Ô.S.; Dos Santos, K.M.O. Effects of tropical fruit pulps and partially hydrolysed galactomannan from Caesalpinia pulcherrima seeds on the dietary fibre content, probiotic viability, texture and sensory features of goat dairy beverages. *LWT* **2014**, *59*, 196–203. [CrossRef]
17. Kalicka, D.; Znamirowska, A.; Pawlos, M.; Szajnar, K. Effect of the type of thickener for selected quality features of peach yoghurts during refrigerated storage. In *Technological Shaping of Food Quality*; Wójciak, K.M., Dolatowski, Z.J., Eds.; Wydawnictwo Naukowe PTTŻ: Kraków, Poland, 2015; pp. 69–78.
18. Gawai, K.M.; Mudgal, S.P.; Prajapati, J.B. Chapter 3—Stabilizers, Colorants, and Exopolysaccharides in Yogurt A2—Shah, Nagendra P. In *Yogurt and Health Disease Prevention*; Academic Press: Cambridge, MA, USA, 2017; pp. 49–68.
19. Gutiérrez, C.; Rubilar, M.; Jara, C.; Verdugo, M.; Sineiro, J.; Shene, C. Flaxseed and flaxseed cake as a source of compounds for food industry. *J. Soil Sci. Plant Nutr.* **2010**, *10*, 454–463. [CrossRef]
20. Kajla, P.; Sharma, A.; Sood, D.R. Flaxseed-a potential functional food source. *J. Food Sci. Technol.* **2015**, *52*, 1857–1871. [CrossRef]
21. Rocha-Mendoza, D.; Kosmerl, E.; Krentz, A.; Zhang, L.; Badiger, S.; Miyagusuku-Cruzado, G.; Mayta-Apaza, A.; Giusti, M.; Jiménez-Flores, R.; García-Cano, I. Invited review: Acid whey trends and health benefits. *J. Dairy Sci.* **2021**, *104*, 1262–1275. [CrossRef]
22. Hu, Y.; Shim, Y.Y.; Reaney, M.J.T. Flaxseed Gum Solution Functional Properties. *Foods* **2020**, *9*, 681. [CrossRef]
23. Guo, Q.; Zhu, X.; Zhen, W.; Li, Z.; Kang, J.; Sun, X.; Wang, S.; Cui, S.W. Rheological properties and stabilizing effects of high-temperature extracted flaxseed gum on oil/water emulsion systems. *Food Hydrocoll.* **2021**, *112*, 106289. [CrossRef]
24. Drozłowska, E.; Bartkowiak, A.; Trocer, P.; Kostek, M.; Tarnowiecka-Kuca, A.; Bienkiewicz, G.; Łopusiewicz, Ł. The influence of flaxseed oil cake extract on oxidative stability of microencapsulated flaxseed oil in spray-dried powders. *Antioxidants* **2021**, *10*, 211. [CrossRef]
25. Drozłowska, E.; Bartkowiak, A.; Trocer, P.; Kostek, M.; Tarnowiecka-Kuca, A.; Łopusiewicz, Ł. Formulation and Evaluation of Spray-Dried Reconstituted Flaxseed Oil-In-Water Emulsions Based on Flaxseed Oil Cake Extract as Emulsifying and Stabilizing Agent. *Foods* **2021**, *10*, 256. [CrossRef]
26. Drozłowska, E.; Bartkowiak, A.; Łopusiewicz, Ł. Characterization of flaxseed oil bimodal emulsions prepared with flaxseed oil cake extract applied as a natural emulsifying agent. *Polymers* **2020**, *12*, 2207. [CrossRef]
27. Khalloufi, S.; Corredig, M.; Goff, H.D.; Alexander, M. Flaxseed gums and their adsorption on whey protein-stabilized oil-in-water emulsions. *Food Hydrocoll.* **2009**, *23*, 611–618. [CrossRef]
28. Drozłowska, E.; Łopusiewicz, Ł.; Mężyńska, M.; Bartkowiak, A. The effect of native and denaturated flaxseed meal extract on physiochemical properties of low fat mayonnaises. *J. Food Meas. Charact.* **2020**, *14*, 1135–1145. [CrossRef]
29. Fedeniuk, R.W.; Biliaderis, C.G. Composition and Physicochemical Properties of Linseed (*Linum usitatissimum* L.) Mucilage. *J. Agric. Food Chem.* **1994**, *42*, 240–247. [CrossRef]
30. Drozłowska, E.; Łopusiewicz, Ł.; Mężyńska, M.; Bartkowiak, A. Valorization of flaxseed oil cake residual from cold-press oil production as a material for preparation of spray-dried functional powders for food applications as emulsion stabilizers. *Biomolecules* **2020**, *10*, 153. [CrossRef]
31. Dmytrów, I.; Szczepanik, G.; Kryza, K.; Mituniewicz-Małek, A.; Lisiecki, S. Impact of polylactic acid packaging on the organoleptic and physicochemical properties of tvarog during storage. *Int. J. Dairy Technol.* **2011**, *64*, 569–577. [CrossRef]
32. Łopusiewicz, Ł.; Drozłowska, E.; Trocer, P.; Kwiatkowski, P.; Bartkowiak, A.; Gefrom, A.; Sienkiewicz, M. The Effect of Fermentation with Kefir Grains on the Physicochemical and Antioxidant Properties of Beverages from Blue Lupin (*Lupinus angustifolius* L.) Seeds. *Molecules* **2020**, *25*, 5791. [CrossRef]
33. AOAC. *Official Methods of Analysis*; Association of Official Analytical Chemists: Gaithersburg, MD, USA, 1997.
34. *PN-ISO 5534:2005*; Determination of Total Dry Matter Content (Reference Method). Polish Committee for Standardization: Warszawa, Poland, 2005. (In Polish)
35. Yu, D.; Kwon, G.; An, J.; Lim, Y.S.; Jhoo, J.W.; Chung, D. Influence of prebiotic biopolymers on physicochemical and sensory characteristics of yoghurt. *Int. Dairy J.* **2021**, *115*, 104915. [CrossRef]
36. Kubo, M.T.K.; Maus, D.; Xavier, A.A.O.; Mercadante, A.Z.; Viotto, W.H. Transference of lutein during cheese making, color stability, and sensory acceptance of Prato cheese. *Food Sci. Technol.* **2013**, *33*, 81–88. [CrossRef]
37. *ISO 22935-2:2013-07*; Milk and Milk Products. Sensory Analysis. Part 2: Recommended Methods of Sensory Evaluation. Polish Committee for Standardization: Warszawa, Poland, 2013. (In Polish)

38.	*PN-ISO 22935-3:2013-07*; Milk and Milk Products. Sensory Analysis. Part 3: Guidelines for Assessing the Conformity of Sensory Properties with Product Specifications Using the Scoring Method. Polish Committee for Standardization: Warszawa, Poland, 2013. (In Polish)
39.	Kiełczewska, K.; Pietrzak-Fiećko, R.; Nalepa, B. Acid whey-based smoothies with probiotic strains. *J. Elem.* **2020**, *25*, 1435–1448. [CrossRef]
40.	AbdulAlim, T.S.; Abeer, F.; Zayan, A.F.; Pedro, H.; Campelo, P.H.; Bakry, A.M. Development of new functional fermented product: Mulberry-whey beverage. *J. Nutr. Food Technol.* **2018**, *1*, 64–69. [CrossRef]
41.	Sady, M.; Najgebauer-Lejko, D.; Domagala, J. The suitability of different probiotic strains for the production of fruit-whey beverages. *Acta Sci. Pol. Technol. Aliment.* **2017**, *16*, 421–429. [CrossRef]
42.	Madhavi, T.V.; Shah, R.K. Microbiological study of synbiotic fermented whey drink. *Res. J. Anim. Husb. Dairy Sci. Res. Pap.* **2017**, *8*, 1–7. [CrossRef]
43.	Hadinezhad, M.; Duc, C.; Han, N.F.; Hosseinian, F. Flaxseed Soluble Dietary Fibre Enhances Lactic Acid Bacterial Survival and Growth in Kefir and Possesses High Antioxidant Capacity. *J. Food Res.* **2013**, *2*, 152. [CrossRef]
44.	Basiri, S.; Haidary, N.; Shekarforoush, S.S.; Niakousari, M. Flaxseed mucilage: A natural stabilizer in stirred yogurt. *Carbohydr. Polym.* **2018**, *187*, 59–65. [CrossRef]
45.	Kailasapathy, K.; Chin, J. Survival and therapeutic potential of probiotic organisms with reference to *Lactobacillus acidophilus* and *Bifidobacterium* spp. *Immunol. Cell Biol.* **2000**, *78*, 80–88. [CrossRef]
46.	Terpou., A.; Bosnea, L.; Kanellaki, M. Effect of Mastic Gum (Pistacia Lentiscus Via Chia) as a Probiotic Cell Encapsulation Carrier for Functional Whey Beverage Production. *SCIOL Biomed.* **2017**, *1*, 1–10.
47.	Farahani, Z.K.; Mousavi, S.M.A.E.; Ardebili, S.M.S.; Bakhoda, H. Modification of sodium alginate by octenyl succinic anhydride to fabricate beads for encapsulating jujube extract. *Curr. Res. Food Sci.* **2022**, *5*, 157. [CrossRef]
48.	Tabibloghmany, F.; Hojjatoleslamy, M.; Farhadian, F.; Ehsandoost, E. Effect of linseed (*Linum usitatissimum* L.) hydrocolloid as edible coating on decreasing oil absorption in potato chips during Deep-fat frying. *Int. J. Agric. Crop Sci.* **2013**, *6*, 63–69.
49.	Aamer, R.A.; El-Kholy, W.M.; Mailam, M.A. Production of Functional Beverages from Whey and Permeate Containing Kumquat Fruit. *Alexandria J. Food Sci. Technol.* **2017**, *14*, 41–56. [CrossRef]
50.	Alane, D.; Raut, N.; Kamble, D.B.; Bhotmange, M. Studies on preparation and storage stability of whey based mango herbal beverage. *Int. J. Chem. Stud.* **2017**, *5*, 237–241.
51.	Ismail, A.E.; Abdelgader, M.O.; Ali, A.A. Microbial and chemical evaluation of Whey-Based Mango beverage. *Adv. J. Food Sci. Technol.* **2011**, *3*, 250–253.
52.	Salaün, F.; Mietton, B.; Gaucheron, F. Buffering capacity of dairy products. *Int. Dairy J.* **2005**, *15*, 95–109. [CrossRef]
53.	Kalyas, A.; Ürkek, B. Effect of Flaxseed Powder on Physicochemical, Rheological, Microbiological and Sensory Properties of Yoghurt. *Braz. Arch. Biol. Technol.* **2022**, *65*, 2022. [CrossRef]
54.	Van Long, N.N.; Rigalma, K.; Coroller, L.; Dadure, R.; Debaets, S.; Mounier, J.; Vasseur, V. Modelling the effect of water activity reduction by sodium chloride or glycerol on conidial germination and radial growth of filamentous fungi encountered in dairy foods. *Food Microbiol.* **2017**, *68*, 7–15. [CrossRef]
55.	Skryplonek, K.; Jasińska, M. The quality of fermented probiotic drinks obtained from frozen acid whey and milk during cooling storage. *Food Sci. Technol. Qual.* **2016**, *1*, 32–44. [CrossRef]
56.	Skryplonek, K.; Jasińska, M. Fermented probiotic beverages based on acid whey. *Acta Sci. Pol. Technol. Aliment.* **2015**, *14*, 397–405. [CrossRef]
57.	Dehghan, B.; Kenari, R.E.; Amiri, Z.R. Stabilization of whey-based pina colada beverage by mixed Iranian native gums: A mixture design approach. *J. Food Meas. Charact.* **2022**, *16*, 171–179. [CrossRef]
58.	Hashemi, F.S.; Pezeshky, A.; Gharedaghi, K.; Javani, R. Effect of Hydrocolloids on Sensory Properties of the Fermented Whey Beverage. In Proceedings of the 1st International Conference on Natural Food Hydrocolloids, Mashhad, Iran, 22–23 October 2014.
59.	Zendeboodi, F.; Yeganehzad, S.; Sadeghian, A. Optimizing the formulation of a natural soft drink based on biophysical properties using mixture design methodology. *J. Food Meas. Charact.* **2018**, *12*, 763–769. [CrossRef]
60.	Liu, J.; Shim, Y.Y.; Tse, T.J.; Wang, Y.; Reaney, M.J.T. Flaxseed gum a versatile natural hydrocolloid for food and non-food applications. *Trends Food Sci. Technol.* **2018**, *75*, 146–157. [CrossRef]
61.	Amiri, M.S.; Mohammadzadeh, V.; Yazdi, M.E.T.; Barani, M.; Rahdar, A.; Kyzas, G.Z. Plant-Based Gums and Mucilages Applications in Pharmacology and Nanomedicine: A Review. *Molecules* **2021**, *26*, 1770. [CrossRef]
62.	Cui, W.; Mazza, G. Physicochemical characteristics of flaxseed gum. *Food Res. Int.* **1996**, *29*, 397–402. [CrossRef]
63.	Chen, H.H.; Xu, S.Y.; Wang, Z. Gelation properties of flaxseed gum. *J. Food Eng.* **2006**, *77*, 295–303. [CrossRef]
64.	Durham, R.J.; Hourigan, J.A. Waste Management and Co-product Recovery in Food Processing. In *Woodhead Publishing Series in Food Science, Technology and Nutrition*; Waldron, K.W., Ed.; Woodhead Publishing: Sawston, UK, 2007; pp. 332–387. ISBN 978-1-84569-025-0.
65.	Uebelacker, M.; Lachenmeier, D.W. Quantitative determination of acetaldehyde in foods using automated digestion with simulated gastric fluid followed by headspace gas chromatography. *J. Autom. Methods Manag. Chem.* **2011**, *2011*, 907317. [CrossRef]
66.	Shepard, L.; Miracle, R.E.; Leksrisompong, P.; Drake, M.A. Relating sensory and chemical properties of sour cream to consumer acceptance. *J. Dairy Sci.* **2013**, *96*, 5435–5454. [CrossRef]

67. Gierczynski, I.; Guichard, E.; Laboure, H. Aroma perception in dairy products: The roles of texture, aroma release and consumer physiology. A review. *Flavour Fragr. J.* **2011**, *26*, 141–152. [CrossRef]
68. Zarei, F.; Nateghi, L.; Eshaghi, M.R.; Abadi, M.E.T. Production of gamma-aminobutyric acid (Gaba) in whey protein drink during fermentation by lactobacillus plantarum. *J. Microbiol. Biotechnol. Food Sci.* **2020**, *9*, 1087–1092. [CrossRef]
69. Keshtkaran, M.; Mohammadifar, M.A.; Asadi, G.H.; Nejad, R.A.; Balaghi, S. Effect of gum tragacanth on rheological and physical properties of a flavored milk drink made with date syrup. *J. Dairy Sci.* **2013**, *96*, 4794–4803. [CrossRef] [PubMed]
70. Behbahani, M.; Abbasi, S. Stabilization of Flixweed (*Descurainia sophia* L.) syrup using native hydrocolloids. *Iran. J. Nutr. Sci. Food Technol.* **2014**, *9*, 31–38.
71. Neis, E.P.J.G.; Dejong, C.H.C.; Rensen, S.S. The role of microbial amino acid metabolism in host metabolism. *Nutrients* **2015**, *7*, 2930–2946. [CrossRef] [PubMed]
72. Ansari, F.; Pourjafar, H.; Bahadori, M.B.; Pimentel, T.C. Effect of microencapsulation on the development of antioxidant activity and viability of lactobacillus acidophilus la5 in whey drink during fermentation. *Biointerface Res. Appl. Chem.* **2021**, *11*, 9762–9771. [CrossRef]
73. Virtanen, T.; Pihlanto, A.; Akkanen, S.; Korhonen, H. Development of antioxidant activity in milk whey during fermentation with lactic acid bacteria. *J. Appl. Microbiol.* **2007**, *102*, 106–115. [CrossRef]

*Article*

# Safety Assessment of Organic High-Protein Bars during Storage at Ambient and Refrigerated Temperatures

Monika Trząskowska [1], Katarzyna Neffe-Skocińska [1,*], Anna Okoń [2], Dorota Zielińska [1], Aleksandra Szydłowska [1], Anna Łepecka [2] and Danuta Kołożyn-Krajewska [1]

[1] Institute of Human Nutrition Sciences, Warsaw University of Life Sciences (WULS), Nowoursynowska St. 159C, 02-776 Warszawa, Poland

[2] Department of Meat and Fat Technology, Prof. Wacław Dąbrowski Institute of Agricultural and Food Biotechnology-State Research Institute, Rakowiecka St. 36, 02-532 Warszawa, Poland

* Correspondence: katarzyna_neffe_skocinska@sggw.edu.pl; Tel.: +48-225937067; Fax: +48-225937068

**Citation:** Trząskowska, M.; Neffe-Skocińska, K.; Okoń, A.; Zielińska, D.; Szydłowska, A.; Łepecka, A.; Kołożyn-Krajewska, D. Safety Assessment of Organic High-Protein Bars during Storage at Ambient and Refrigerated Temperatures. *Appl. Sci.* **2022**, *12*, 8454. https://doi.org/10.3390/app12178454

Academic Editors: Joanna Trafialek and Mariusz Szymczak

Received: 21 July 2022
Accepted: 22 August 2022
Published: 24 August 2022

**Publisher's Note:** MDPI stays neutral with regard to jurisdictional claims in published maps and institutional affiliations.

**Abstract:** This study aimed to assess the safety characteristics of organic high-protein bars (HPB) during storage at ambient and refrigerated temperatures based on selected microbiological and chemical indicators. After production, the total number of microorganisms ranged from 3.90–4.26 log CFU/g;. The *Enterobacteriaceae* family was present at 2.81–3.32 log CFU/g, and the count of yeasts and moulds was 2.61–3.99 log CFU/g. No *Salmonella* sp. was found in 25 g of the product. *Bacillus cereus* was present in samples B1 and B2. *Staphylococcus aureus* was presented in samples below the detection limit (<2 log CFU/g). During the storage of products, the number of microorganisms varied. After production and storage, in all samples of HPB, the amount of mycotoxins was below the detection limit. The presence of histamine and tryptamine was not found in the HPB throughout the study period. Regarding TBARS, it can be concluded that the use of prunes and oat flakes (B2 bar composition) in the production of organic bars, and refrigerated storage, reduces the degree of fat oxidation. Among the tested variants, the composition of the B3 bar seemed to be the safest and worth further research, mainly due to the lower frequency of undesirable microorganisms. The protective antioxidative effect of prunes and oat flakes in bars stored at 22 °C indicates the value of the composition of bar B2. The appropriate composition modifications and the use of heat treatment proved to be effective in improving the safety characteristics of HPB. Relying on the results it is possible to store HPB for at least 3 months. Next to standard safety parameters, the unique and effective to increase the safety of HPB is controlling the presence of *B. cereus* and other low water activity (aw) resistant microorganisms.

**Keywords:** biogenic amines; fat oxidation; food safety; mycotoxin; organic bar; meal replacement

## 1. Introduction

Nowadays, one of the major health goals is to achieve and maintain a desirable body weight, and this is accompanied by high awareness of the types of diets used for weight control. Meal replacements are substitutes for a wholesome dish in the form of a snack, bar or powder for making a drink or soup. They should be palatable, portion-controlled, exclude the need to choose and prepare foods, and eliminate food varieties that can spur over-consumption of energy [1,2]. Incorporating meal replacements (1–2 per day) into traditional lifestyle interventions is effective in treating overweight or obese patients [3,4]. Among other food trends, organic food products are considered to have higher nutritional value, compared to conventional ones. As well, organic plant raw materials contain fewer nitrates and pesticide residues [5].

A high-protein bar is an example of meal replacement. Its production should include the use of high-quality raw materials that bring biologically active substances, i.e., the basic nutrients, as well as non-nutritive compounds naturally occurring in the raw material or in the product subjected to the technological process that affects physiological and metabolic functions of the body [6]. Previous research conducted by Szydłowska et al. [7] concerned the development of bar recipes made from organic plant-origin material (cereal, dried fruits, nuts, chocolate and others), as well as whey protein concentrate (WPC). The protein content was dependent on the product composition. Muesli and coconut bars contained a high amount of proteins in total (at least 20% energy from proteins according to Regulation (EC) No 1924/2006 [8]. All bars were characterized by a high scavenging capacity of DPPH (2,2-diphenyl-1-picryl-hydrazyl-hydrate) free radicals. However, some organic ingredients were contaminated with *Bacillus cereus*, and these bacteria were also found in muesli and pumpkin bars.

Although these organic bars had low water activity (aw) (0.63–0.74), which has clear advantages concerning controlling the growth of foodborne pathogens, there are nevertheless some major concerns. Desiccation stress is a common event in natural and food industry environments that significantly affects bacterial viability, but they can resist this [9,10].

The study aimed to assess the safety characteristics of organic high-protein during storage at ambient and refrigerated temperatures. Moreover, an attempt was made to define the date of minimum durability and the storage conditions of the bars. Selected microbiological and chemical indicators were evaluated.

## 2. Materials and Methods

### 2.1. Preparation of High-Protein Bar

The organic bars were produced under industrial conditions. The ingredients used in the production of the 100 g bar are listed in Table 1. The manufacture of organic bars was as follows. Pumpkin seeds, spelt, oat flakes and coconut shreds when added were roasted and grounded. Prunes and/or dried apricots were mixed with boiling water and allowed to soak for an hour, then the water was discarded, and the fruits were mixed until smooth. Then, water, sunflower oil, honey, and inulin were added and mixed. Dried cherries, freeze-dried raspberries were mixed, and all ingredients were blended. Subsequently, part of the WPC was added, mixed thoroughly with the remaining ingredients and the bar was formed. Next, chocolate (organic chocolate, 70% cacao content, organic cane sugar) was liquefied and the unused amount of WPC was added (proportionally for every 100 g of chocolate to 10 g of WPC). Bars were covered with chocolate glaze, packed in the metalized polyester polyethylene (MP) and sealed. Three different bars were prepared and named B1, B2, and B3. Bar B1 consisted of an equal amount of prune and dried apricots with no addition of oat flakes but coconut shreds. Bar B2 contained only prunes and B3 only dried apricots, and instead of coconut shreds they were made from oat flakes as well.

Bars were stored at refrigerated (4 °C) and ambient temperature (22 °C). The measurements of microbiological quality and water activity (aw) were carried out after production (day 0) and in the 1st, 2nd and 3rd months of storage. The presence of selected biogenic amines and selected mycotoxins were measured on day 0 and after 3 months of storage.

### 2.2. Microbiological Quality Evaluation

To evaluate the microbiological quality of the organic high-protein bars, the spread plate technique on an appropriate agar medium was performed, determining the total viable count (TVC) following ISO 4833-1:2013 [11] on nutrient agar (LabM, Heywood, UK). The number of rods from the *Enterobacteriaceae* family (ENT) by ISO 21528-2:2017 [12] on MacConkey agar No. 3 (LabM, Heywood, UK), and the number of yeasts and moulds (TYMC), following ISO 21527-1:2008 [13] on YGC agar (Sabouraud Dextrose with chloramphenicol lab Agar, Biomaxima, Lublin, Poland). 10 g of samples B1, B2 or B3 were used for evaluation. The measurements for each bar sample (B1, B2 or B3) were performed in triplicate.

**Table 1.** Variants of study beer beverages and their composition.

| Ingredient (g) | Bar Symbol | | |
|---|---|---|---|
| | **B1** | **B2** | **B3** |
| Whey protein concentrate (WPC) | 15.9 | 16.7 | 16.3 |
| Pumpkin seeds | 14.3 | 15.0 | 14.6 |
| Spelt flakes | 11.9 | 12.5 | 12.2 |
| Prune | 7.9 | 16.7 | 0.0 |
| Dried apricots | 7.9 | 0.0 | 16.3 |
| Oat flakes | 0.0 | 8.3 | 8.1 |
| Coconut shreds | 7.9 | 0.0 | 0.0 |
| Honey | 3.2 | 3.3 | 3.3 |
| Sunflower oil | 2.4 | 2.5 | 2.4 |
| Inulin | 2.4 | 2.5 | 2.4 |
| Dried cherries | 1.6 | 1.7 | 1.6 |
| Freeze-dried raspberries | 0.8 | 0.8 | 0.8 |
| Water | 7.9 | 3.3 | 5.7 |
| Chocolate | 15.9 | 16.7 | 16.3 |
| Sum | 100 | 100 | 100 |

The presence of pathogenic bacteria was determined using the enrichment culture method with media indicated in the standards, such as XLD agar (Xylose Lysine Deoxycholate Agar, LabM, Heywood, UK) and RAPID'Salmonella agar (Bio-Rad, Hercules, USA) to determine the presence of *Salmonella* spp., following ISO 6579-1:2017 [14]. To determine the presence of *Bacillus cereus* agar PEMBA (LabM, Heywood, UK) was used following ISO 21871:2006 [15]. The presence of *Staphylococcus aureus* was detected using Baird-Parker agar (LabM, Heywood, UK) according to PN-EN ISO 6888-1:2001 [16]. Samples (25 g) of B1, B2 or B3 were used for evaluation. The measurements for each bar sample (B1, B2 and B3) were performed in triplicate.

### 2.3. Water Activity Evaluation

The measurement of water activity (aw) was performed using an AQUALAB 4TE Series Water Activity Meter (METER Group, Inc., Pullman, WA, USA) according to ISO 21807:2004 [17]. The water activity of the sample was measured with a photoelectric sensor using the dew point detection method. According to the producers' instructions, approx. 5 g of samples B1, B2 or B3 were placed in the test container and then in the measuring chamber of the apparatus. The mean value of three replicates was assumed as the final result of the water activity of the tested samples.

### 2.4. Mycotoxin Evaluation

Mycotoxin determinations were carried out by an accredited external Hamilton laboratory (https://hamilton.com.pl/en/ (accessed on 21 August 2022)) according to the methodology described in the standards authorized by the Polish Committee for Standardization (PKN), the sum of aflatoxins according to PN-EN 14123:2008 [18], and ochratoxin A according to PN-EN 14132:2010 [19]. The content of zearalenone deoxynivalenol was determined by high-performance liquid chromatography according to the J.S. Hamilton's internal procedures [20,21]. The test was carried out after production and after the 3rd month of storage.

### 2.5. Biogenic Amines Content

The analysed amines were 2-phenylethyl amine, Cadaverine, Histamine, Putrescine, Spermidine, Spermine, Tryptamine, Tyramine. The content of biogenic amines in organic high-protein bars was determined by an accredited external Hamilton laboratory (https://hamilton.com.pl/en/ (accessed on 21 August 2022)) and immediately after production in industrial conditions, as well as after three months of storage at two temperatures. The presence of biogenic amines was determined using liquid chromatographic analyses.

Biogenic amines were separated using a liquid chromatograph consisting of a quaternary pump, a vacuum degasser, an autosampler, a LC-UV/DAD detector with variable wavelength, and a fluorescence detector. The methodology of the analyses was according to Smělá et al. [22].

### 2.6. Analysis of Lipid Oxidation TBARS

The determination was performed according to the modified Salih method according to Pikul et al. [23]. The absorbance was measured at a wavelength of 532 nm against a blank containing 5 cm$^3$ 4% perchloric acid and 5 cm$^3$ TBA reagent.

### 2.7. Presence of Pests

The assessment of the presence of warehouse pests consisted of an inspection of the packaging of B1, B2 and B3 bars immediately after production and during storage, after checking that the packaging is undamaged. Whole pests, fragments of their bodies, droppings, fume and other traces of presence were sought. After the first inspection, the bars were placed in an airtight container. Sticky traps were placed at the place of storage. Observation of pests was carried out after the 1st, 2nd and 3rd months of storage.

### 2.8. Statistical Analysis

One-way or multiple factor analysis of variance (ANOVA) with a linear model, as well as post hoc Bonferroni test, were applied to the statistical analysis of the data. Statistical significance was recognized when $p < 0.05$. Tests were conducted using Statistica 13 (StatSoft Inc., Tulsa, OK, USA).

## 3. Results

### 3.1. Microbiological Quality Evaluation

The results of the average number of selected groups of microorganisms in the tested variants of high-protein bars are shown in Figure 1.

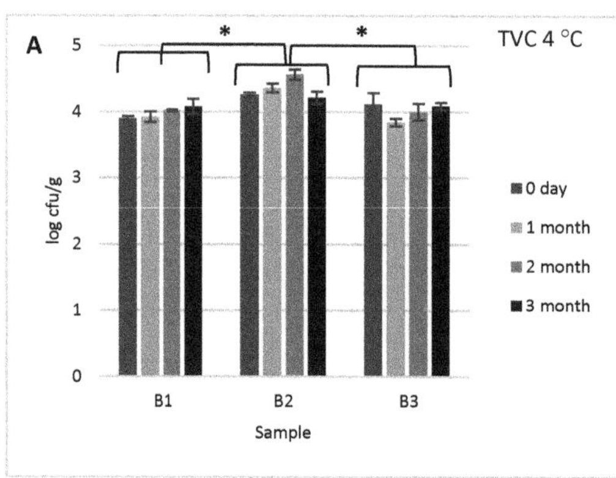


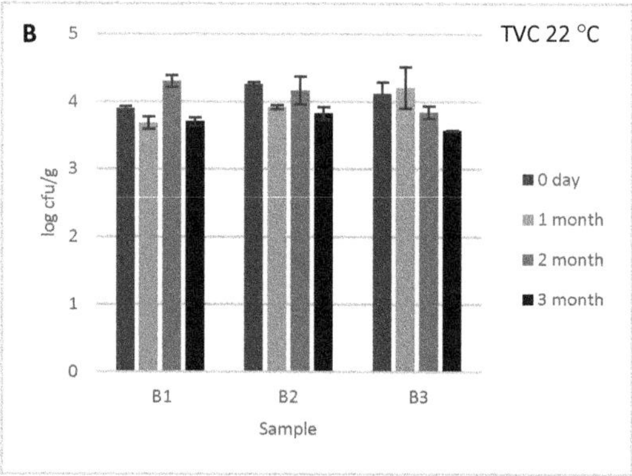


**Figure 1.** *Cont.*

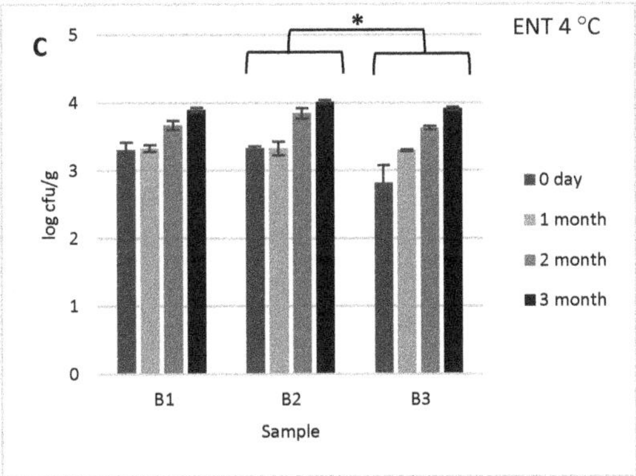


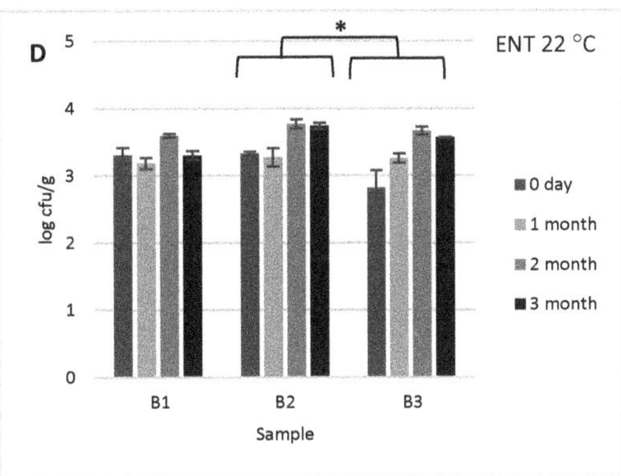


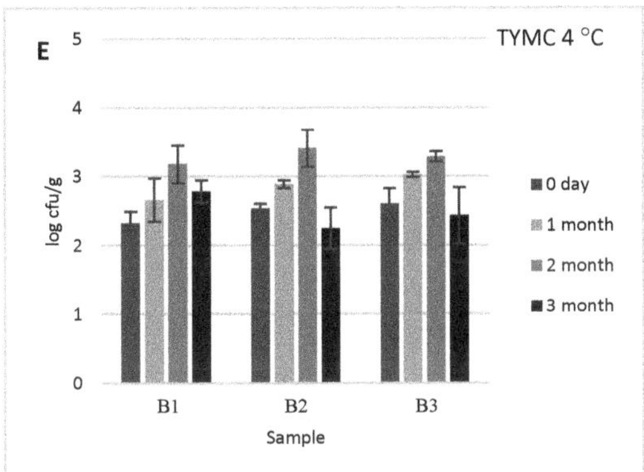


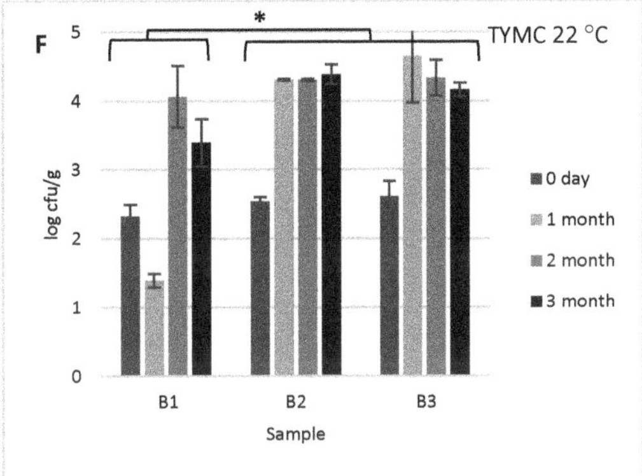


**Figure 1.** Mean number of selected microorganisms in the organic high-protein bars samples (B1, B2 and B3) stored for 3 months at 4 or 22 °C. (**A**,**B**) Total number of microorganisms (TVC); (**C**,**D**) number of Enterobacteriaceae (ENT); (**E**,**F**) count of yeasts and moulds (TYMC). $n = 3$. Values of the variants of the product (B1, B2, B3) marked with an asterisk differ significantly according to the post hoc Bonferroni test ($p < 0.05$). Bars represent mean standard error.

Immediately after production, the microbiological quality of bars differed, although at a similar level. The total number of microorganisms (TVC) ranged from 3.90–4.26 log CFU/g. The results of the ANOVA analysis revealed that during the storage of products at 4 °C, the TVC did not change significantly ($p > 0.05$). However, the results of the post hoc Bonferroni test indicated that sample B2 contained considerably more TVC versus B1 and B3 bars ($p < 0.05$), whereas at 22 °C, all the values were slightly different (Figure 1A,B).

The number of Enterobacteriaceae (ENT) family after production was in the range of 2.81–3.32 log CFU/g. Regardless of the product variant and the storage temperature, the number of these microorganisms increased significantly after 3 months of storage ($p < 0.05$) (Figure 1C,D).

The count of yeasts and moulds (TYMC) directly after production was in the range of 2.61–3.99 log CFU/g. After a month of storage at 4 °C, in sample B1 a decrease in TYMC to approx. 1.32 log CFU/g was observed. However, in subsequent cold storage, the number of these microorganisms increased. In the other variants, the TYMC increased, when in the samples stored at 22 °C the increase was more dynamic and higher by about 1 logarithmic order compared to the samples stored at 4 °C (Figure 1E,F). On some bars, aerial mould growth was observed after 3 months of storage.

In the tested variants of high-protein bars, no pathogenic bacteria *Salmonella* sp. was found in 25 g of the product. Pathogenic bacteria of the species *Bacillus cereus* were present in samples B1 and B2 immediately after production and in the following months of storage at 4 and 22 °C. The presence of *B. cereus* was also found in the B3 high-protein bar variant, with two exceptions (after production and one month of storage at 22 °C) (Table 2).

**Table 2.** Evaluation of the presence of pathogenic bacteria in the tested products.

| Time (Day/Month) | Temperature of Storage (°C) | Pathogen Sample | B. cereus B1 | B2 | B3 | Salmonella sp. B1 | B2 | B3 | S. aureus B1 | B2 | B3 |
|---|---|---|---|---|---|---|---|---|---|---|---|
| | | | | (Presence) | | | | | | (log CFU/g) | |
| 0 | | | + | + | − | − | − | − | <2.00 | <2.00 | <2.00 |
| 1 | 4 | | + | + | + | − | − | − | 3.05 | 3.06 | 2.50 |
| | 22 | | + | + | − | − | − | − | 3.05 | 2.36 | <2.00 |
| 2 | 4 | | + | + | + | − | − | − | 2.83 | 3.04 | <2.00 |
| | 22 | | + | − | + | − | − | − | <2.00 | <2.00 | <2.00 |
| 3 | 4 | | + | + | + | − | − | − | 2.80 | 2.06 | <2.00 |
| | 22 | | + | + | + | − | − | − | <2.00 | <2.00 | <2.00 |

Explanatory notes: + present; − not present.

Staphylococcus aureus was presented in the samples after the production of the bars below the detection limit, i.e., <2 log CFU/g. After one month of storage at 4 or 22 °C, the S. aureus number was in the range of 2.36–3.07 log CFU/g. In subsequent research periods (2 and 3 months) of storage at 4 °C, S. aureus was detected in samples B1 and B2, but not in the variant B3. All tested variants (B1, B2 and B3) stored for 2 and 3 months at 22 °C contained S. aureus below the detection limit (Table 2).

### 3.2. Water Activity Evaluation

Figure 2 shows changes in water activity in the samples of organic bars with higher protein content during storage at different temperature conditions. The initial value was in the range of 0.64–0.68, the lowest aw was in sample B3, and the highest was in sample B2. There was no statistical difference between the tested variants (B1, B2 and B3) stored in both studied conditions regarding aw ($p > 0.05$). However, during 3 months of storage, the value significantly increased in all tested organic high-protein bars ($p < 0.05$). At this point at 4 °C, bar B2 had the highest aw d (0.81) and t bar B3 had the lowest aw (0.75) while at 22 °C the water activity of B3 bars was 0.77 versus B2 with 0.74. To sum up, bar B3 most often had the lowest aw.

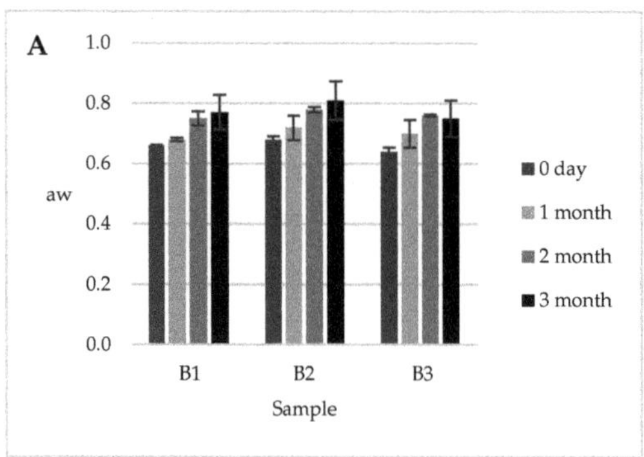


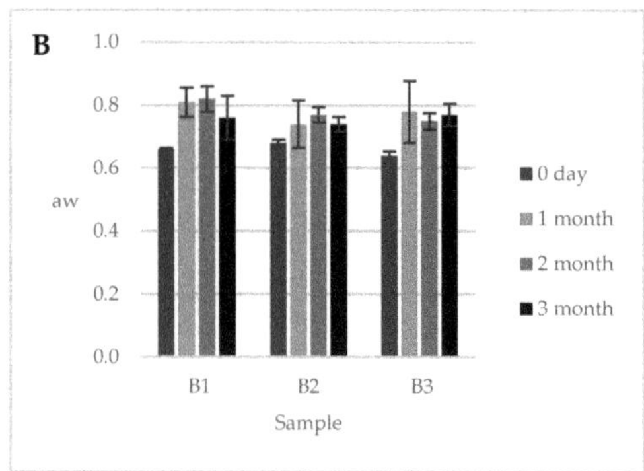


**Figure 2.** Water activity (aw) of tested products during storage (**A**) at 4 °C and (**B**) at 22 °C; $n = 3$. Bars represent mean standard error.

### 3.3. Mycotoxin Content

Table 3 presents detailed results of the content of selected mycotoxins measured in tested variants of the organic high-protein bars. Immediately after production and upon 3 months of storage, in all tested samples of high-protein bars, the amount of mycotoxin was below the detection limit of analytical methods.

**Table 3.** Presence of selected mycotoxins in HPB bars immediately after production (time 0) and after 3 months of storage at 4 °C and 22 °C ($n = 3$).

| Mycotoxin (µg/kg) | Products at Day 0 | | | Products after Storage at 4 °C | | | Products after Storage at 22 °C | | |
|---|---|---|---|---|---|---|---|---|---|
| | B1 | B2 | B3 | B1/4 | B2/4 | B3/4 | B1/22 | B2/22 | B3/22 |
| Aflatoxin B1 | <0.10 | <0.10 | <0.10 | <0.10 | <0.10 | <0.10 | <0.10 | <0.10 | <0.10 |
| Aflatoxin B2 | <0.05 | <0.05 | <0.05 | <0.05 | <0.05 | <0.05 | <0.05 | <0.05 | <0.05 |
| Aflatoxin G1 | <0.10 | <0.10 | <0.10 | <0.10 | <0.10 | <0.10 | <0.10 | <0.10 | <0.10 |
| Aflatoxin G2 | <0.05 | <0.05 | <0.05 | <0.05 | <0.05 | <0.05 | <0.05 | <0.05 | <0.05 |
| Sum of aflatoxin B1, B2, G1, G2 | <0.10 | <0.10 | <0.10 | <0.10 | <0.10 | <0.10 | <0.10 | <0.10 | <0.10 |
| Ochratoxin A | <0.25 | <0.25 | <0.25 | <0.25 | <0.25 | <0.25 | <0.25 | <0.25 | <0.25 |
| Zearalenone | <10 | <10 | <10 | <10 | <10 | <10 | <10 | <10 | <10 |
| Deoxynivalenol | <100 | <100 | <100 | <100 | <100 | <100 | <100 | <100 | <100 |

### 3.4. Biogenic Amines Content

The content of biogenic amines in the organic high-protein bars immediately after production in industrial conditions and after a three-month storage period in two temperatures (4 and 22 °C) is shown in Table 4. The presence of biogenic amines, such as histamine and tryptamine, was not found in the study products throughout the research period. There were no statistically significant differences between the three variants compared in a given period ($p > 0.05$). However, changes in the content of biogenic amines under the influence of temperature and storage time were shown. A statistically significant decrease in the content of biogenic amines was observed in the case of putrescine and spermidine in all three tested bar variants ($p < 0.05$).

### 3.5. Analysis of Lipid Oxidation as TBARS

The lowest value of the TBARS index during 0, 1, and 2 months was found in the B2 bar at 4 °C at 0.38, 0.32 and 0.36 mg MDA/kg of the product, respectively. Among all experimental trials, the greatest changes in fat oxidation were observed in the B1 bar from 0.47 at day 0 to 1.96 mg MDA/kg after 3 months.

### 3.6. Presence of Pests

After nspection, no damage was found to the packaging in which bars B1, B2 and B3 were placed. The tested products did not contain live pests, parts of their bodies or other traces of their activity. During storage, no traces of the presence of pests were found in the tested products.

**Table 4.** Presence of selected biogenic amines in HPB bars immediately after production (time 0) and after 3 months of storage at 4 °C and 22 °C.

| Biogenic Amines (mg/kg) | Products at Day 0 | | | Products after Storage at 4 °C | | | Products after Storage at 22 °C | | |
|---|---|---|---|---|---|---|---|---|---|
| | B1 | B2 | B3 | B1/4 | B2/4 | B3/4 | B1/22 | B2/22 | B3/22 |
| 2-phenylethyl amine | 1.24 ± 0.13 aA | 1.31 ± 0.1 aA | 1.14 ± 0.12 aA | n.d. | 2.95 ± 0.65 aA | 1.66 ± 0.37 aA | 1.24 ± 0.27 aA | 1.41 ± 0.31 aA | 1.35 ± 0.30 aA |
| Cadaverine | 2.09 ± 0.24 aA | 2.68 ± 0.21 aA | 2.05 ± 0.27 aA | n.d. | 2.47 ± 0.35 aA | 2.99 ± 0.42 aA | 3.30 ± 0.46 aA | 3.94 ± 0.55 aA | 3.09 ± 0.31 aA |
| Histamine | <1.00 | <1.00 | <1.00 | n.d. | <1.00 | <1.00 | <1.00 | <1.00 | <1.00 |
| Putrescine | 11.8 ± 0.31 aA | 12.2 ± 0.27 aA | 12.7 ± 0.27 aA | n.d. | 14.9 ± 3.9 aA | 13.8 ± 3.60 aA | 8.63 ± 2.24 aB | 7.38 ± 1.92 aB | 8.08 ± 2.10 aB |
| Spermidine | 7.61 ± 0.21 aA | 9.49 ± 0.29 aA | 7.67 ± 0.25 aA | n.d. | 7.25 ± 1.02 aA | 6.8 ± 0.96 aA | 4.03 ± 0.56 aB | 5.86 ± 0.82 aB | 5.19 ± 0.73 aB |
| Spermine | 3.71 ± 0.15 aA | 4.53 ± 0.17 aA | 3.67 ± 0.26 aA | n.d. | 3.84 ± 0.85 aA | 3.44 ± 0.76 aA | 1.74 ± 0.38 aA | 2.77 ± 0.61 aA | 2.35 ± 0.52 aA |
| Tryptamine | <5.00 | <5.00 | <5.00 | n.d. | <5.00 | <5.00 | <5.00 | <5.00 | <5.00 |
| Tyramine | 2.47 ± 0.35 aA | 2.55 ± 0.31 aA | 2.29 ± 0.32 aA | n.d. | 2.24 ± 0.18 aA | 2.16 ± 0.17 aA | 2.17 ± 0,17 aA | 1.66 ± 0.13 aA | 2.01 ± 0.16 aA |

Explanatory notes: n.d.-no data. Data are expressed as mean ± SD. Means in the same row within the same column followed by different lowercase letters represent significant differences ($p < 0.05$). Means in the same row between different columns followed by different capital letters represent significant differences ($p < 0.05$).

## 4. Discussion

Consumer interest in food quality and safety is considered to be one of the main reasons for the growing demand for organically grown food, which consumers perceive as healthier and safer [24]. Previous research proved that it is possible to apply organic raw materials to develop the product intended for physically active people (high protein content) with high nutritional value and acceptable, high sensory quality [7]. However, a disturbing presence of *B. cereus* has been observed during storage. Organic products seem to be susceptible to microbial contamination compared to conventional products on the same level [25]. Therefore, broader research was needed to evaluate the safety of the product and to set possible storage times. For this reason, the presented study was planned to investigate more deeply the safety characteristic of high-protein bars made from organic raw materials.

The composition of high-protein bars determined by Szydłowska et al. [7] was modified to combat a few safety dilemmas revealed during the experiment and for wider use of raw materials native to Poland. The reformulation consisted of removing coconut flour and replacing it with oat flakes. Dried dates were replaced with prunes and/or apricots. Cane sugar was replaced with honey. Due to the microbiological contamination of raw materials, as in pumpkin seeds among others, the production process was modified. The change concerned the use of heat treatment, i.e., roasting of pumpkin seeds, oat and spelt flakes as well as soaking prunes and/or apricots in hot water to reduce the number of microorganisms in the finished product.

Insects present in food can have both direct and indirect effects on human health. The direct effect is the contamination of food with arthropod fragments as well as related contaminants e.g., allergenic, carcinogenic or microorganisms. The most significant indirect effect is that their presence can change the storage microenvironment, making durable products suitable for the rapid growth of fungi and other microorganisms present in the environment [26]. The tested products during storage were not infested with pests and it can be concluded that all food safety characteristics were connected with the natural microbial content and its activity or chemical changes during storage.

Regarding microbiological quality, the changes implemented in the bar's production did not considerably lower the TVC in the reformulated bars. However, in comparison with the pumpkin bars used in the Szydłowska et al. [7] study (TVC approx. 4.6 log CFU/g), the determined value in the present study was lower (TVC approx. 4.06 log CFU/g) and justifies the introduction of thermal processes during the manufacture of the bars. It is worth noting that products manufactured under industrial conditions usually contain a higher number of microorganisms compared to laboratory-scale production, due to the more diverse microbial environment and a greater number of people involved in the production. The proof of that is the higher number of Enterobacteriaceae family in tested products (about 4 log CFU/g) (Figure 1) whereas in the bars obtained from laboratory production the count was around 2 log CFU/g [7]. Other researchers found varied data regarding the TVC number in similar products i.e., in bars made from oat and corn flakes with brown rice and dried fruits detected the TVC on the level of about 2.08 log CFU/g [27]. Bhakha et al. [28] counted aerobes on the level of 8.9 log CFU/g in the pulse-based snack bars with cereals and dried fruits addition.

Fungi cause major spoilage of foods and feedstuffs. Certain environmental factors can induce the production of secondary metabolites i.e., mycotoxins. Contamination of food and feed by mycotoxins is considered one of the most serious food safety problems in the world because these fungal metabolites can be teratogenic, mutagenic, carcinogenic and immunosuppressive, and can cause serious damage to animal and human health [29,30]. In the study of Szydłowska et al. [7], just after production, the TYMC was below the detection threshold, but during one month of cold storage (4 °C), substantial growth was observed to average 5.5 log CFU/g, whereas in the modified product the TYMC was approximately about 3 log CFU/g during the whole 3 months cold storage period. A similar count of the TYMC was identified in the study of Bhakha et al. [28], where pulse-based snack bars contained yeasts and moulds at around 5.1 log CFU/g during storage at room temperature but 4.3 log CFU/g stored at 4 °C. The research of Munshi et al. [27] revealed approx. 0.78 log CFU/g in cereals-based bars, whereas in popped pearl millet bars during storage at ambient temperature the TYMC was below the detection limit [31]. Nevertheless, after analysing the presence of yeasts and moulds it can be stated that applied changes positively influenced this safety characteristic mainly because selected mycotoxins were not detected (Table 3), which means that moulds did not find the proper condition for toxin production.

Regarding food pathogen occurrence, the tested high-protein bars were free from Salmonella spp. but Staphylococcus aureus and *B. cereus* could be possible hazards in this kind of product. Although the number of S. aureus did not exceed 3 log CFU/g, and the possibility of toxin production was low, additional preventive measures should be implemented to protect the product more efficiently. The more serious risk arises from the occurrence of *B. cereus*. Under EU food law, the limits are set only for dried infant formulas and dried dietary foods for special medical purposes intended for infants below six months of age, and allow the presence of *B. cereus* up to 500 CFU/g [32]. Most cases of food-borne outbreaks caused by the *B. cereus* group have been associated with concentrations above 5 log CFU/g. Nevertheless, cases of both emetic and diarrheal illness have been reported involving lower levels of *B. cereus*. At postharvest, the main management option for controlling *B. cereus* group strains in the food chain is to maintain the foods refrigerated at $\leq 7$ °C. Other efficient control measures include heat treatment, high hydrostatic pressure, pulsed light, irradiation and chemical sanitisers. Because some of these may fail to inactivate spores, effective methods should be established, for example, a combination of the above-mentioned procedures [33]. For example, Hamanaka et al. [34] found infrared radiation (IR) effective in the inactivation of Bacillus subtilis spores, and the combination of IR heating with UV irradiation markedly accelerated the killing efficiency of microorganisms on fresh figs [35]. The results of Singh et al. [31] suggest that proper packaging material may improve the microbial stability of snack bars. In this research, the product packaged in metalized polyester polyethylene (MP) contained about 1 log CFU/g less the TVC than wrapped in high-density polyethylene (HDPE) foil.

The iroasting step of spelt flakes was not effective enough to inactivate all *B. cereus* cells or spores. The presence of *B. cereus* was found in all tested variants. It is worth noting that the B3 bar was the least contaminated with pathogenic spores, so perhaps its composition played a role in the dynamics of pathogen growth.

Water activity (aw) as an indicator of water availability allows prediction of the course of biological processes, especially the growth and development of microorganisms. A higher aw value indicates a better condition for microbial growth. Generally, the minimum aw at which microorganisms can grow is 0.60, but these numbers are variable. The minimum aw of most bacteria is approximately 0.87. Survival of only a few cells of some foodborne pathogens may be sufficient to cause disease. In a desiccated state, metabolism is greatly reduced, i.e., growth does not occur, but vegetative cells and spores may remain viable for several months or even years [10]. Measured values were approx. 0.65 at the beginning of the storage and increased to approx. 0.79 after 3 months of storage (Figure 2). Even though aw was in the range considered safe, the growth of some pathogens (Table 2) suggests that further action is needed to protect the product fully for example the addition of ingredients that have strong antimicrobial activity. Another possibility is the application selected thermal processes.

Biogenic amines (BAs) are naturally occurring ingredients in food products and influence the sensory properties of foodstuffs. In low concentrations, they do not pose a threat to health, and their increased content may be the result of the activity of endogenous enzymes contained in raw materials used for food production or microbial decarboxylation of amino acids taking place [36]. The toxic effects of BAs depend on individual sensitivity, lifestyle and diet, including consumption of ethanol, as well as on the kind of biogenic amine and its concentration in food. BAs such as histamine, tyramine, cadaverine, putrescine and spermidine or spermine are involved in several pathogenic syndromes, representing a risk for consumer health [36–38]. The study bars with increased protein content we analysed to assess the potential presence of biogenic amines as a significant threat to human health. The content of biogenic amines in the study bars did not indicate high levels of these compounds. In food, the presence of biogenic amines such as histamine and tyramine is a public health concern, since these are the most notorious foodborne toxins [38]. The toxic concentration of tyramine, a biogenic amine also found in fruit, especially prunes, is 100–800 mg/kg. The values in the study bars did not exceed this range. In the case of histamine, the maximum concentration of this compound in food products should not exceed 200 mg/kg of the product [39]. In the investigated bars, this amount did not exceed 1 mg/kg of the product. It should be noted that very low amounts of histamine and tyramine, considered to be one of the most toxic biogenic amines, were found in the tested products. Despite the toxic influence of biogenic amines on food quality and human health, there is no specific regulation regarding BA content in food, except for the European Food Safety Authority, which developed a qualitative risk assessment concerning biogenic amine presence in fermented foods and concentrations that can induce adverse effects in consumers [40]. In out study of non-fermented, organic high-protein bars, no high concentration of biogenic amines was found that would pose a risk of food poisoning.

Fat oxidation is one of the significant problems in the production and storage of food products, as oxidative rancidity causes quality deterioration, loss of nutritional value, and the formation of toxic compounds. Factors that accelerate these changes include oxygen, light, and high temperatures. Based on the results of the research, it was found that the applied storage conditions of the bars under refrigerated conditions (4 °C) had an inhibitory effect on lipid oxidative changes, as evidenced by the results of the TBARS index (Table 5). The TBARS values of bars stored at 4 °C were lower than the rancidity period (1–2 mg MDA/kg sample) for food products and proved their good physicochemical quality [41]. The use of prune and oat flakes in B2 bars had a significant ($p < 0.05$) effect on the inhibition of oxidative changes in fat. The protective effect of prunes and oat flakes was particularly evident in bars stored at 22 °C. The antioxidant properties of prunes and oat flakes were also confirmed in their research by Vinson et al. [42], and Ryan et al. [43]. According to the values of TBARS, it can be concluded that the use of prunes and oat flakes (B2 bar composition) in the production of organic bars and refrigerated storage reduces the degree of fat oxidation.

**Table 5.** The value of the bar TBARS index (mg MDA/kg).

| Time and Temperature of Storage (Month/°C) | Bar Symbol | | |
|---|---|---|---|
| | B1 | B2 | B3 |
| 0 | 0.47 ± 0.03 aB | 0.38 ± 0.03 aA | 0.48 ± 0.02 abB |
| 1/4 | 0.59 ± 0.03 abC | 0.32 ± 0.02 aA | 0.47 ± 0.02 aB |
| 2/4 | 0.53 ± 0.06 abB | 0.36 ± 0.03 aA | 0.45 ± 0.07 aAB |
| 3/4 | 0.63 ± 0.05 bAB | 0.75 ± 0.06 bB | 0.61 ± 0.06 bA |
| 1/22 | 0.90 ± 0.03 cB | 0.54 ± 0.05 abA | 1.01 ± 0.05 dB |
| 2/22 | 0.81 ± 0.03 cB | 0.45 ± 0.02 abA | 0.74 ± 0.04 cB |
| 3/22 | 1.96 ± 0.17 dA | 1.33 ± 0.41 cA | 1.57 ± 0.11 eA |

Explanatory notes: data are expressed as mean ± SD. Means in the same row within the same column followed by different lowercase letters represent significant differences ($p < 0.05$). Means in the same row between different columns followed by different capital letters represent significant differences ($p < 0.05$).

## 5. Conclusions

The nutritional quality and safety of organic products should be evaluated simultaneously to deliver scientific evidence of the quality of an organic product. The results of this research provide unique knowledge about a product that modern consumers and food producers are interested in.

Manufactured in industrial conditions, the organic high-protein bars were of good but not the best microbiological quality, with only slight heat treatment as preservation.

Nevertheless, based on the presented results it is possible to store the organic high-protein bar for at least 3 months.

There is the possibility of further modifications to achieve microbial stability at an ambient storage temperature; for example, selecting ingredients that contain fewer microorganisms and/or are rich in antimicrobial compounds. In addition, choosing proper packaging materials and/or storage temperature may improve the safety characteristics of organic high-protein bars.

It can be concluded that the composition of the organic high-protein bars and their microbiological and physicochemical stability indicate the safety of this innovative product. Among the tested variants, the composition of the B3 bar seemed to be the safest and worth further research, mainly due to the lower frequency of undesirable microorganisms. On the other hand, the protective antioxidative effect of prunes and oat flakes in bars stored at 22 °C indicates the value of the composition of bar B2.

Next to standard safety parameters, increasing the safety of HPB involves controlling the presence of *B. cereus* and other low-aw-resistant microorganisms.

**Author Contributions:** Conceptualization, M.T., D.Z., A.S. and D.K.-K.; methodology, K.N.-S., A.O. and A.Ł.; software, M.T.; validation, K.N.-S. and D.Z.; formal analysis, M.T. and A.S.; investigation, M.T. and D.Z.; resources, K.N.-S. and A.O.; data curation, M.T.; writing—original draft preparation, M.T.; writing—review and editing, D.Z., A.S. and K.N.-S.; visualization, M.T. and K.N.-S.; supervision, D.K.-K. and D.Z.; project administration, D.K.-K.; funding acquisition, D.K.-K. All authors have read and agreed to the published version of the manuscript.

**Funding:** This research was funded by the Ministry of Agriculture and Rural Development (decision No. HOR.re.027.6.2018) from a subsidy for organic farming research entitled: Processing of plant and animal products with organic methods: Research on optimization and development of innovative processing solutions to increase the value of healthy organic products.

**Institutional Review Board Statement:** The study was conducted following the Declaration of Helsinki (World Medical Association. Declaration of Helsinki: Ethical principles for medical research involving human subjects. JAMA J. Am. Med. Assoc. 2013, 310, 2191–2194).

**Informed Consent Statement:** Not applicable.

**Data Availability Statement:** Not applicable.

**Conflicts of Interest:** The authors declare no conflict of interest.

## References

1. Annunziato, R.A.; Timko, C.A.; Crerand, C.E.; Didie, E.R.; Bellace, D.L.; Phelan, S.; Kerzhnerman, I.; Lowe, M.R. A randomized trial examining differential meal replacement adherence in a weight loss maintenance program after one-year follow-up. *Eat. Behav.* **2009**, *10*, 176–183. [CrossRef] [PubMed]
2. Miraballes, M.; Fiszman, S.; Gámbaro, A.; Varela, P. Consumer perceptions of satiating and meal replacement bars, built up from cues in packaging information, health claims and nutritional claims. *Food Res. Int.* **2014**, *64*, 456–464. [CrossRef]
3. Ashley, J.M.; Herzog, H.; Clodfelter, S.; Bovee, V.; Schrage, J.; Pritsos, C. Nutrient adequacy during weight loss interventions: A randomized study in women comparing the dietary intake in a meal replacement group with a traditional food group. *Nutr. J.* **2007**, *6*, 12. [CrossRef] [PubMed]
4. Levitsky, D.A.; Pacanowski, C. Losing weight without dieting. Use of commercial foods as meal replacements for lunch produces an extended energy deficit. *Appetite* **2011**, *57*, 311–317. [CrossRef] [PubMed]
5. Matt, D.; Rembiałkowska, E.; Luik, A.; Peetsmann, E.; Pehme, S. *Quality of Organic vs. Conventional Food and Effects on Health: Report*; Estonian University of Life Sciences: Tartu, Estonia, 2011.
6. Parn, O.J.; Bhat, R.; Yeoh, T.K.; Al-Hassan, A.A. Development of novel fruit bars by utilizing date paste. *Food Biosci.* **2015**, *9*, 20–27. [CrossRef]
7. Szydłowska, A.; Zielińska, D.; Łepecka, A.; Trząskowska, M.; Neffe-Skocińska, K.; Kołożyn-Krajewska, D. Development of functional high-protein organic bars with the addition of whey protein concentrate and bioactive ingredients. *Agriculture* **2020**, *10*, 390. [CrossRef]
8. *Regulation (EC) No 1924/2006 of the European Parliament and of the Council of 20 December 2006 on Nutrition and Health Claims Made on Foods 2006*; Publications Office of the European Union: Luxembourg, 2016.
9. Fong, K.; Wang, S. Strain-specific survival of *Salmonella* enterica in peanut oil, peanut shell, and chia seeds. *J. Food Prot.* **2016**, *79*, 361–368. [CrossRef]

10. Beuchat, L.R.; Komitopoulou, E.; Beckers, H.; Betts, R.P.; Bourdichon, F.; Fanning, S.; Joosten, H.M.; Ter Kuile, B.H. Low-water activity foods: Increased concern as vehicles of foodborne pathogens. *J. Food Prot.* **2013**, *76*, 150–172. [CrossRef]

11. *ISO 4833-1*; Microbiology of the Food Chain-Horizontal Method for the Enumeration of Microorganisms-Part 1: Colony Count at 30 Degrees C by the Pour Plate Technique. ISO: Geneva, Switzerland, 2013.

12. *ISO 21528-2*; Microbiology of Food and Feeding Stuffs-Horizontal Method for the Detection and Enumeration of Enterobacteriaceae-Part 2: Colony Count Method. ISO: Geneva, Switzerland, 2017.

13. *ISO 21527-1*; Microbiology of Food and Animal Feeding Stuffs-Horizontal Method for the Enumeration of Yeasts and Moulds-Part 1: Colony Count Technique in Products with Water Activity Greater Than 0,95. ISO: Geneva, Switzerland, 2008.

14. *ISO 6579-1*; Microbiology of the Food Chain-Horizontal Method for the Detection, Enumeration and Serotyping of *Salmonella*-Part 1: Detection of *Salmonella* spp. ISO: Geneva, Switzerland, 2017.

15. *ISO 21871*; Microbiology of Food and Animal Feeding Stuffs-Horizontal Method for the Determination of Low Numbers of Presumptive *Bacillus Cereus*-Most Probable Number Technique and Detection Method. ISO: Geneva, Switzerland, 2006.

16. *PN-EN ISO 6888-1*; Food and Feed Microbiology-Horizontal Method for Determining the Number of Coagulase-Positive Staphylococci (*Staphylococcus Aureus* and Other Species)-Part 1: Method Using Baird-Parker Agar Medium. PKN: Warsaw, Poland, 2001.

17. *ISO 21807:2004*; Microbiology of Food and Animal Feeding Stuffs—Determination of Water Activity. ISO: Geneva, Switzerland, 2004.

18. *PN-EN 14123*; Foodstuffs-Determination of Aflatoxin B1 and the Sum of Aflatoxin B1, B2, G1 and G2 in Hazelnuts, Peanuts, Pistachios, Figs and Paprika Powder-High Performance Liquid Chromatographic Method with Post-Column Derivatisation and Immunoaffinity Column Cleanup. PKN: Warsaw, Poland, 2008.

19. *PN-EN 14132*; Foodstuffs-Determination of Ochratoxin A in Barley and Roasted Coffe-HPLC Method with Immunoaffinity Column Clean-Up. PKN: Warsaw, Poland, 2010.

20. J.S. Hamilton Laboratory Determination of Zearalenone Content; Range: 10÷4000 µg/kg by High-Performance Liquid Chromatography-PB-44/HPLC wyd. III z dn. 28.02.2009. Available online: https://hamilton.com.pl/wp-content/uploads/2017/03/Zatwierdzony_przez_MON_Zakres_Akredytacji_OiB_20.10.2016.pdf (accessed on 21 August 2022).

21. J.S. Hamilton Laboratory Determination of Deoxynivalenol Content; 100÷20 000 µg/kg by High-Performance Liquid Chromatography-PB-226/LC wyd. III z dn. 02.01.2015. Available online: https://hamilton.com.pl/wp-content/uploads/2017/0 3/Zatwierdzony_przez_MON_Zakres_Akredytacji_OiB_20.10.2016.pdf (accessed on 21 August 2022).

22. Smělá, D.; Pechová, P.; Komprda, T.; Klejdus, B.; Kubáň, V. Liquid chromatographic determination of biogenic amines in a meat product during fermentation and long-term storage. *Czech J. Food Sci.* **2003**, *21*, 167–175. [CrossRef]

23. Pikul, J.; Leszczynski, D.E.; Kummerow, F.A. Evaluation of three modified TBA methods for measuring lipid oxidation in chicken meat. *J. Agric. Food Chem.* **1989**, *37*, 1309–1313. [CrossRef]

24. Magkos, F.; Arvaniti, F.; Zampelas, A. Organic food: Buying more safety or just peace of mind? A critical review of the literature. *Crit. Rev. Food Sci. Nutr.* **2006**, *46*, 23–56. [CrossRef] [PubMed]

25. Gomiero, T. Food quality assessment in organic vs. conventional agricultural produce: Findings and issues. *Appl. Soil Ecol.* **2018**, *123*, 714–728. [CrossRef]

26. Hubert, J.; Stejskal, V.; Athanassiou, C.G.; Throne, J.E. Health hazards associated with arthropod infestation of stored products. *Annu. Rev. Entomol.* **2018**, *63*, 553–573. [CrossRef] [PubMed]

27. Munshi, R.; Kochhar, A.; Kaur, A. Nutrient selection and optimization to formulate a nutrient bar stable on storage and specific to women at risk of osteoporosis. *J. Food Sci. Technol.* **2020**, *57*, 3099–3107. [CrossRef] [PubMed]

28. Bhakha, T.; Ramasawmy, B.; Toorabally, Z.; Neetoo, H.; Bhakha, T.; Ramasawmy, B.; Toorabally, Z.; Neetoo, H. Development, characterization and shelf-life testing of a novel pulse-based snack bar. *AIMS Agric. Food* **2019**, *4*, 756–777. [CrossRef]

29. Adeyeye, S.A.O. Fungal mycotoxins in foods: A review. *Cogent Food Agric.* **2016**, *2*, 1213127. [CrossRef]

30. Jin, J.; Beekmann, K.; Ringø, E.; Rietjens, I.M.C.M.; Xing, F. Interaction between food-borne mycotoxins and gut microbiota: A review. *Food Control* **2021**, *126*, 107998. [CrossRef]

31. Singh, R.; Singh, K.; Nain, M.S. Nutritional evaluation and storage stability of popped pearl millet bar. *Curr. Sci.* **2021**, *120*, 1374–1381. [CrossRef]

32. *Regulation (EU) No 1441/Commission Regulation (EC) No 1441/2007 of 5 December 2007 Amending Regulation (EC) No 2073/2005 on Microbiological Criteria for Foodstuffs 2007*; Publications Office of the European Union: Luxembourg, 2017.

33. EFSA Panel on Biological Hazards. Risks for public health related to the presence of *Bacillus cereus* and other *Bacillus* spp. including *Bacillus thuringiensis* in foodstuffs. *EFSA J.* **2016**, *14*, e04524. [CrossRef]

34. Hamanaka, D.; Uchino, T.; Hu, W.; Yasunaga, E. Effects of infrared radiation on inactivation of *Bacillus subtilis* spore and *Aspergillus niger* spore. *J. Jpn. Soc. Agric. Mach.* **2002**, *64*, 69–75. [CrossRef]

35. Hamanaka, D.; Norimura, N.; Baba, N.; Mano, K.; Kakiuchi, M.; Tanaka, F.; Uchino, T. Surface decontamination of fig fruit by combination of infrared radiation heating with ultraviolet irradiation. *Food Control* **2011**, *22*, 375–380. [CrossRef]

36. Tabanelli, G. Biogenic amines and food quality: Emerging challenges and public health concerns. *Foods* **2020**, *9*, 859. [CrossRef] [PubMed]

37.  Costa, M.P.; Rodrigues, B.L.; Frasao, B.S.; Conte-Junior, C.A. Biogenic amines as food quality index and chemical risk for human consumption. In *Food Quality: Balancing Health and Disease*; Handbook of Food Bioengineering; Holban, A.M., Grumezescu, A.M., Eds.; Academic Press: Cambridge, MA, USA, 2018; pp. 75–108. ISBN 978-0-12-811442-1.

38.  Ruiz-Capillas, C.; Herrero, A.M. Impact of biogenic amines on food quality and safety. *Foods* **2019**, *8*, 62. [CrossRef]

39.  Waqas, A.; Mohammed, G.I.; Al-Eryani, D.A.; Saigl, Z.M.; Alyoubi, A.O.; Alwael, H.; Bashammakh, A.S.; O'Sullivan, C.K.; El-Shahawi, M.S. Biogenic amines formation mechanism and determination strategies: Future challenges and limitations. *Crit. Rev. Anal. Chem.* **2020**, *50*, 485–500. [CrossRef]

40.  EFSA Panel on Biological Hazards. Scientific opinion on risk based control of biogenic amine formation in fermented foods. *EFSA J.* **2011**, *9*, 2393. [CrossRef]

41.  Etemadian, Y.; Shabanpour, B.; Ramzanpour, Z.; Shaviklo, A.R.; Kordjazi, M. Production of the corn snack seasoned with brown seaweeds and their characteristics. *J. Food Meas. Charact.* **2018**, *12*, 2068–2079. [CrossRef]

42.  Vinson, J.A.; Zubik, L.; Bose, P.; Samman, N.; Proch, J. Dried fruits: Excellent in vitro and in vivo antioxidants. *J. Am. Coll. Nutr.* **2005**, *24*, 44–50. [CrossRef]

43.  Ryan, L.; Thondre, P.S.; Henry, C.J.K. Oat-based breakfast cereals are a rich source of polyphenols and high in antioxidant potential. *J. Food Compos. Anal.* **2011**, *24*, 929–934. [CrossRef]

 *applied sciences*

*Article*

# The Effect of Cold Plasma on Selected Parameters of Bovine Colostrum

Elżbieta Bogusławska-Wąs [1], Alicja Dłubała [1], Wojciech Sawicki [1], Małgorzata Ożgo [2,*] and Adam Lepczyński [2]

[1] Department of Applied Microbiology and Human Nutrition Physiology, Faculty of Food Sciences and Fisheries, West Pomeranian University of Technology Szczecin, Papieża Pawła VI 3, 71-459 Szczecin, Poland; elzbieta.boguslawska-was@zut.edu.pl (E.B.-W.); alicja.dlubala@zut.edu.pl (A.D.); wojciech.sawicki@zut.edu.pl (W.S.)

[2] Department of Physiology, Cytobiology and Proteomics, Faculty of Biotechnology and Animal Husbandry, West Pomeranian University of Technology Szczecin, Klemensa Janickiego 29, 71-270 Szczecin, Poland; adam.lepczynski@zut.edu.pl

* Correspondence: malgorzata.ozgo@zut.edu.pl; Tel.: +48-91-449-6774

**Abstract:** The main problem in processing bovine colostrum is preserving as many beneficial compounds as possible, most of which have low thermal stability. The present study evaluates the possibility of using cold plasma (CP) as a decontamination technology and its effect on selected biologically active fractions of freeze-dried bovine colostrum. The plasma process was carried out in air, nitrogen, and oxygen environments. The results revealed that the sterilization process using CP caused slight changes in the colour of the samples expressed by the attributes ΔC, ΔL, Δh and ΔE. The decontamination effect depended on the gas used and the type of microorganism. The highest decontamination effects were gained under oxygen conditions, where reductions were obtained for total psychrophilic bacteria (THPC) by log 1.24, mesophilic bacteria (THMC) by log 1.02, *Enterobacteriaceae* by log 1.16, *E. coli* by log 0.96, yeast (TYMC) by log 0.92. A significantly lower decontaminating effect was obtained for Gram-positive bacteria and sporophytic forms. Additionally, the application of CP, regardless of the gas used, affected the modification of protein structure and reduction of immunoglobulin concentration. as proven by proteomics analyses (1-DE, 2-DE, MALDI–TOF MS). The same applied to β-lactoglobulin in air and oxygen and BSA in nitrogen and air.

**Keywords:** cold plasma; colostrum; proteome; decontamination technology; microbiological safety

**Citation:** Bogusławska-Wąs, E.; Dłubała, A.; Sawicki, W.; Ożgo, M.; Lepczyński, A. The Effect of Cold Plasma on Selected Parameters of Bovine Colostrum. *Appl. Sci.* **2023**, *13*, 5490. https://doi.org/10.3390/app13095490

Academic Editor: Emilio Martines

Received: 6 April 2023
Revised: 25 April 2023
Accepted: 26 April 2023
Published: 28 April 2023

## 1. Introduction

Colostrum is the first milk for newborn mammals. It is a valuable source of nutrients and contains numerous bioactive compounds crucial for mammal growth and development, such as immunoglobulins, lactoferrin, lactoperoxidase, lysozyme, lactalbumin, hormones, enzymes, and whey proteins [1]. Immunostimulators: lactoferrin, colostrin, cytokines, and leukocytes are particularly important in colostrum as they modulate the newborn immune system [2,3]. The composition of colostrum is highly variable in terms of protein, fat, cytokines, growth factors, carbohydrates, vitamins, and minerals content. It depends on many factors, such as the breed and age of the animal, prepartum nutrition, and the course of birth.

The functional features of colostrum proteins are determined by the properties and structure of their molecules. Their size, shape, spatial arrangement, ability to create conglomerates, susceptibility to denaturation, and amino acid sequences are among the most important features. Colostrum contains more than 30 different proteins, which can be divided into two groups: casein (75–80%) and whey proteins (up to 20%) [4]. Nineteen amino acids have been determined in colostrum proteins, including all essential amino acids [5].

The most critical group of colostrum proteins are whey globular proteins. In bovine colostrum, there are $\alpha$-lactalbumin (15–20%) and $\beta$-lactoglobulin (55–65%) fractions, as well as immunoglobulins (9%), bovine serum albumin (BSA) (5.5%), lactoferrin, lactoperoxidase, phospholipoproteins, bioactive factors and enzymes [6]. In addition to their nutritional value, they also exhibit functional activities, including, most notably, immunomodulatory and antioxidant activity. They also regulate fatty acid synthesis in the liver, mitigating oxidative damage to DNA and oxidation of LDL [7].

Bovine colostrum has recently been used as a functional food, i.e., food of natural origin that benefits health. Unfortunately, most of the biologically active substances found in colostrum are thermolabile. Pasteurization degrades most desirable substances, including lactoferrin, depriving colostrum of its antigenic properties and ability to bind iron. In addition, spray drying of colostrum degrades lysozyme [8], and this technology does not guarantee the preservation of the health-promoting components of colostrum due to high temperatures. Therefore, freeze-drying is significantly more commonly used to stabilize colostrum.

The abundance of nutrients and biologically active substances in colostrum, especially proteins with defensive properties for newborns and therapeutic and prophylactic properties for older animals and humans, has prompted the development of new non-thermal and highly energy-efficient techniques that can be used to effectively reduce microbial contaminants in food [9–11] while maintaining the stability of biological components. Cold plasma is a technology with a high potential for application in the food processing sector.

Despite the fact it is essential from the quality and health and safety risk assessment perspective, there is a lack of information on the microbiological content of colostrum. According to Fecteau et al. [12], the microorganisms isolated from colostrum can be divided into four groups: (1) pathogens from the mammary gland, which include: *S. uberis*, *S. dysgalactiae* and *S.aureus*, (2) physiological flora of bovine skin and mucous membranes, e.g., *A. pyogenes*, *Corynebacterium* spp., *Pasteurella* spp., *Streptococcus* spp., *Staphylococcus* spp., and yeast. (3) faecal contaminants, e.g., *Enterococcus* spp., *E. coli* and other coliform bacteria (*Klebsiella* spp., *Enterobacter* spp.) and (4) environmental contaminants, e.g., *Micrococcus* spp., *Bacillus* spp., *Acinetobacter* spp., *Pseudomonas* spp., *Flavobacterium* spp. Therefore, how colostrum is sourced and handled after milking should be performed in a way that reduces contamination and inhibits and eliminates further growth of microorganisms without lowering its biological properties.

Several green technologies have been used in the food industry to decontaminate production lines and products from microorganisms that could degrade product quality, causing recalls and the emergence of food-borne illnesses. Over the past decade, the use of plasma has gained widespread application in the food industry as a relatively new and promising non-thermal decontamination technology [13]. Cold plasma (CP) is the application of plasma as ionized or partially ionized gases to inactivate bacterial cells. The plasma is created either by heating the gas in a closed chamber under deep vacuum conditions or by using radiofrequency or microwave energy to excite the gas molecules to produce free radicals and electrons, which are the main components of the plasma [14]. These components have a destructive effect on microorganisms, including bacterial spores, yeast, moulds, and viruses.

The antimicrobial effect of plasma is multimodal and includes both physical and chemical effects. However, the predominant mode of action is unknown. In addition, the impact of plasma on a microbial cell varies depending on the type of microorganisms (bacteria, fungi), working gas, kind of plasma, process parameters, substrate, condition of the substrate, and surface roughness [14]. For example, argon with oxygen and nitrogen emits four times more UV photons than pure argon [15]. Hence the mode of action could be significantly different in both cases. The advantage of cold plasma sterilization is that its components involved in destroying microorganisms are perishable. Therefore, there is no danger of them being present in the finished food.

Given recent developments in plasma technology, many distinct plasma systems have been designed. Among them can be distinguished low-temperature atmospheric-pressure plasma (APP) [16] or microwave-driven discharge (MWD). The dielectric barrier discharge (DBD) plasma system, which is widely considered the most commonly used APP system [17], generates plasma between two electrodes that are covered with dielectric layers [18].

Therefore, this study aims to evaluate the suitability of low-temperature plasma for inactivating microorganisms present in freeze-dried bovine colostrum while preserving the biological value of the raw material.

## 2. Materials and Methods

### 2.1. Materials

A total of 25 freeze-dried samples of bovine colostrum provided by a single manufacturer of bovine colostrum supplements were tested. The samples represented six batches. Each batch was sourced from a different farm according to the manufacturer's declaration.

### 2.2. Microbiological Analysis

All microbiological analyses were performed in accordance with accepted standards ISO 93/A-86034/02 [19]. To determine the quantitative and qualitative diversity of microbiota of freeze-dried bovine colostrum samples (before and after plasma decontamination), 10 g weights were prepared and homogenized in 90 mL of buffered peptone water (Oxoid, Basingstoke, UK). Decimal dilutions were performed according to EN ISO ISO 6887-5: 2020-10 [20]. The following ISO standards were used to determine specific groups of microorganisms: total bacterial count (THMC and THPC) EN ISO 4833-2:2013-12 [21], total yeast and mould count (TYMC and TMMC) ISO 21527-1: 2009 [22] using DRBC (Oxoid, UK), the total count of *Enterobacteriaceae* (TEb) and *E. coli* (Ec) ISO 21528-2: 2017-08 [23] using ChromaCult coli® (Millipore, Burlington, MA, USA), total count of *Staphylococcus* spp. (TSt) ISO 6888-2:2001+A1: 200 [24], using Baird-Parker Agar (Oxoid, UK) aerobic spore-forming bacteria (TCBac) ISO 7932: 2005/A1:2020-09 [25] using MYP Agar (Oxoid, UK), the total count of *Enterococcus* spp. (TEcc) and total count of *Streptococcus* spp. (TStr) ISO 15788:2009 [26] using BEA (Oxoid, UK) as a confirmation medium, lactic acid bacteria (LAB) ISO 15214:2002 [27] using MRS Agar (Scharlau, Sentmenat, Spain).

### 2.3. Measurement of pH

The pH of colostrum samples was measured using a pH meter (AD 12, Adwa, Romania) in the mixture obtained by adding 10 mL of redistilled water to 10 g of sample and homogenizing for 30 s.

### 2.4. Colour Analysis

The colour of test samples was assessed with an objective method using colourimeter WR 18 (FRU®, Shenzhen Wave Optoelectronics Technology Co., Ltd., Shenzhen, China), based on white standard tile ($L^*$ = +92.4; $a^*$ = −0.04; $b^*$ = +1.9) and CIE $L^* a^* b^*$, illuminant D65, observer 10°, illumination mode d/8 and calibre 8 mm. The hue (h) and (C) chroma of colour were calculated according to Dmytrów et al. [28]. The colour difference ($\Delta$E) was calculated according to Pankaj et al. [29]. The values obtained were classified as very distinct ($\Delta$E > 3), distinct (1.5 < $\Delta$E < 3) and small difference (1.5 < $\Delta$E) following Pankaj et al. [29].

### 2.5. Sensory Evaluation

Thirty participants took part in the sensory evaluation. Taste, odour, and texture were evaluated in accordance with ISO 22935-2:2013-07 [30] and ISO 22935-3:2013-07 [31]. The sensory characteristics of dried dairy products were included in the evaluation of each variant, as indicated by Drake et al. [32] and Abdalla et al. [33]. Samples of 10 g were placed in closed Petri dishes and served in random order. Distilled water (~20 °C) was used as a taste neutralizer between each sample. The results for each descriptor were expressed as an arithmetic mean and reported in conventional units on a scale of 1–5.

### 2.6. Cold Plasma Treatment

A cold vacuum, low-pressure plasma generator-Tetra 100 RF-PC-D (Diener Electronics-Plasma Surface Technology, Germany), was sterilized. The variable factors in the plasma process were the gases used: air, oxygen, and nitrogen. Before decontamination, the freeze-dried colostrum samples (10 g each) were packed in Tyvek® designed for safe cold plasma decontamination (DuPont, Poland, Warsaw) and sealed. All 25 samples were treated with cold plasma one at a time. Packages were sterilized from each batch in five independent replicates. The results obtained were averaged. Gas flow rate, chamber temperature and time to achieve vacuum were automated and resulted from keeping plasma conditions constant. Sterilization conditions are presented in Table 1.

**Table 1.** Conditions for generating and maintaining cold plasma depend on the ionizing gas used.

| Sample | Gas Flow | Time to Obtain Vacuum | Temperature | Power | Process Time | Pressure in the Chamber |
|---|---|---|---|---|---|---|
|  | [cm$^3$/min] | [min] | [°C] | [W] | [min] | [mbar] |
| Air | 20–50 | 60 | 27–30 |  |  |  |
| Nitrogen | 20–48 | 60 | 28–32 | 300 | 60 | 0.6 |
| Oxygen | 20–50 | 70 | 26–30 |  |  |  |

### 2.7. Protein Extract Preparation

Samples of freeze-dried bovine colostrum were defatted by double centrifugation at 4 °C, 4500× *g* (Centrifuge MPW 351R. MPW Med. Instruments, Poland, Warsaw) first for 20 min, then again for 10 min. The lipid layer was removed each time. The next step was to precipitate the casein with 50% acetic acid. After precipitation, the samples were centrifuged at 4 °C, 3380× *g* for 25 min, and the obtained supernatant was transferred to Eppendorf tubes. The protein concentration in the samples was measured using the Protein Assay (Bio-Rad, Hercules, CA, USA) before proteomic analyses. Samples were stored at a temperature of −35 °C before further analysis.

### 2.8. Protein Profiling Using SDS-PAGE (1-DE)

The colostrum proteins were analyzed using 4% (*w/v*) stacking gel and 12% separating gels. To prepare the 12% separation gel solution, the following was added: 10 mL 30% acrylamide/Bis Solution 37.5: 1 (Bio-Rad, Hercules, CA, USA), 6.25 mL 1.5 M Tris HCl (pH 8.8) (Bio-Rad, Hercules, CA, USA), 250 µL SDS 10% (*w/v*) (Sigma-Aldrich, Co., St. Louis, MO, USA), 8.375 mL deionized water, 125 µL 10% APS (ammonium persulfate) (Sigma -Aldrich, Co., St. Louis, MO, USA), 127.5 µL N, N, N′, N′-Tetramethylenediamine (Sigma-Aldrich, Co., St. Louis, MO, USA). To prepare a 4% concentration gel solution, 1.65 mL of 30% acrylamide/Bis Solution 37.5:1 (Bio-Rad, Hercules, CA, USA), 3.15 mL of 0.5 M Tris HCl (pH 6.8) (Bio-Rad), 125 µL of 10% SDS solution (*w/v*) (Sigma-Aldrich, Co., St. Louis, MO, USA), 7.5 mL of deionized water, 62.5 µL of 10% APS (Sigma-Aldrich, Co., St. Louis, MO, USA), 12.5 µL of TEMED (S Sigma-Aldrich, Co., St. Louis, MO, USA) were added. The gels were stained with Coomassie brilliant blue R250 to visualize the protein bands. Finally, gels were photographed using GelDoc Go Gel Imaging System (Bio-Rad, Hercules, CA, USA). Computerized gel analysis was performed using KTE Gel Scan (Kucharczyk Electrophoretic Techniques, Poland, Warsaw). Molecular weight determination was performed by comparison of peptides against the protein molecular weight marker Precision Plus Protein Standards 10–250 kD (Bio-Rad, Hercules, CA, USA).

### 2.9. Two-Dimensional Electrophoresis (2-DE)

Bovine colostrum whey protein samples were separated using 2-DE electrophoresis in accordance with the protocol previously described by Lepczyński et al. [34]. Briefly, after the protein quantification using the Bradford method (Bio-Rad Protein Assay, Bio-Rad), the protein samples were combined with the sample buffer (7 M urea, 2 M thiourea, 4% *w/v* CHAPS, 0.2% *w/v* 3–10 carrier ampholytes, 1% *w/v* 1,4-dithiothreitol) to obtain 600 μg of total protein in the volume of 250 μL. The first dimension of the protein separation (IEF–isoelectric focusing) was run in 11 cm length IPG strips with nonlinear pH gradient 3–10 (Bio-Rad, Hercules, CA, USA) using Protean i12® IEF Cell (Bio-Rad, Hercules, CA, USA). Before direct, IEF protein samples were loaded into the strips using a combined rehydration method (6 h, 20 °C–passive; 50 V, 12 h, 20 °C–active). The IEF was run for a total of 35 kVh. After the separation, the IPG strips were equilibrated for 15 min in basal buffer (6 M urea, 0.5 M Tris/HCl, pH 6.8, 2% *w/v* SDS, 30% *w/v* glycerol) with 1% DTT addition (15 min). Next, the strips were washed in the basal buffer with a 2.5% iodoacetamide addition. The second-dimension separation (SDS-PAGE) was processed in 12% polyacrylamide gels using Protean Plus™ Dodeca Cell™ electrophoretic chamber (Bio-Rad, Hercules, CA, USA) under the following condition: 40 V and subsequently at 100 V for 2.5 h and 16 h, respectively (20 °C).

After the 2-DE separation in gels, protein detection was performed with CBB G-250 according to the procedure described by Lepczyński et al. [35]. First, the gels were fixed for 3 h in a buffer containing: 50% *v/v* ethanol and 5% *v/v* phosphoric acid in ddH2O. Then, the gels were stained in Bio-Rad Protein Assay (Bio-Rad, Bio-Rad, CA, USA) 20-fold diluted in ddH2O (3 h). After that, the excess of the unbound dye was washed out in the ddH2O (3 × 15 min). After washing, gel images were equalized using GS-800™ Calibrated Densitometer (Bio-Rad, Hercules, CA, USA). Finally, densitometric analysis of the gels representing the samples was performed using PDQuest 8.0.1 advanced (Bio-Rad, Hercules, CA, USA) software.

### 2.10. MALDI—TOF (Matrix Assisted Laser Desorption/Ionisation-Time of Flight) Mass Spectrometry

After manual picking of all visualized spots from the 2-D gels, their proteins were identification identified by the aid of (MALDI-TOF) mass spectrometric analysis (Microflex (Bruker Daltonics, Bremen, Germany) mass spectrometer, with MALDI ionization type and ToF analyser according to the procedure as previously described by Ożgo et al. [36].

### 2.11. Statistical Analysis

All analyses were repeated three times. Significant differences ($p < 0.05$) were determined by using analysis of variance (ANOVA) with Tukey's tests were carried out using Statistica 13.0 (StatSoft, Kraków, Poland), and $p$-value $< 0.05$ are considered to be statistically significant.

## 3. Results and Discussion

As the first secretion of the mammary gland, colostrum is characterized by many nutrients and biologically active compounds. However, it is also a source of microorganisms that can adversely affect its quality and consumer safety. Therefore, it has been assumed that the number of microorganisms in raw colostrum samples should not exceed 5 log cfu/mL for the total bacterial count and 4 log cfu/mL for coliforms [37].

In a study by Godden et al. [37] in heat-treated samples, these values were as high as 4.3 log cfu/mL and 2.0 log cfu/mL. This may pose a significant risk of pathogen transmission. Furthermore, the presence of microorganisms also decreases immunoglobulin activity due to the binding of bacteria to free immunoglobulins in the small intestine or blocking the transport of IgG molecules by intestinal epithelial cells [38]. As a result, the absorption of colostrum components is impaired, and the transfer of passive immunity is reduced. Therefore, bacterial contamination, together with Ig concentration (>50 mg/mL), is the basic criteria for assessing colostrum quality, i.e., its nutritional value and microbiological safety [37].

Microbiological analysis of freeze-dried colostrum samples showed the absence of pathogenic bacteria: *Salmonella* spp., *S. aureus* (coagulase positive), *L. monocytogenes* and *Cronobacter* spp. However, normal inhabitants of bovine skin and mucosa (NI), environmental contaminants (EC) and faecal contaminants (FC) were present. Total bacteria count (TCB), including individual samples, was determined in the log at 3.91–5.54 cfu/g. Twenty percent of the samples were above the accepted levels of microbial contamination, as defined by Godden et al. [35]. According to Table 2, the highest differences in the tested colostrum samples were determined for microorganisms included in the NI group. In individual microbial groups (THMC including TSt, TStr or LAB and TYMC), statistically significant differences ($p \leq 0.05$) were confirmed, depending on the batch.

**Table 2.** Microbiotic diversity of freeze-dried bovine colostrum samples.

| Bacteria | Average | LL-UL | SEM | *p*-Value |
|---|---|---|---|---|
| Normal inhabitants of bovine skin and mucosa (NI) | log cfu/g | | | |
| Mesophilic bacteria (THMC) | 4.70 * | 3.21–4.86 | 0.082 | 0.001 |
| *Staphylococcus* spp. (TSt) | 1.89 * | 0.5–3.29 | 0.148 | 0.031 |
| *Streptococcus* spp. (TStr) | 1.87 * | 1.00–2.30 | 0.135 | 0.049 |
| Yeast (TYMC) | 2.95 * | 1.61–3.25 | 0.019 | 0.046 |
| LAB | 2.94 * | 1.29–3.72 | 0.006 | 0.001 |
| Environmental contaminants (EC) | | | | |
| Psychrotrophic bacteria (THPC) | 5.24 | 4.56–5.78 | 0.366 | 0.897 |
| *Bacillus* spp. (TCBac) | 2.96 | 2.48–3.02 | 0.313 | 0.604 |
| Moulds (TMMC) | 1.97 | 1.74–2.24 | 0.185 | 0.091 |
| Faecal contaminants (FC) | | | | |
| *Enterobacteriacea* (TEb) | 2.43 * | 1.18–2.98 | 0.267 | 0.049 |
| *E. coli* (Ec) | 1.38 * | 0.00–2.00 | 0.348 | 0.048 |
| *Enterococcus* spp. (TEcc) | 2.37 | 1.69–2.59 | 0.134 | 0.167 |

*—statistically significant differences between samples, LL—lower limit, UL—upper limit.

The differences in the FC group counts could affect the THMC count. In this case, the statistical significance of the results was determined for *Enterobacteriaceae* (TEb), including *E. coli*. In contrast to TEb, which was determined in all colostrum samples, the detection of *E. coli* was incidental. The counts of faecal coliforms (Ecc) were noticeable but not statistically significant. The comparable numbers of identified microorganisms were present in all samples regardless of their batch in the EC group.

In the present study, the application of cold plasma air, nitrogen and oxygen as the working gases resulted in a noticeable reduction in the total microbial count (TBc). Unfortunately, the differences were not statistically significant. Depending on the gas used, a decrease of 0.68 log cfu/g (air), 0.83 log cfu/g (nitrogen) and 0.92 log cfu/g was achieved. It is confirmed that the level of microbial inactivation relies on the mixture of different reactive forms of ions, free electrons and molecules generated in the plasma and reactive nitrogen species (RNS), oxygen species (ROS) and hydrogen species (RHS) are the most important active plasma agents formed. Their number and type depend on the plasmonizer system and the technological parameters, including the ionizing gas [39].

In the present study, the greatest decontamination effects were obtained in oxygen as the working gas. In the case of TBC, a minimum reduction of 1 log was not achieved. These effects of plasma treatment were confirmed for THMC (1.02 log reduction) and THPC (1.24 log reduction). Similar results were observed for the reduction of *Enterobacteriaceae*-by 1.16 log and slightly lower for *E. coli*-by 0.96 log (Table 3). Due to their cell wall structure, Gram-negative rods are more sensitive to reactive nitrogen and oxygen species (RNOS) than Gram-positive bacteria [40–42]. Our study confirms these findings. Inactivation of *Enterococcus* spp., *Staphylococcus* spp., and *Streptococcus* spp. was no greater than 0.78 logs

(Tst), and the results were comparable and not dependent on the working gas used. Only for LAB, among which representatives of *Lactococcus* sp. and *Lactobacillus* sp. were determined, a gas-dependent level of cfu reduction was established. In this case, the plasma process was most effective (reduction of 0.88 logs) in the experiment with oxygen (Table 3). Lactic acid bacteria (LAB) are classified as EC group, and their natural occurrence in colostrum is usually considered a source of strains with probiotic functions [43]. However, the nutrients of colostrum available to LAB can promote the proliferation of these bacteria, leading to a drop in colostrum pH caused by the active fermentation process. The consequence is a reduction of the nutritional values of colostrum [44].

**Table 3.** The level of microorganisms' reduction after plasma-induced freeze-dried colostrum samples.

| | Normal Inhabitants of Bovine Skin and Mucosa | | | | | Environmental Contaminants | | | Faecal Contaminants | | |
|---|---|---|---|---|---|---|---|---|---|---|---|
| Sample | THMC | TSt | TStr | TYMC | LAB | THPC | TCBac | TMMC | TEb | Ec | TEcc |
| | | | | | log cfu/g | | | | | | |
| Air | 0.72 | 0.61 | 0.54 | 0.91 | 0.16 a | 0.73 | 0.76 | 0.45 | 0.68 a | 0.64 | 0.79 |
| Nitrogen | 1.19 | 0.78 | 0.65 | 0.62 | 0.56 | 1.09 | 0.53 | 0.49 | 1.14 | 0.88 | 0.75 |
| Oxygen | 1.02 | 0.73 | 0.75 | 0.92 | 0.88 b | 1.24 | 0.81 | 0.45 | 1.16 b | 0.96 | 0.74 |
| SEM | 0.127 | 0.149 | 0.158 | 0.174 | 0.084 | 0.124 | 0.369 | 0.681 | 0.079 | 0.079 | 0.019 |
| *p*-value | 0.101 | 0.122 | 0.326 | 0.094 | 0.032 | 0.701 | 0.112 | 0.843 | 0.043 | 0.133 | 0.052 |

a,b—statistically significant differences. THMC—total count of mesophilic bacteria, THPC—total count of psychrortophic bacteria, TYMC and TMMC—total count of yeast and mould, TEb—total count of *Enterobacteriaceae*, Ec—total count of *E. coli*, TSt—total count of *Staphylococcus* spp., TCBac—total count of aerobic spore-forming bacteria, TEcc—total count of *Enterococcus* spp., TStr—total count of *Streptococcus* spp., LAB—total count of lactic acid bacteria (LAB).

As it has already been established, the reaction of microorganisms to cold plasma varies. The single or synergistic interaction of reactive plasma components activates a series of successive mechanisms of cell inactivation [40,42,45]. The adsorption of reactive plasma molecules to the cell surface results in the formation of chemical compounds that generate changes in the cell membrane. This affects the regulation of transport in the cell and intracellular changes in pH [44]. There are three primary mechanisms: oxidative damage to cell proteins and cell membrane lipids and damage to DNA for RNOS action [40,46,47]. In addition, reactive oxygen and nitrogen forms can be transported across the cell membrane into the cell and oxidate intracellular components [48]. Therefore, as in Gram-negative bacteria, a thinner cell membrane structure results in a less pronounced resistance against these plasma stress factors. This is confirmed by the results of our study and Fröhling et al. [41], who showed significantly lower sensitivity of *L. innocua* than *E. coli*. The faintest bacterial decontamination effect in the colostrum samples tested was obtained in sporulating bacteria and moulds. Regardless of the gas used, bacterial counts were lower by 0.50–0.67 log and filamentous fungi by 0.45–0.49 log (Table 3).

Bacterial and mould spores are characterized by high resistance to destructive environmental factors, including those generated by CP. The multilayered structure of the envelopes and the presence of dipicolinic acid (DPA) and SASPs (small amide-soluble proteins) reduce their sensitivity to CP-generated factors [49]. Understanding the mechanisms of cold plasma action requires further research. According to Deng et al. [50], the destruction of spores, like vegetative forms, results from a violation of the integrity of the spore and leakage of cytoplasm. Reineke et al. [15] show that UV radiation plays an important function in spores' inactivation. UV photons emitted during the plasma process cause dimerization of thymine and cytosine bases, which results in reduced bacterial reproduction. However, due to the UV-absorbing abilities of the outer structures of the spores, the DNA of bacterial spores is not degraded but only partially fragmented [51]. According to Wang et al. [52], proteins associated with bacterial germination, which are inactivated under the influence of plasma-generated components, are of greater importance in the process of inactivating spores.

One of the factors regulating changes in the colostrum is pH value. Any drastic change can affect undesirable changes in its taste, texture, odour, or shelf life. Despite differences in the colour parameters of the control samples used in our study, their pH was not affected and was determined within the range (6.61–6.71). The plasma processes, regardless of the gas used, did not affect pH changes (Table 4). Such an insignificant effect of plasma treatment on pH was also observed in other studies [53–55]. Reactive plasma radicals react only with surface water in solid and dry products, forming acidic ions or acids in limited amounts (depending on the working gas used). The observed changes in pH and acidity depend on the matrix type and usually apply to liquid environments [56,57].

**Table 4.** Changes in pH and colour parameters of freeze-dried bovine colostrum after plasma process.

| Samples | pH | a*(-) | b* | L* | h (-) | C | ΔE |
|---|---|---|---|---|---|---|---|
| Control | 6.656 | 1.94 | 18.29 | 82.84 [a] | 83.92 | 18.40 [a] | - |
| Air | 6.663 | 2.31 | 18.43 | 82.11 | 82.87 | 18.57 | 0.82 [A] |
| Nitrogen | 6.654 | 2.04 | 19.64 | 81.49 [b] | 84.18 | 19.74 [b] | 1.90 [B,C] |
| Oxygen | 6.667 | 1.87 | 19.27 | 82.49 | 84.46 | 19.36 | 1.04 [D] |
| SEM | 0.033 | 0.039 | 0.554 | 0.166 | 0.855 | 0.338 | 0.159 |
| *p*-value | 0.992 | 0.056 | 0.082 | 0.012 | 0.055 | 0.000 | 0.000 |

[a,b] statistically significant differences among the control and plasma samples, [A,B,C,D] statistically significant differences among plasma samples.

Colostrum colour is considered one of the parameters indicating its quality. A paler colour is associated with a lower quality in terms of overall composition compared to more yellowish and darker colours [58]. Significant differences (at $p \leq 0.05$) between the tested colostrum samples were already established in preliminary tests (Table 4). Depending on their origin (batch), they were characterized by a different colour profile. This was indicated by the ranges of marked colour attributes, i.e., brightness (L: 82.357–83.403), saturations (C: 17.180–19.474) and hue (h: 82.84–84.534). The plasma process carried out solely with nitrogen contributed to the statistically significant increase in the C parameter. In addition, lower values of L* (81.49) were recorded, with a simultaneous slight increase in b* (19.64) and reduction in a* (−2.04) compared to the untreated CP sample (Table 4). The cumulative effect of the colour parameters: brightness, redness/greenness (a*) and yellowness/blueness (b*) determines the total colour difference (ΔE) of the samples tested. It was found that using air and oxygen as working gases in the plasma process had no significant effect on the colour change of the colostrum samples (ΔE $\leq$ 1.5), obtaining ΔE values of 0.82 and 1.04, respectively.

In the experiments with nitrogen, the obtained value of ΔE: 1.90 indicated colour differences at a noticeable (1.5 < ΔE < 3) and statistically significant level (Table 4). Our results confirmed no significant changes in the colour of milk plasma treatment in air and nitrogen, as previously reported by Gurol et al. [59] and Yong et al. [60] and Chen et al. [6]. However, it should be noted that a different type of plasma was used in our experiment, and the exposure of the matrix to the plasma was much longer, which, according to Nikmaram et al. [61], may impact results.

Determined differences in the values of basic colour parameters in nitrogen may indicate non-enzymatic reactions leading to colour changes in the analysed samples. According to Hertwig et al. [42] this is likely due to CP-induced oxidation processes. The study by Segat et al. [62] found an effect of cold plasma on the oxidation of milk proteins resulting in slight colour changes, in this case, yellowing.

The sensory evaluation of plasma-treated products also included taste, aroma, and texture. The overall sensory results do not indicate that the plasma process affected the sensory characteristics of freeze-dried colostrum at a statistically significant level (Table 5). However, in the panelists' evaluation, the highest overall sensory score was obtained for samples of plasma treatment in nitrogen. In contrast, the lowest rating was for samples of plasma treatment in oxygen, primarily due to an intense grassy and fishy odor and an unacceptable rancid, bitter aftertaste. Similar taste and odour sensations were also indicated for air-treated colostrum, but they obtained higher notes. The perceptible changes in the tested samples may be due to the reaction of reactive oxygen species formed during the plasma process, mainly with fatty acids. Free radicals, precursors of lipid hydroperoxides, cause their oxidation and breakdown primarily to aldehydes.

**Table 5.** Effect of cold plasma on sensory qualities of freeze-dried bovine colostrum.

| Sample | Smell | Taste | Structure | Overall Sensory Quality |
|---|---|---|---|---|
| Control | 3.954 | 3.944 | 3.978 | 3.972 |
| Air | 3.833 | 3.899 | 3.944 | 3.745 |
| Nitrogen | 4.012 | 4.051 | 3.816 | 4.034 |
| Oxygen | 3.675 | 3.749 | 3.801 | 3.933 |
| SEM | 0.450 | 0.801 | 0.394 | 0.654 |
| *p*-value | 0.178 | 0.450 | 0.765 | 0.125 |

The accumulation of secondary oxidation products is the cause of the release of odors described by the panelists as fishy, rancid, and oxidized, which was also observed in their study by Yong et al. [60]. In addition, lipid peroxidation increased with higher voltage and longer treatment time [63]. According to the panelists, a caramel smell was perceptible in all plasma-treated samples, which was considered desirable.

Based on the results, the bands from the separation 1-DE of bovine colostrum proteins were identified as lactoperoxidase (94 kDa), lactotransferrin (76 kDa), and serum albumin (62 kDa). Immunoglobulin G was identified as two different protein bands, namely IgG heavy chain and IgG light chain, with molecular weights of about 48 kDa and 24 kDa, respectively. B-lactoglobulin and α-lactalbumin were identified as protein bands with molecular weights of 18 and 15 kD, respectively (Figure 1).

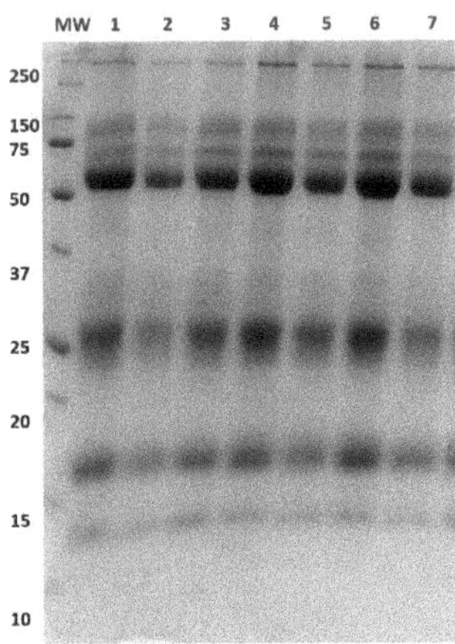


**Figure 1.** Results of 1D electrophoretic separation of samples and bovine colostrum; MW Molecular weight 10–250 kD, lane 1 N$_2$ cold plasma, lane 2 O$_2$ cold plasma, lane three air cold plasma, lanes 4–7 control condition.

2-DE and MALDI TOF-based analysis created a protein profile for lyophilised bovine colostrum. We have generated a 2-D map of this medium protein in the range of isoelectric points (pI) from 3 to 10 and molecular masses (Mr) from 250 to less than 20 kDa. The protein profile comprised ca. 90 distinct protein spots, among which we successfully identified 54 representing 15 gene expression products. The representative 2-D map of lyophilized bovine colostrum whey protein is presented in Figure 2. Most of these proteins are characteristic of domestic cattle (Bos taurus). In addition, nearly 50% of gene products were identified as peptides representing immunoglobulins. Among the other identified gene products, most are prevalent as colostrum/milk proteins: serum albumin, transferrin, β-lactoglobulin, casein S1 and prepronociceptin. The list of identified proteins, including detailed information on the identification parameters of each protein spot, is given in Table 6.

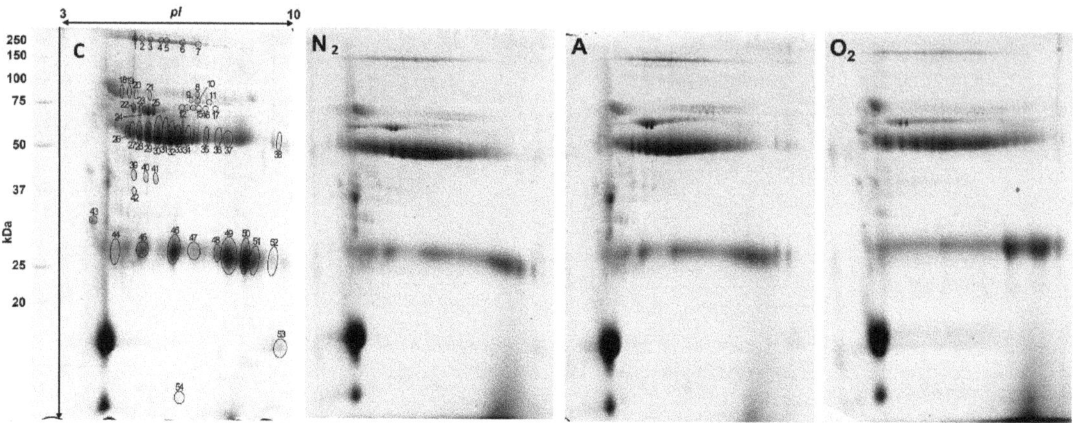

**Figure 2.** 2-DE map of the differentially expressed protein spots found in the colostrum spot numbers corresponds to those in Table 6. Control condition (C) nitrogen cold plasma (N₂), air (A) and oxygen (O₂).

Food processing technologies have a significant impact not only on food safety but also on the structure of proteins and their interactions [64]. For example, studies show that cold plasma induces changes in the structure and functionality of various proteins.

As a part of this study, a comparative densitometric analysis of proteome maps of lyophilized bovine colostrum whey protein after its treatment with cold plasma in different atmospheric environments in contrast to the untreated sample was also performed. The analysed protein profile representatives are shown in Figure 2. The untreated sample is shown on panel C, and the cold plasma-treated samples in the presence of nitrogen, oxygen and atmospheric air are presented on panels N₂, O₂ and A, respectively. In addition, the comparative analysis was performed for the identified protein spots; its results are shown in Table 6.

Densitometric analysis indicated changes occurring in various fractions regardless of the ionizing gas used compared to control samples. Immunoglobulins were noticeable in all samples subjected to plasma sterilization. However, a reduction in the intensity of these spots was observed for all test variants, indicating a reduction in the amount of this fraction.

Samples of freeze-dried bovine colostrum treated with plasma sterilization showed reduced β-lactoglobulin. Plasma sterilization using air, nitrogen and oxygen as ionizing gas had a negative effect on serum albumin fractions. Similar results for β-lactoglobulin and serum albumin bands were obtained by Ng et al. [65]. They attributed the disappearance of protein bands to forming new aggregates involving disulfide bonds and hydrophobic and electrostatic interactions. It has been suggested that cold plasma treatment causes aggregation of whey proteins [62]. The results indicated that the ionizing gases used in our study had a negative effect on the fraction of α-lactalbumin and serotransferrin.

**Table 6.** Proteins identified in the freeze-drying of bovine colostrum with their abundance in the control condition (C) and treated with cold plasma in the presence of nitrogen ($N_2$), air (A), and oxygen ($O_2$). Spot number indicates the number labelling the spots in Figure 2. For each identified protein spot: the protein name, gene name, and accession number are shown in Uniprot/NCBI database. Also, the identification parameters using MALDI-ToF MS are given: several peptides matched per gene product (Pep. Mach.), percentage of the sequence coverage (Seq. %), the Mascot score (score) and predicted pI and Mr.

| No. | Protein Name | Gene Name | Accession Number | C | $N_2$ | $N_2$/C | A | A/C | $O_2$ | $O_2$/C | Pep. Mach | Seq.% /Score | pI/Mr pH/kDa | Sp. |
|---|---|---|---|---|---|---|---|---|---|---|---|---|---|---|
| 1 | Immunoglobulin gamma heavy chain | IGHG | AQT27060 | 251.6 | 25.8 | 0.1 | 14.9 | 0.06 | 51 | 0.2 | 28 | 65/82 | 6.21/51.09 | B |
| 2 | | | | 144.1 | 9.6 | 0.07 | 7.3 | 0.05 | 31.7 | 0.22 | 18 | 42/87 | | |
| 3 | | | | 174.4 | 146 | 0.84 | 118.9 | 0.68 | 50.4 | 0.29 | 18 | 42/100 | | |
| 4 | | | | 173.6 | 82.6 | 0.48 | 31.2 | 0.18 | 64.7 | 0.37 | 16 | 42/82 | | |
| 5 | | | | 152.1 | 121.4 | 0.8 | 257.9 | 1.7 | 52.3 | 0.34 | 15 | 41/84 | | |
| 6 | | | | 57.9 | 125.4 | 2.17 | 203.3 | 3.51 | 89.8 | 1.55 | 14 | 41/84 | | |
| 7 | | | | 43.8 | 7.8 | 0.18 | 133.6 | 3.05 | 34.5 | 0.79 | 14 | 41/86 | | |
| 8 | l-lactate dehydrogenase A chain | LDHA | P00339 | 58.6 | 22.8 | 0.39 | 38.2 | 0.65 | 23.4 | 0.4 | 13 | 58/64 | | S |
| 9 | Sclerostin domain-containing protein 1 | SOSTDC1 | Q9CQN4 | 28 | 16.9 | 0.61 | 41.3 | 1.48 | 16.5 | 0.59 | 7 | 33/63 | 9.87/23.73 | M |
| 10 | Serotransferrin | TRF | Q299443 | 15.9 | 4.4 | 0.28 | 11.4 | 0.72 | 0.3 | 0.02 | 39 | 63/78 | 6.75/79.87 | B |
| 11 | | | | 24.7 | 24.3 | 0.98 | 50.5 | 2.04 | 19.8 | 0.8 | 40 | 62/63 | | |
| 12 | | | | 42 | 25.2 | 0.6 | 85.7 | 2.04 | 116.9 | 2.78 | 41 | 65/66 | | |
| 13 | | | | 31.2 | 40.7 | 1.3 | 62.5 | 2 | 57.2 | 1.83 | 41 | 66/99 | | |
| 14 | | | | 38 | 22.1 | 0.58 | 28.8 | 0.76 | 2.5 | 0.06 | 17 | 33/61 | | |
| 15 | | | | 23.7 | 24.2 | 1.02 | 12.3 | 0.52 | 15.4 | 0.65 | 45 | 68/81 | | |
| 16 | | | | 10.2 | 18.1 | 1.78 | 23.1 | 2.27 | 13.6 | 1.34 | 43 | 68/120 | | |
| 17 | | | | 16 | - | - | - | - | - | - | 36 | 62/82 | | |
| 18 | Secreted immunoglobulin mu 2 heavy chain constant region, partial | IGMG | ANN46371 | 87.4 | 17.9 | 0.2 | 73.1 | 0.84 | 23.4 | 0.27 | 23 | 74/93 | 5.48/50.30 | B |
| 19 | Immunoglobulin mu heavy chain constant region, partial | IGMG | AAO37096 | 99.6 | 63.2 | 0.63 | 114.5 | 1.15 | 77 | 0.77 | 22 | 69/88 | 5.49/49.85 | B |
| 20 | Secreted immunoglobulin mu 2 heavy chain constant region, partial | IGMG | ANN46371 | 49.7 | 20.8 | 0.42 | 0.3 | 0.01 | 14.1 | 0.28 | 19 | 66/71 | 5.48/50.30 | B |
| 21 | Serum albumin precursor | ALB | NP851335 | 30.1 | 4.2 | 0.14 | 14.4 | 0.48 | 2.6 | 0.09 | 24 | 43/98 | 5.82/71.27 | B |
| 22 | Serum albumin | ALB | P02769 | 379.9 | 62 | 0.16 | 54.5 | 0.14 | 71.7 | 0.19 | 17 | 34/75 | 5.82/71.24 | B |
| 23 | | | | 465.7 | 248.2 | 0.53 | 334.9 | 0.72 | 206.5 | 0.44 | 17 | 31/75 | | |
| 24 | | | | 351.2 | 264.6 | 0.75 | 738.1 | 2.1 | 328.3 | 0.93 | 21 | 41/87 | | |
| 25 | | | | 604.4 | 69.7 | 0.12 | 99.9 | 0.17 | 162.6 | 0.27 | 18 | 34/76 | | |
| 26 | Immunoglobulin gamma heavy chain | IGHG | AQT27060 | 596.1 | 363.9 | 0.61 | 475.3 | 0.8 | 111.5 | 0.19 | 16 | 48/22 | 6.21/51.09 | B |
| 27 | | | | 866.9 | 105.1 | 0.12 | 390.6 | 0.45 | 1891 | 2.18 | 24 | 53/80 | | |

**Table 6.** *Cont.*

| No. | Protein Name | Gene Name | Accession Number | C | N₂ | N₂/C | A | A/C | O₂ | O₂/C | Pep. Mach | Seq.% /Score | pI/Mr pH/kDa | Sp. |
|---|---|---|---|---|---|---|---|---|---|---|---|---|---|---|
| 28 | Secreted immunoglobulin gamma1 heavy chain constant region, partial | IGHG | ANN46377 | 2237 | 2301 | 1.03 | 2428 | 1.09 | 287.2 | 0.13 | 14 | 60/79 | 6.07/36.32 | B |
| 29 | Immunoglobulin gamma heavy chain | IGHG | AQT27060 | 2275 | 1017 | 0.45 | 1618 | 0.71 | 2225 | 0.98 | 26 | 56/89 | 6.21/50.09 | B |
| 30 | | | | 5708 | 4441 | 0.78 | 4624 | 0.81 | 3248 | 0.57 | 15 | 48/82 | 6.21/50.09 | B |
| 31 | | | | 69.1 | 759.5 | 11 | 160.6 | 2.33 | 1212 | 17.6 | 23 | 53/91 | | |
| 32 | | | | 1777 | 657.4 | 0.37 | 5782 | 3.25 | 627 | 0.35 | 16 | 49/109 | | |
| 33 | | | | 94 | 759.5 | 8.08 | 160.6 | 1.71 | 1212 | 12.9 | 17 | 50/107 | | |
| 34 | | | | 698.5 | 990.2 | 1.42 | 1057 | 1.51 | 779.5 | 1.12 | 16 | 47/87 | | |
| 35 | | | | 1087 | 566.7 | 0.52 | 1282 | 1.18 | 1709 | 1.57 | 15 | 41/81 | | |
| 36 | | | | 952.9 | 2962 | 3.11 | 413.2 | 0.43 | 1967 | 2.06 | 15 | 48/80 | | |
| 37 | | | | 1122 | 100.3 | 0.09 | 219.6 | 0.2 | 494.7 | 0.44 | 16 | 53/90 | | |
| 38 | t-RNA specific adenosine deaminase 2 | ADAT2 | Q5E9J7 | 55.8 | 24.9 | 0.45 | 49.7 | 0.89 | 41.3 | 0.74 | 14 | 61/61 | 6.33/21.57 | B |
| 39 | | | | 29.4 | 11.5 | 0.39 | 14.9 | 0.51 | 17.9 | 0.61 | 14 | 61/58 | | |
| 40 | Immunoglobulin gamma heavy chain | IGHG | AQT27060 | 43.2 | 18.5 | 0.43 | 35.8 | 0.83 | 12.3 | 0.28 | 26 | 58/83 | 6.21/50.09 | B |
| 41 | | | | 51.4 | 21.8 | 0.42 | 30.3 | 0.59 | 10.8 | 0.21 | 15 | 43/91 | | |
| 42 | IgG heavy chain precursor (B/MT.4A.17.H5.A5) | IGHG | CAA44699 | 25.7 | 18.7 | 0.73 | 21.5 | 0.83 | 5.6 | 0.22 | 22 | 58/91 | 6.10/51.39 | B |
| 43 | Alpha-S1-casein | CSN1S1 | P02662 | 283.9 | 58.1 | 0.2 | 45.7 | 0.16 | 34.6 | 0.12 | 6 | 35/59 | 4.98/24.57 | B |
| 44 | T-complex protein 1 subunit epsilon isoform d | TCP1 | NP_001293084 | 668.8 | 643.3 | 0.96 | 2000 | 2.99 | 863.8 | 1.29 | 29 | 72/81 | 5.86/49.95 | H |
| 45 | Unnamed protein product, partial | CAF90119 | CAF90119 | 1558 | 1783 | 1.14 | 1941 | 1.25 | 710 | 0.46 | 12 | 64/95 | 9.78/13.15 | T |
| 46 | PREDICTED: nck-associated protein 5 isoform X5 | | XP_012901706 | 4206 | 1057 | 0.25 | 1757 | 0.42 | 1033 | 0.25 | 95 | 45/83 | 8.25/22.35 | MP |
| 47 | hypothetical protein EI555_020003, partial | TKC51465 | TKC51465 | 1185 | 191.5 | 0.16 | 489.4 | 0.41 | 840.7 | 0.71 | 21 | 61/82 | 10.5/46.66 | MM |
| 48 | Kv channel-interacting protein 1 | KCIP1 | Q9NZI2 | 1325 | 50 | 0.04 | 253.6 | 0.19 | 433.9 | 0.33 | 26 | 81/66 | 5.10/26.97 | H |
| 49 | Coiled-coil domain-containing protein 183 isoform X2 | CCDC183 | XP_030896878 | 3331 | 1156 | 0.35 | 1247 | 0.37 | 3818 | 1.15 | 38 | 67/85 | 9.90/32.80 | L |
| 50 | | | | 2019 | 749.5 | 0.37 | 1594 | 0.79 | 2021 | 1 | 37 | 68/86 | | |
| 51 | | | | 4315 | 2094 | 0.49 | 144.7 | 0.03 | 2743 | 0.64 | 41 | 71/89 | | |
| 52 | | | | 44.6 | 181.5 | 4.07 | 187.3 | 4.2 | 928.3 | 20.8 | 35 | 76/82 | | |
| 53 | Beta-lactoglobulin | LGB | P02754 | 155.1 | 22.3 | 0.14 | 21.7 | 0.14 | 6.9 | 0.04 | 23 | 96/62 | 4.93/20.27 | B |
| 54 | Prepronociceptin | PNOC | O62647 | 31 | 4.2 | 0.14 | 10.9 | 0.35 | 0.2 | 0.01 | 5 | 46/56 | 9.67/20.64 | B |

Pep. Mach.–peptides matched, Seq %–sequence coverage %, score–Mascot score value, pI/Mr–isoelectric point/molecular range, Sp.–species, B-Bos taurus; S–sus scrofa, M–Mus musculus, H-Homo sapiens, T-Tetraodon nigroviridis, MP-Mustela putorius furo, MM-Monodon Monoceros, L-Leptonychotes weddellii.

In the 1-DE gels, bands of protein fractions were presently smeared and wider compared to the control sample. This may be caused by some cold plasma-induced structural modification resulting in molecular weight changes. These changes are most likely caused by reactive oxygen and nitrogen species and UV radiation, which can potentially modify protein structures [66]. This is supported by the study of Zhou et al. [67], in which modification of amino acids by cold plasma-induced reactions, e.g., oxidation, hydroxylation, dehydrogenation, sulfonation and dimerization, were found.

In addition, literature data [68–70] indicate protein aggregation as a consequence of plasma treatment, which can lead to the disappearance of protein bands in SDS-PAGE. Oxidation induces the unfolding of proteins, which increases their surface hydrophobicity, and causes protein aggregation and polymerization due to the formation of disulfide bonds, dityrosine bonds and other intermolecular bonds [71].

Segat et al. [62] treated whey proteins with dielectric barrier discharge (DBD) plasma and showed an increase in carbonyl groups, surface hydrophobicity, and aperiodic structures. This resulted in a decrease in free SH groups and improved foaming and emulsifying ability.

## 4. Conclusions

Cold plasma is a technology with an extensive range of applications. The variety of molecules activated during plasma generation, the near-zero electrical charge (net), and the low temperature-usually below 40 °C—are only some factors influenced by the type of gas supply and the configuration of the device. Many variables involved in the plasma process and the kind of matrix affect the decontamination results. The differences in parameters in the plasma technology limit the ability to compare the results, especially in the case of colostrum, which until now has not been widely used as the matrix in the plasma process.

As a food supplement, colostrum is available in powder form produced by freeze-drying or spray-drying. The technologies are not intended to eliminate microbial contaminants but mainly to increase their stability, especially during storage. Yet the nutritional values of colostrum and its basic parameters depend on the level of microbial contamination. The application of CP has the potential to inactivate microorganisms while keeping biologically active compounds of colostrum intact. However, our study's results on the decontamination efficiency of plasma are similar to those described in other publications. It could be explained by the fact that product-plasma interactions were limited to the surface of the matrix due to the powder–like state of the product. Our results reflected the dependence of the limited penetration depth of plasma components on the effectiveness of microbial inactivation throughout the sample volume. In addition, reactive plasma molecules in contact with a moist environment produce more oxidizing radicals (ex. OH*, H*, $H_2O_2$, $O_3$) by which the level of inactivation of microorganisms such as bacteria, yeast, and molds increases. Our studies indicate that using oxygen and nitrogen as working gases generates conditions conducive to the highest reduction of the determined microorganisms. Unfortunately, they negatively affected the sensory characteristics of the colostrum samples. The plasma also induced changes in protein fractions.

In general, cold plasma is an emerging and highly promising technology for dairy product processing, including colostrum, with minimal change in quality. However, further studies are required to prove if, by adjusting cold plasma treatment conditions, including the type of gas, voltage, treatment time and plasma source, non-thermal inactivation of microorganisms in commercial colostrum preparations can be achieved without affecting the sensory properties of the product.

**Author Contributions:** Conceptualization, E.B.-W.; methodology, E.B.-W., A.D., A.L., M.O. and W.S.; formal analysis, E.B.-W., A.D., M.O. and A.L.; writing—original draft preparation, review and editing E.B.-W., A.D., A.L. and W.S.; visualization E.B.-W. All authors have read and agreed to the published version of the manuscript.

**Funding:** This research was funded by Intelligent Development Operational Program POIR 02.03.02–30–0065-20-00.

**Institutional Review Board Statement:** Not applicable.

**Informed Consent Statement:** Not applicable.

**Data Availability Statement:** Any data or material supporting this study's findings can be made available by the corresponding author upon request.

**Acknowledgments:** The authors are grateful to Dagmara Mędrala-Klein, for her kindness and English editing.

**Conflicts of Interest:** The authors declare no conflict of interest.

## References

1. Brian, A.M.; Patric, F.F.; Paul, S.; Alan, K. Composition and properties of bovine colostrum: A review. *Dairy Sci. Technol.* **2016**, *96*, 133–158.
2. Pakkanen, R.; Aalto, J. Growth factors and antimicrobial factors of bovine colostrum. *Int. Dairy J.* **1997**, *7*, 258–297. [CrossRef]
3. Shen, R.L.; Thymann, T.; Østergaard, M.V.; Støy, A.C.; Krych, Ł.; Nielsen, D.S.; Lauridsen, C.; Hartmann, B.; Holst, J.J.; Burrin, D.G.; et al. Early gradual feeding with bovine colostrum improves gut function and NEC resistance relative to infant formula in preterm pigs. *Am. J. Physiol. Gastrointest. Liver. Physiol* **2015**, *309*, 310–320. [CrossRef]
4. Playford, R.J.; Weiser, M.J. Bovine Colostrum: Its Constituents and Uses. *Nutrients* **2021**, *13*, 265. [CrossRef]
5. Płusa, T. Immunomodulatory proteins in colostrum. *Pol. Merkur. Lekarski.* **2009**, *26*, 8–10.
6. Chen, D.; Peng, P.; Zhou, N.; Cheng, Y.; Min, M.; Ma, Y.; Mao, Q.; Chen, P.; Chen, C.; Ruan, R. Evaluation of Cronobacter sakazakii inactivation and physicochemical property changes of non-fat dry milk powder by cold atmospheric plasma. *Food Chem.* **2019**, *290*, 270–276. [CrossRef]
7. Chiang, S.H.; Chang, C.Y. Antioxidant properties of caseins and whey proteins from colostrums. *J. Food Drug. Anal.* **2005**, *13*, 57–63. [CrossRef]
8. Wieczorek—Dąbrowska, M.; Wójcik, P.; Malinowski, E. Importance of cow colostrum and factors determining its quality. *Balice Przegląd Hod.* **2013**, *81*, 9–10.
9. Kim, J.S.; Lee, E.J.; Choi, E.H.; Kim, Y.J. Inactivation of Staphylococcus aureus on the beef jerky by radio-frequency atmospheric pressure plasma discharge treatment. *Innov. Food Sci. Emerg. Technol.* **2014**, *22*, 124–130. [CrossRef]
10. Toepfl, S.; Mathys, A.; Heinz, V.; Knorr, D. Review: Potential of High Hydrostatic Pressure and Pulsed Electric Fields for Energy Efficient and Environmentally Friendly Food Processing. *Food Rev. Int.* **2006**, *22*, 405–423. [CrossRef]
11. van Boekel, M.; Fogliano, V.; Pellegrini, N.; Stanton, C.; Scholz, G.; Lalljie, S.; Somoza, V.; Knorr, D.; Jasti, P.R.; Eisenbrand, G. A review on the beneficial aspects of food processing. *Mol. Nutr. Food Res.* **2010**, *54*, 1215–1247. [CrossRef]
12. Fecteau, G.; Baillargeon, P.; Higgins, R.; Paré, J.; Fortin, M. Bacterial contamination of colostrum fed to newborn calves in Québec dairy herds. *Can. Vet. J.* **2002**, *43*, 523–527.
13. Nwabor, O.F.; Onyeaka, H.; Miri, T.; Obileke, K.; Anumudu, C.; Hart, A. A Cold Plasma Technology for Ensuring the Microbiological Safety and Quality of Foods. *Food Eng. Rev.* **2022**, *14*, 535–554. [CrossRef]
14. Yepez, X.V.; Misra, N.N.; Keener, K.M. Nonthermal Plasma Technology. In *Food Safety Engineering. Food Engineering Series*; Demirci, A., Feng, H., Krishnamurthy, K., Eds.; Springer: New York, NY, USA, 2020; pp. 607–628.
15. Reineke, K.; Langer, K.; Hertwig, C.; Ehlbeck, J.; Schlüter, O. The impact of different process gas compositions on the inactivation effect of an atmospheric pressure plasma jet on Bacillus spores. *Innov. Food Sci. Emerg. Technol.* **2015**, *30*, 112–118. [CrossRef]
16. Kim, H.-J.; Yong, H.I.; Park, S.; Choe, W.; Jo, C. Effects of dielectric barrier discharge plasma on pathogen inactivation and the physicochemical and sensory characteristics of pork loin. *Curr. Appl. Phys.* **2013**, *13*, 1420–1425. [CrossRef]
17. Lee, H.J.; Jung, S.; Jung, H.; Park, S.; Choe, W.; Ham, J.S.; Jo, C. Evalution of dielectric barier discharge plasma system for inactivatating pathogens on cheese slices. *J. Anim. Sci. Technol.* **2012**, *54*, 191–198. [CrossRef]
18. Moreau, M.; Orange, N.; Feuilloley, M.G.J. Non-thermal plasma technologies: New tools for bio-decontamination. *Biotechnol. Adv.* **2008**, *26*, 610–617. [CrossRef]
19. *PN—93/A-86034/02*; Milk and Dairy Products. Microbiological Testing. General Principles of Testing. Polish Committee for Standardization: Warsaw, Poland, 2018.
20. *ISO 6887-5:2020-10*; Microbiology of the Food Chain—Preparation of Samples, Stock Suspension and Decimal Dilutions for Microbiological Testing—Part 5: Specific Rules for the Preparation of Milk and Milk Products. Polish Committee for Standardization: Warsaw, Poland, 2020.
21. *ISO 4833-2:2013-12*; Microbiology of the Food Chain—Horizontal Method for the Determination of Microbial Counts—Part 2: Determination of Counts by Surface Culture at 30 Degrees C. Polish Committee for Standardization: Warsaw, Poland, 2013.
22. *ISO 21527-1:2009*; Microbiology of Food and Feed—Horizontal Method for the Determination of Yeast and Mold Counts—Part 1: Method for Counting Colonies in Products with Water Activity Higher than 0.95. Polish Committee for Standardization: Warsaw, Poland, 2009.
23. *ISO 21528-2:2017-08*; Food Chain Microbiology—Horizontal Method for Detection and Enumeration of Enterobacteriaceae—Part 2: Colony Counting Method. Polish Committee for Standardization: Warsaw, Poland, 2017.

24. *ISO 6888-2:2001+A1:200*; Microbiology of Food and Feed—Horizontal Method for the Determination of Coagulase-Positive Staphylococci (Staphylococcus Aureus and Other Species)—Part 2: Method Using Rabbit Plasma Agar Medium and Fibrynogen. Polish Committee for Standardization: Warsaw, Poland, 2004.
25. *ISO 7932:2005/A1:2020-09*; Microbiology of Food and Feed—Horizontal Method for Determining the Number of Putative Bacillus Cereus—Method of Counting Colonies at 30 Degrees C. Polish Committee for Standardization: Warsaw, Poland, 2020.
26. *PN-EN 15788:2009*; Detection and Enumeration of Enterococcus (E. faecium) spp. Polish Committee for Standardization: Warsaw, Poland, 2009.
27. *ISO 15214:2002*; Microbiology of Food and Feed—Horizontal Method for the Determination of the Number of Mesophilic Lactic Fermentation Bacteria—Plate Method at 30 Degrees C. Polish Committee for Standardization: Warsaw, Poland, 2002.
28. Dmytrów, I.; Szymczak, M.; Szkolnicka, K.; Kamiński, P. Development of Functional Acid Curd Cheese (Tvarog) with Antioxidant Activity Containing Astaxanthin from Shrimp Shells Preliminary Experiment. *Foods* **2021**, *10*, 895. [CrossRef]
29. Pankaj, S.K.; Bueno-Ferrer, C.; Misra, N.N.; O'Neill, L.; Tiwari, B.K.; Bourke, P.; Cullen, P.J. Physicochemical characterization of plasma-treated sodium caseinate film. *Food Res. Int.* **2014**, *66*, 438–444. [CrossRef]
30. *ISO 22935-2:2013-07*; Milk and Dairy Products—Sensory Analysis—Part 2: Recommended Methods for Sensory Evaluation. Polish Committee for Standardization: Warsaw, Poland, 2013.
31. *ISO 22935-3:2013-07*; Milk and Dairy Products—Sensory Analysis—Part 3: Guidelines for Evaluating the Conformity of the Properties of Sensory Attributes with Product Specifications Using the Scoring Method. Polish Committee for Standardization: Warsaw, Poland, 2013.
32. Drake, M.A.; Karagul-Yuceer, Y.; Cadwallader, K.R.; Civtlle, G.V.; Tong, P.S. Determination of the sensory attributes of dried milk powders and dairy ingredients. *J. Sens. Stud.* **2003**, *18*, 199–2016. [CrossRef]
33. Abdalla, A.K.; Smith, J.K.; Lucey, K. Physical Properties of Nonfat Dry Milk and Skim Milk Powder. *Int. J. Dairy Sci.* **2017**, *12*, 149–154. [CrossRef]
34. Lepczyński, A.; Ożgo, M.; Dratwa-Chałupnik, A.; Robak, P.; Pyć, A.; Zaborski, D.; Herosimczyk, A. An update on medium- and low-abundant blood plasma proteome of horse. *Animal* **2018**, *12*, 76–87. [CrossRef]
35. Lepczyński, A.; Ożgo, M.; Michałek, K.; Dratwa-Chałupnik, A.; Grabowska, M.; Herosimczyk, A.; Liput, K.P.; Poławska, E.; Kram, A.; Pierzchała, M. Effects of three-month feeding high fat diets with different fatty acid composition on myocardial proteome in mice. *Nutrients* **2021**, *13*, 330. [CrossRef]
36. Ozgo, M.; Lepczynski, A.; Herosimczyk, A. Two-dimensional gel-based serum protein profile of growing piglets. *Turk. J. Biol.* **2015**, *39*, 320–327. [CrossRef]
37. Godden, S.M.; Lombard, J.E.; Woolums, A.R. Colostrum Management for Dairy Calves. *Vet. Clin. North. Am. Food. Anim. Pract.* **2019**, *35*, 535–556. [CrossRef]
38. Lorenz, I.; Mee, J.F.; Earley, B.; More, S.J. Calf health from birth to weaning. General aspects of disease prevention. *Ir. Vet. J.* **2011**, *16*, 3–8. [CrossRef]
39. Bourke, P.; Ziuzina, D.; Han, L.; Cullen, P.J.; Gilmore, B.F. Microbiological interactions with cold plasma. *J. Appl. Microbiol.* **2017**, *123*, 308–324. [CrossRef]
40. Stoffels, E.; Sakiyama, Y.; Graves, D.B. Cold atmospheric plasma: Charged species and their interactions with cells and tissues. *IEEE Trans. Plasma Sci.* **2008**, *36*, 1441–1457. [CrossRef]
41. Fröhling, A.; Baier, M.; Ehlbeck, J.; Knorr, D.; Schlüter, O. Atmospheric pressure plasma treatment of *Listeria innocua* and *Escherichia coli* at polysaccharide surfaces: Inactivation kinetics and flow cytometric characterization. *Innov. Food Sci. Emerg. Technol.* **2012**, *13*, 142–150. [CrossRef]
42. Hertwig, C.; Reineke, K.; Ehlbeck, J.; Erdoğdu, B.; Rauh, C.; Schlüter, O. Impact of remote plasma treatment on natural microbial load and quality parameters of selected herbs and spices. *J. Food Eng.* **2015**, *167*, 12–17. [CrossRef]
43. Cummins, C.; Lorenz, I.; Kennedy, E. Short communication: The effect of storage conditions over time on bovine colostral immunoglobulin G concentration, bacteria, and pH. *J. Dairy Sci.* **2016**, *99*, 4857–4863. [CrossRef] [PubMed]
44. Fasse, S.; Jarmo, A.; Björn, F.; Gun, W. Bovine Colostrum for Human Consumption—Improving Microbial Quality and Maintaining Bioactive Characteristics through Processing. *Dairy* **2021**, *2*, 556–575. [CrossRef]
45. Moisan, M.; Barbeau, J.; Moreau, S.; Pelletier, J.; Tabrizian, M.; Yahia, L.H. Low-temperature sterilization using gas plasmas: A review of the experiments and an analysis of the inactivation mechanisms. *Int. J. Pharm.* **2001**, *226*, 1–21. [CrossRef]
46. Schlüter, O.; Fröhling, A. *Non-Thermal Processing Cold Plasma for Bioefficient Food Processing, Reference Module in Food Science Encyclopedia of Food Microbiology*, 2nd ed.; Elsevier: New York, NY, USA, 2014; pp. 948–953.
47. Li, J.; Sakai, N.; Watanabe, M.; Hotta, E.; Wachi, M. Study on plasma agent effect of a direct-current atmospheric pressure oxygen-plasma jet on inactivation of *E. coli* using bacterial mutants. *IEEE Trans. Plasma Sci.* **2013**, *41*, 935–941. [CrossRef]
48. Joshi, S.G.; Cooper, M.; Yost, A.; Paff, M.; Ercan, U.K.; Fridman, G.; Friedman, G.; Fridman, A.; Brooks, A.D. Nonthermal dielectric-barrier discharge plasma-induced inactivation involves oxidative DNA damage and membrane lipid peroxidation in Escherichia coli. *Antimicrob. Agents. Chemother.* **2011**, *55*, 1053–1062. [CrossRef]
49. Hertwig, C.; Meneses, N.; Mathys, A. Cold atmospheric pressure plasma and low energy electron beam as alternative nonthermal decontamination technologies for dry food surfaces: A review. *Trends Food Sci. Technol.* **2018**, *77*, 131–142. [CrossRef]
50. Deng, X.; Shi, J.; Kong, M.G. Physical mechanisms of inactivation of Bacillus subtilis spores using cold atmospheric plasmas. *IEEE Trans. Plasma Sci.* **2006**, *34*, 1310–1316. [CrossRef]

51. Tseng, S.; Abramzon, N.; Jackson, J.O.; Lin, W.J. Gas discharge plasmas are effective in inactivating Bacillus and Clostridium spores. *Appl. Microbiol. Biotechnol.* **2012**, *93*, 2563–2570. [CrossRef]
52. Wang, S.; Doona, C.J.; Setlow, P.; Li, Y.Q. Use of Raman Spectroscopy and Phase-Contrast Microscopy to Characterize Cold Atmospheric Plasma Inactivation of Individual Bacterial Spores. *Appl. Environ. Microbiol.* **2016**, *82*, 5775–5784. [CrossRef]
53. Segat, A.; Misra, N.N.; Cullen, P.J.; Innocente, N. Effect of atmospheric pressure cold plasma (ACP) on activity and structure of alkaline phosphatase. *Food Bioprod. Process.* **2016**, *98*, 181–188. [CrossRef]
54. Manoharan, D.; Stephen, J.; Radhakrishnan, M. Study on low-pressure plasma system for continuous decontamination of milk and its quality evaluation. *J. Food Process. Preserv.* **2020**, *45*, 234–245. [CrossRef]
55. Wan, Z.; Misra, N.N.; Li, G.; Keener, K.M. High voltage atmospheric cold plasma treatment of Listeria innocua and Escherichia coli K-12 on Queso Fresco (fresh cheese). *LWT* **2021**, *146*, 111406. [CrossRef]
56. Oehmigen, K.; Hahnel, M.; Brandenburg, R.; Wilke, C.; Weltmann, K.D.; von Woedtke, T. The Role of Acidification for Antimicrobial Activity of Atmospheric Pressure Plasma in Liquids. *Plasma Process. Polym.* **2010**, *7*, 250–257. [CrossRef]
57. Traylor, M.J.; Pavlovich, M.J.; Karim, S.; Hait, P.; Sakiyama, Y.; Clark, D.S.; Graves, D.B. Long-term antibacterial efficacy of air plasma-activated water. *J. Phys. Appl. Phys.* **2011**, *44*, 134–145. [CrossRef]
58. Gross, J.J.; Kessler, E.C.; Bruckmaier, R.M. Colour measurement of colostrum for estimation of colostral IgG and colostrum composition in dairy cows. *J. Dairy. Res.* **2014**, *81*, 440–444. [CrossRef] [PubMed]
59. Gurol, C.; Ekinci, F.Y.; Aslan, N.; Korachi, M. Low Temperature Plasma for decontamination of *E. coli* in milk. *Int. J. Food. Microbiol.* **2012**, *157*, 1–5. [CrossRef]
60. Yong, H.I.; Kim, H.-J.; Park, S.; Kim, K.; Choe, W.; Yoo, S.J.; Jo, C. Pathogen inactivation and quality changes in sliced cheddar cheese treated using flexible thin-layer dielectric barrier discharge plasma. *Food Res. Int.* **2015**, *69*, 57–63. [CrossRef]
61. Nikmaram, N.; Keener, K.M. The effects of cold plasma technology on physical, nutritional, and sensory properties of milk and milk products. *LWT* **2022**, *154*, 112729. [CrossRef]
62. Segat, A.; Misra, N.N.; Cullen, P.J.; Innocente, N. Atmospheric pressure cold plasma (ACP) treatment of whey protein isolate model solution. *Innov. Food Sci. Emerg. Technol.* **2015**, *29*, 247–254. [CrossRef]
63. Wu, X.; Luo, Y.; Zhao, F.; Murad, M.S.; Mu, G. Influence of dielectric barrier discharge cold plasma on physicochemical property of milk for sterilization. *Plasma Process. Polym.* **2021**, *18*, 245–256. [CrossRef]
64. Singh, P.K.; Huppertz, T. Chapter 8—Effect of Nonthermal Processing on Milk Protein Interactions and Functionality. In *Milk Proteins*; Boland, M., Singh, H., Eds.; Academic Press: Toronto, ON, Canada, 2020; pp. 293–324.
65. Ng, S.W.; Lu, P.; Rulikowska, A.; Boehm, D.; O'Neill, G.; Bourke, P. The effect of atmospheric cold plasma treatment on the antigenic properties of bovine milk casein and whey proteins. *Food Chem.* **2021**, *342*, 128283. [CrossRef] [PubMed]
66. Ramazzina, I.; Tappi, S.; Rocculi, P.; Sacchetti, G.; Berardinelli, A.; Marseglia, A.; Rizzi, F. Effect of Cold Plasma Treatment on the Functional Properties of Fresh-Cut Apples. *J. Agric. Food. Chem.* **2016**, *64*, 8010–8018. [CrossRef] [PubMed]
67. Zhou, R.; Zhuang, J.; Zong, Z.; Zhang, X.; Liu, D.; Bazaka, K.; Ostrikov, K. Interaction of Atmospheric-Pressure Air Microplasmas with Amino Acids as Fundamental Processes in Aqueous Solution. *PLoS ONE* **2016**, *11*, 134–148. [CrossRef]
68. Held, S.; Tyl, C.E.; Annor, G.A. Effect of radio frequency cold plasma treatment on intermediate wheatgrass (Thinopyrum intermedium) flour and dough properties in comparison to hard and soft wheat (*Triticum aestivum* L.). *J. Food. Qual.* **2019**, *34*, 156–176. [CrossRef]
69. Ji, H.; Dong, S.; Han, F.; Li, Y.; Chen, G.; Li, L.; Chen, Y. Effects of Dielectric Barrier Discharge (DBD) Cold Plasma Treatment on Physicochemical and Functional Properties of Peanut Protein. *Food Bioprocess Technol.* **2018**, *11*, 344–354. [CrossRef]
70. Zhang, S.; Huang, W.; Feizollahi, E.; Roopesh, M.S.; Chen, L. Improvement of pea protein gelation at reduced temperature by atmospheric cold plasma and the gelling mechanism study. *Innov. Food Sci. Emerg. Technol.* **2021**, *67*, 102–112. [CrossRef]
71. Zhang, W.; Xiao, S.; Ahn, D.U. Protein oxidation: Basic principles and implications for meat quality. *Crit. Rev. Food. Sci. Nutr.* **2013**, *53*, 1191–1201. [CrossRef]

*Article*

# The Antibacterial Effect of the Films Coated with the Layers Based on *Uncaria tomentosa* and *Formitopsis betulina* Extracts and ZnO Nanoparticles and Their Influence on the Secondary Shelf-Life of Sliced Cooked Ham

Magdalena Ordon [1], Weronika Burdajewicz [1,2], Józef Sternal [2], Marcin Okręglicki [2] and Małgorzata Mizielińska [1,*]

[1] Center of Bioimmobilisation and Innovative Packaging Materials, Faculty of Food Sciences and Fisheries, West Pomeranian University of Technology Szczecin, Janickiego 35, 71-270 Szczecin, Poland; magdalena.ordon@zut.edu.pl (M.O.); w.burdajewicz@onet.pl (W.B.)

[2] MarDruk Opakowania sp. z o.o., sp.k., Ul. Krakowska 83c, 34-120 Andrychów, Poland; technolog@mardrukopakowania.pl (J.S.); marcin@mardrukopakowania.pl (M.O.)

* Correspondence: malgorzata.mizielinska@zut.edu.pl

**Abstract:** Microbial surface contamination of the cooked ham sliced by shop assistants has an influence on its potential secondary shelf-life. Active packaging may be the solution. The first goal of the work was to obtain coatings based on *Formitopsis betulina* and *Uncaria tomentosa* extracts (separately), with the addition of nanoparticles of ZnO (as a synergetic agent to increase plant extracts' antimicrobial effectiveness), that could be active against selected bacterial strains. Another aim of the work was to determine the influence of obtained active layers on microbial purity and texture of the samples. The results of the study demonstrated that the layers containing *F. betulina* or *U. tomentosa* extracts with the addition of ZnO nanoparticles inhibited *Staphylococcus aureus*, *Bacillus atrophaeus*, and *Escherichia coli* growth completely and decreased the number of *Pseudomonas syringae* cells. Analyzing microbial purity of sliced cooked ham portions after storage, it was observed that the active packaging based on *U. tomentosa* extracts had a greater influence on the total count than the coatings containing *F. betulina* extracts and that both active packaging materials affected the microbial purity of the cooked ham. Analyzing the textural parameter changes after 48 h of storage, a decrease in the gumminess, chewiness and cohesiveness values was noted, but only for the ham portions stored in bags covered with the coating based on *U. tomentosa*. The effectiveness of the active packaging on textural parameters after 96 h of storage was not noticed.

**Keywords:** active packaging; cooked ham; ham storage; *Uncaria tomentosa* extract; *Formitopsis betulina* extract; antibacterial properties

Citation: Ordon, M.; Burdajewicz, W.; Sternal, J.; Okręglicki, M.; Mizielińska, M. The Antibacterial Effect of the Films Coated with the Layers Based on *Uncaria tomentosa* and *Formitopsis betulina* Extracts and ZnO Nanoparticles and Their Influence on the Secondary Shelf-Life of Sliced Cooked Ham. *Appl. Sci.* **2023**, *13*, 8853. https://doi.org/10.3390/app13158853

Academic Editor: Monica Gallo

Received: 27 June 2023
Revised: 20 July 2023
Accepted: 30 July 2023
Published: 31 July 2023

## 1. Introduction

With the increasing demand for ready-to-eat food products and globalization, concerns for food quality and safety are expanding. Meat-based perishable food products are subjected to bacterial contamination causing undesirable reactions [1]. The sliced meat products are rich in fats, proteins, and some (added) carbohydrates that may be metabolized by spoilage microorganisms, resulting in off-odors, off-flavors, gas formation, discoloration, changes in texture, and slime-formation. These changes make the product unsuitable for consumption. Although cooking eliminates the majority of the microorganisms, products may be contaminated during handling operations after the cooking process, for instance, during packaging or slicing by shop assistants [2,3]. Microbial surface contamination of cooked meat products has an influence on the potential shelf-life stability, creating important food preservation challenges [2]. Vacuum and modified-atmosphere packaging (MAP) allows for an extension of the shelf-life period of meat-based products with minimal changes in the chemical and physical properties of the products, such as sliced cooked ham [2,3]. In order to inhibit the growth of microorganisms responsible for food spoilage in ready-to-eat sliced ham, bioactive compounds can be incorporated into product formulation, coated on its surface, or incorporated into the packaging material [1]. Cooked ham packaged in a modified atmosphere is a very popular product, subjected to abundant microbial contamination throughout its shelf life that can lead to deterioration of both textural properties and microbial purity [3,4]. Consumers are happy to purchase sliced ham packed in a vacuum or in MAP. However, many of them decide to buy cooked ham that is sliced in shops, and there can be situations where it is contaminated by the staff while being cut [3]. The assessment of appropriate secondary shelf life, defined as the time after package opening or after buying a meat product sliced in a butcher shop by a shop assistant during which the food product retains a required level of quality, is pivotal for reducing domestic food waste [4]. The need to produce safe and healthy meat-based products not based on chemical additives has resulted in the control of both spoilage and pathogenic microorganisms. This requirement has been led by consumer demand for safe food products while, in addition, leading to a huge decrease in processing, cutting out many preservatives, and an extension in shelf-life [5]. Active food packaging plays an important role in the preservation of the quality of food products and ensuring their safety [6]. Packaging materials covered with active coatings are designed to release active substances into the product and, by doing so, can extend the shelf life of the food while maintaining its safety and quality [6,7]. The fabrication of active films, which are based on a surface covering method, contains a number of biobased carriers and antibacterial substances, including those based on plant extracts to create active layers, that can result in an increase in both the primary or secondary shelf-life of foodstuffs [3–5].

*Uncaria tomentosa*, known as a cat's claw, is a medicinal plant used as complementary medicine to treat many diseases. Extensive research carried out over many years has demonstrated its antibacterial, antiviral, anti-inflammatory, antioxidant, anticonceptive, and immunostimulant effects, as well as not being toxic [8,9]. *Fomitopsis betulina* is a plant known as a birch polypore. It is a medicinal and edible mushroom (when young). The effectiveness of *F. betulina* originates mostly from the triterpenoids, especially lanostane derivatives present in the fungus [9,10]. The results of the previous work [9] have shown that *Fomitopsis betulina* extract was unable to inhibit *S. aureus* growth. A lower than 1-log reduction in the number of bacterial cells was noted for 5% and 10% plant extracts. A greater than 2-log reduction in the number of this strain was only observed in the case of a 50% extract of *F. betulina*. It should be underlined that *U. tomentosa* extract had a much lower influence on the *S. aureus* cells than the *F. betulina* extract. When analyzing the effect of *F. betulina* and *U. tomentosa* extracts on *B. subtilis* cells, it was demonstrated that all of the extracts showed lower than a 2-log reduction but higher than a 1-log reduction in the number of bacterial cells. The results of the previous study have also confirmed that *F. betulina* and *U. tomentosa* extracts had no activity against *E. coli* cells [9]. An excellent example is an earlier work carried out by the authors [11], who analyzed the antibacterial effect of a methyl-hydroxy-propyl-cellulose (MHPC) active layer (G) containing 5% of geraniol in the coating carrier. The authors observed a 3-log reduction in the number of *S. aureus* cells, a 4-log reduction in the number of *E. coli* cells, and a 2-log reduction in the number of *P. syringae* cells. It should also be underlined that a characteristic aroma of geraniol in the active layer was noticeable. The findings of the previous work demonstrated that a decrease in the amount of geraniol (by 99.75%) in the MHPC carrier (G1) had a significant effect on the activity of the coating G1 against Gram-positive and Gram-negative strains. This coating slightly decreased the number of *S. aureus* strains (1 log). The G1 layer was found not to be active against *E. coli* and *P. syringae* cells; however, a lowering of the scent of the coating G1 compared to G was observed. The authors confirmed [12] that the coating containing 0.082 g of ZnO nanoparticles in 100 mL of MHPC carrier inhibited the growth of both Gram-positive and Gram-negative bacteria. The results of an earlier study [11] confirmed that the addition of the 0.041 g of nanoparticles (decreased amount) to the 100 mL of MHPC carrier containing 0.0125 g of geraniol (decreased amount compared to the MHPC containing 5% of geraniol) led to obtaining the active layer, which inhibited the growth of *S. aureus* cells. Additionally, a 4-log reduction in the number of *E. coli* and *P. syringae* cells was noticed.

Zinc oxide nanoparticles are widely used as antimicrobial agents. They have ideal properties such as biocompatibility, high safety, long-term efficiency, stability, and an uncomplicated production process, making them a proper option for active coatings/packaging applications. Additionally, the US Food and Drug Administration (FDA) has affirmed the safety of these nanoparticles [12,13]. The antibacterial activity of ZnO NPs against different bacterial strains such as Gram-negative *Klebsiella* sp., *E. coli*, and Gram-positive *S. aureus*, including its AMR variant (MRSA), vancomycin-resistant enterococci, and *B. subtilis*, was confirmed [12–19]. The various advantages of ZnO NPs may be attributed to their unique features, including their semiconducting behavior, high transparency in the visible light range, and high UV absorption capabilities. Ultraviolet radiation can lead to a degradation in the antimicrobial activity of packaging coverings by inactivation of the active agent introduced into the coating. The addition of an active agent which is resistant to UV in an active layer, or the addition of a substance with shielding properties, may prevent inactivation of the antimicrobial layer after UV irradiation [12,14,17–19].

As an assumption [11,12], it may be concluded that the addition of ZnO nanoparticles into the *F. betulina* or *U. tomentosa* (separately) extracts can lead to a synergistic effect between these plant extracts and nanoparticles, and as a result, it may increase their antibacterial activity. The coatings containing nano ZnO as an additive will be resistant to ultraviolet irradiation due to their UV-shielding properties; they will maintain the plant extract's active compound effectiveness [12,17–19].

The first purpose of the work was to obtain coatings based on *F. betulina* and *U. tomentosa* extracts with the addition of a decreased number of nanoparticles of ZnO (to 0.021 g per 100 mL of coating carrier) that could be active against selected Gram-positive and Gram-negative bacterial strains. The ZnO nanoparticles were used as a synergetic agent to increase plant extracts' antimicrobial activity and to prevent their inactivation when added to the coatings. Another aim of the study was to determine the influence of packaging covered with the obtained active layers on the microbial purity and texture of the sliced cooked ham.

## 2. Materials and Methods

### 2.1. Materials

The microorganisms used to verify antibacterial properties of coatings in this research were purchased from a collection from the Leibniz Institute DSMZ (Deutsche Sammlung von Mikroorganismen und Zellkulturen). There was a *Bacillus atrophaeus* DSM 675 IZT, a *Staphylococcus aureus* DSMZ 346, a *Pseudomonas syringae* van Hall 1902 DSM 21482, and an *Escherichia coli* DSMZ 498 strain.

Polypropylene films (A4, 20 μm) were obtained from a MarDruk company (Andrychów, Poland). The solvent dispersion 70 GU279686 (Hubergroup, Warsaw, Poland) was used as coating carrier (CC), as well as a Zinc Oxide AA 44899, (~70 nm) powder (Thermo Fisher GmbH, Kandel Germany), *Uncaria tomentosa*, and *Fomitopsis betulina* (Planteon, Borków Stary, Poland), which were used as antibacterial substances. The 99.8% ethanol (EU-ROCHEM BGD Sp. z o.o. Tarnów, Poland) was used as a plant solvent extract. Agar-agar, TSB (Triptic Soy Broth), TSA (Triptic Soy Agar), and MacConkey agar (Merck, Darmstadt, Germany) were used to determine the antimicrobial properties of all analyzed coatings. To carry out the microbial analysis of sliced, cooked ham pre- and post-storage, a number of chosen tests were performed. Samples were tested to verify the coliform bacteria count through the use of Violet Red Bile Glucose Agar (VRBG) (Merck KGAA, Darmstadt, Germany). As a verification of the total bacterial count and the *S. aureus* count of the sliced cooked ham, PPS (PPS: 0.85% m/v NaCl, 0.1% m/v peptone), PCA (BTL, Łódź, Poland) and Baird-Parker mediums (Merck KGAA, Darmstadt, Germany) were used. These mediums were made in accordance with instructions set by the manufacturer. All mediums, barring

VRBG agar, were then weighed and later suspended in 1 L of distilled water; this was then autoclaved at 121 °C for 15 min. After being weighed, the VRBG agar was then suspended in 1 L of distilled water and heated to a resulting boiling temperature.

### 2.2. Extracts Preparation

The dry plants *Uncaria tomentosa* and *Fomitopsis betulina* were introduced (separately) into a TM6 Thermomix (VORWERK, Wrocław, Poland). The dry plants were ground to a powder (7600 rpm, 20 s). As next step, 100 g of each plant powder was introduced (separately) into 100 mL of 99.8% ethanol. The ethanol solutions containing the *U. tomentosa* and *F. betulina* powders were then introduced into a Microwave (Amica, Wronki, Poland) for 5 min at 70 °C. Next, the samples were introduced into a shaker (Ika, Staufen im Breisgau, Germany) and extracted for 1 h at 70 °C (150 rpm). The extraction process was carried out according to the methods described by the authors [9,20,21]. After the extraction, the plants were separated from the extracts using Büchner funnel, then the samples were filtered through a 0.2 µm filter and evaporated to obtain the dry mass of the samples 13.74% (*F. betulina*) and 21.72% (*U. tomentosa*). Then the extracts were used in the next stage of the experiments.

### 2.3. Coating Preparation

(1. Coating Zn) A total of 0.06 g of ZnO nanoparticles was introduced into 50 mL of ethanol. As a first step, the dispersion was mixed for 1 h using a magnetic stirrer (500 rpm, Ika, Warsaw, Poland). Next, it was sonicated for 30 min. (sonication parameters: amplitude: 20%; cycle: 0.5). A ZnO nanoparticle dispersion was introduced into the 50 mL of the coating carrier (CC) and mixed for 10 min using a magnetic stirrer (500 rpm).

(2. Coatings Ut and Fb) A total of 0.03 g of ZnO nanoparticles was introduced into 45 mL of ethanol. Initially, the dispersion was mixed for 1 h using a magnetic stirrer (500 rpm); the dispersion was then sonicated for 30 min (sonication parameters: amplitude: 20%; cycle: 0.5), while at the same time, the second nano ZnO dispersion was prepared as described above. Then 5 mL of *U. tomentosa* extract (Ut) was introduced into the first dispersion, and 5 mL of *F. betulina* (Fb) extract was introduced into the second dispersion. As next step of the experiment, the dispersions were mixed for 15 min using a magnetic stirrer (500 rpm). A 50 mL of the ZnO nanoparticle dispersions containing Ut and Fb extracts (separately) was introduced (separately) into the 50 mL of the coating carrier (CC) and mixed for 10 min using a magnetic stirrer (500 rpm).

Polypropylene (PP) films were coated using an Unicoater 409 (Erichsen, Hemer, Germany) at a temperature of 24 °C with a 40 µm diameter roller. The coatings were dried for 10 min at a temperature of 50 °C. PP films that were not coated were control samples (PP). Having been cut into square shapes (3 cm × 3 cm), the film samples were tested for their antibacterial properties.

### 2.4. Antibacterial Analysis

The antibacterial activity of the layers containing the Ut or Fb extracts with the addition of ZnO nanoparticles against Gram-negative bacteria: *E. coli P. syringae* and Gram-positive *B. atrophaeus, S. aureus* strains, compared to the non-coated PP film samples was performed according to the ASTM E 2180-01 standard [22]. As first step, agar slurry was prepared by dissolving 0.85 g of NaCl and 0.3 g of agar-agar in 100 mL of deionized water (one agar slurry was prepared for each analyzed microbial strain). Agar slurries were sterilized by autoclaving for 15 min at 121 °C and equilibrated at 45 °C. Then 1 mL of bacterial broth culture ($1 \times 10^8$ cells/mL; for each strain separately) was placed into 100 mL of the molten agar slurry to obtain the final concentration of microorganisms ($1 \times 10^6$ cells/mL). Square samples of the PP film (as control samples) and the samples covered with the UT and Fb coatings (test samples) were placed into sterile Petri dishes (separately). Then 1 mL of inoculated agar slurry was pipetted onto the control and test samples (in triplicates) to obtain a 1 mm film (in-depth). The samples were incubated for 24 h at 37 °C (*E. coli*,

*S. aureus, B. atrophaeus*) and at 28 °C (*P. syringae*). The relative humidity within the incubator (Memmert GmbH, Schwabach, Germany) was 80%. After incubation, the samples were treated from Petri dishes to the sterile bags containing 100 mL of TSB and introduced into the BagMixer®400 VW (Interscience, Saint Nom la Bretêche, France). The samples were then mixed for 2 min. After mixing, serial dilutions of the initial TSB mediums containing control or test samples (separately) were performed. Each dilution was spread in duplicates on agar mediums as follows: (a) the dilutions of the samples containing *B. atrophaeus* and *P. syringae* were spread into TSA agar; (b) the dilutions of the samples containing *S. aureus* were spread into TSA agar containing 9% of NaCl; (c) the dilutions of the samples containing *E. coli* were spread into MacConkey agar. The plates were incubated for 48 h at 37 °C (*E. coli, S. aureus, B. atrophaeus*) and at 28 °C (*P. syringae*).

### 2.5. SEM Analysis of the Coatings

The PP films and PP coated with the active layers Zn (containing ZnO nanoparticles as active substance), Ut (containing *U. tomentosa* extract and ZnO nanoparticles as active agents), and Fb (containing *F. betulina* extract and ZnO nanoparticles as active compounds) were analyzed using a scanning electron microscope (SEM). As first step, the samples were placed on pin stubs and coated with a thin layer of gold in a sputter coater at room temperature (Quorum Technologies Q150R S, Laughton, East Sussex, UK). Microscopic analysis was carried out by the use of a Vega 3 LMU microscope (Tescan, Brno-Kohoutovice, Czech Republic), described earlier. These tests were of vital importance as they would clearly show if the PP films had been thoroughly and homogeneously coated with the active layers. Analysis took place at 24 °C through the use of a tungsten filament with an accelerating voltage of 10 kV being used to create SEM images of the non-active (uncoated) and active (covered) layers that had been put on PP foil. The specimens were then examined from above.

### 2.6. Bags and Spacers Preparation

PP film was covered with the active layers (Ut or Fb) containing plant extracts with the addition of ZnO nanoparticles (separately) from one side for the creation of bags while taken from both sides for the creation of spacers. A control sample (C) was created from PP film that was not coated with any layer. The films that included both the coated and non-coated films were then cut into pieces to prepare bags and for the creation of foil spacers which were square shaped. For the creation of the bags, both coated and non-coated films were welded together (HSE-3, RDM Test Equipment, Hertfordshire, Great Britain) in air conditions deemed normal. The welding parameters were set at pressure—4 kN; temperature—117 °C; time—4 s.

### 2.7. Packaging and Storage

Two bars of cooked ham with the same expiration date (±2 days) were purchased in a local market (Szczecin, Poland). The producer suggested ham consumption within 2–3 days after package opening. The package was opened, and cooked ham was cut into 20 mm slices and 3 mm slices by a shop assistant; the portions were packed into polyethylene bags and immediately transported to the Center of Bioimmobilisation and Innovative Packaging Materials (CBIMO). The sliced ham was then aseptically introduced into bags. All of the cooked ham slices were separated by the use of a square spacer (Figure 1a) and placed into bags, which included (Figure 1b,c):

a. Six PP bags (C) (control samples), these sliced pieces of ham were separated by un-coated PP spacers;
b. Six PP bags coated with Ut coating, these sliced pieces of ham were separated by PP spacers, coated with the Ut layer on each side;
c. Six PP bags covered with Fb coating; these sliced pieces of ham were separated by PP spacers that were coated with the Fb layer on each side.

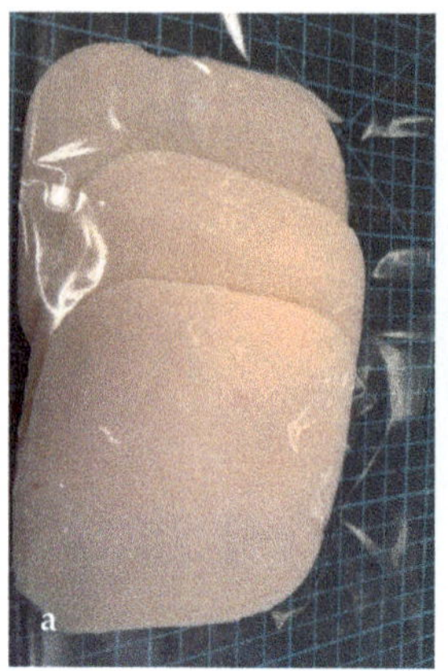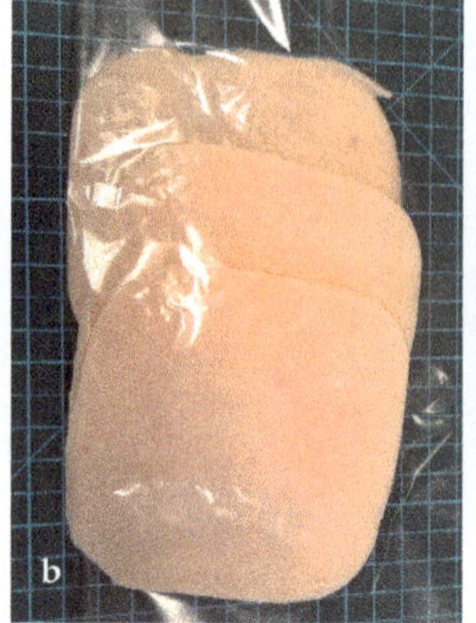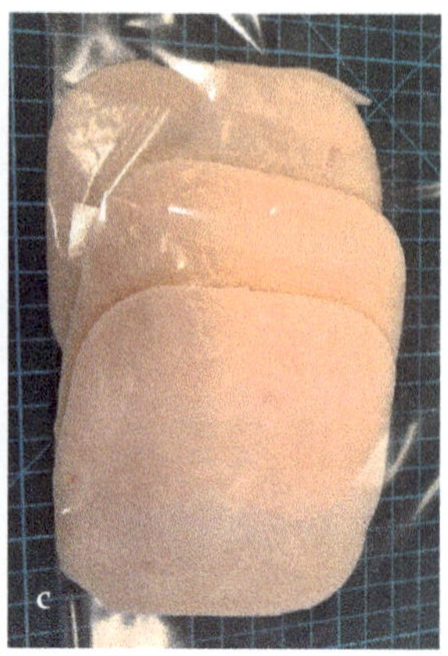

**Figure 1.** The sliced cooked ham: (**a**) in uncoated bag, separated by uncoated PP spacers; (**b**) in bag covered with the Ut coating, separated by PP spacers, coated with the Ut layer on each side; (**c**) in bag covered with the Fb coating, separated by PP spacers that were coated with the Fb layer on each side.

The precooked ham slices were placed in contact with the active layers on each side.

Both the active and non-active bags were welded and sealed (HSE-3, RDM Test Equipment, Hertfordshire, Great Britain) in air conditions deemed normal. Welding parameters were set at pressure—4 kN; temperature –117 °C; time—4 s.

The 18 bags (six of each packaging material) containing sliced cooked ham were then refrigerated at 5 °C. The 9 samples were later examined on 48 h (three bags of each packaging material) and 96 h of storage (three bags of each packaging material). The 20 mm slices were used for textural analysis. The 3 mm slices were used for dry mass analysis, microbiological analysis, SEM, and L*a*b* tests.

### 2.8. The Microbiological Analysis

Microbiological analysis was performed for sliced cooked ham before and after storage (in active and non-active packaging). For each microbiological analysis, $10 \pm 0.1$ g of individual cooked ham slices were aseptically introduced into a sterile stomacher bag in physiological saline peptone solution (PPS: 0.85% m/v NaCl; 0.1% m/v peptone). Then ham slices were homogenized in a Bag Mixer (Interscience, Saint-Nom-la-Brèteche, France) for 120 s, and appropriate decimal dilutions were prepared in PPS. The total count was determined according to PN-EN ISO 4833-2:2013-12 [23] as follows: serial dilutions of the initial, homogenized ham samples (in PPS) were performed; then the 0.1 mL of the initial inoculum and its dilutions were carefully, uniformly spread over the surface of the PCA medium (in duplicates, 90 mm Petri dishes). The plates were left for 15 min for the inoculum to be absorbed into the agar. Then the plates were inverted and introduced into the incubator (Memmert GmbH, Schwabach, Germany). The samples were then incubated for 72 h at 30 °C. After incubation, agar plates with fewer than 300 colonies were selected, and the colonies were counted. The *S. aureus* count analysis was performed according to PN-EN ISO 6888-1 [24] as follows: serial dilutions of the initial homogenized ham samples (in PPS) were performed; then the 1 mL of the initial inoculum and its dilutions were uniformly spread over the surface of the Baird-Parker agar medium (in duplicates, 90 mm Petri dishes). The plates were left for 15 min for the inoculum to be absorbed into the agar. Then the plates were incubated for 48 h at 37 °C. The total coliforms bacteria count was carried out according to PN-ISO 4832:2007 [25] as follows: serial dilutions of the initial homogenized ham samples were performed; then the 1 mL of the initial ham sample and its dilutions were pipetted from test tubes to the Petri dishes; VRBG medium was weighted, suspended in 1 L of distilled water, heated to a resulting boiling temperature (it was very important not to overheat the medium), and equilibrated at 45 °C. Then 15 mL of VRBG medium was pipetted to the Petri dishes containing the samples and their dilutions and mixed carefully. The mediums were left to solidify. As next step, 4 mL of VRBG medium was pipetted on the surface of each sample and left to solidify. The samples were incubated for 24 h at 37 °C.

### 2.9. The Textural Analysis

An analysis of the texture of the cooked ham (20 mm slices) took place in accordance with the PN-ISO 11036:1999 standard: "Sensory analysis. Methodology. Texture profiling" [26]. Zwick/Roell Z 2.5 was used (Wrocław, Poland). An analysis was carried out of the ham samples after delivery from a retail point and after our own storage tests.

### 2.10. Dry Mass Tests

The dry mass was noted in the case of the cooked ham both before being placed in bags and on 48 h and 96 h storage time. A Weight Dryer (Radwag, Warsaw, Poland) was used to ascertain dry mass. The samples from each bag were tested in duplicates.

### 2.11. L*a*b* Tests

The color of the cooked ham was determined through the average of 9 individual measurements from ham slice spots randomly selected by the use of a colorimeter (NR 20 XE, EnviSense) with other related data software. The measurement of color was carried out through an aperture (8 mm diameter) using a CIE L*a*b* color space which utilized a standard 10 observer and Illuminant D65. The selected parameters (that would account for the results) were $\Delta L$ (darkness and lightness difference) and $\Delta E_{lab}$ (total color aberration). The EnviSense protocol was used to calculate parameters.

### 2.12. SEM Analysis of the Sliced Ham

Before being placed into bags and after 48 h and 96 h storage, an SEM analysis was carried out on the cooked ham. In addition, SEM micrographs were taken to visualize the cooked ham micro-texture. The samples of ham were set at 18 h at 4 °C (2% glutaraldehyde in a 0.1 M sodium cacodylate, pH 7.4). Further, the samples were washed with 0.1 M sodium cacodylate and later dehydrated in serial concentrations of ice-cold (–20 °C) methanol (20%, 40%, 60%, 80%, and 100%) at 2 h intervals. The samples of ham were then put in a Petri dish for 5 min. The food samples were then placed on pin stubs, and a thin layer of gold in a sputter coater was used to coat them at room temperature (Quorum Technologies Q150R S, Laughton, East Sussex, UK). The cooked ham samples were analyzed with a scanning electron microscope (SEM). The microscopic analysis of the samples was carried out through the use of a Vega 3 LMU microscope (Tescan, Brno-Kohoutovice, Czech Republic) already described above.

### 2.13. Statistical Analysis

The texture and microbiological analysis results were checked for statistical significance through the use of an analysis of variance (ANOVA) followed by an ANOVA test (one-way). Where $p < 0.05$, the values were seen as significantly different. All analyses were carried out through the use of GraphPad Prism 8 (GraphPad Software, Version 9, San Diego, CA, USA).

## 3. Results

### 3.1. Antibacterial Analysis

The results of the current work demonstrated that the coating (Zn) containing 0.06 g of ZnO nanoparticles in 100 mL of the coating carrier inhibited *S. aureus*, *B. atrophaeus*, and *E. coli* growth completely. The coatings Ut (containing 0.03 g of ZnO nanoparticles and 5% of *U. tomentosa* extract into 100 mL of the coating carrier) and Fb (containing 0.03 g of ZnO nanoparticles and 5% of *F. betulina* extract into 100 mL of the coating carrier) also had a bacteriolytic effect on these strains (Table 1). Unfortunately, neither Zn coating nor Ut and Fb layers inhibited the growth of *P. syringae* cells. It was noted that Zn coating was not effective against the *P. syringae* strain, but the number of these bacteria was decreased by Ut and Fb coatings. Statistical analysis confirmed that the changes in the number of bacterial cells were found to be significant (Figure 2).

**Table 1.** The influence of coatings on *S. aureus*, *E. coli* and *B. syringae* growth.

| The Sample | The Number of Bacterial Cells [CFU/g] | | |
|---|---|---|---|
| | *S. aureus* | *E. coli* | *B. atrophaeus* |
| PP | $4.05 \times 10^6 \pm 4.87 \times 10^5$ * | $9.05 \times 10^7 \pm 1.6 \times 10^6$ * | $7.73 \times 10^5 \pm 1.14 \times 10^5$ * |
| Zn | 0 **** | 0 **** | 0 **** |
| Ut | 0 **** | 0 **** | 0 **** |
| Fb | 0 **** | 0 **** | 0 **** |

*—Standard Deviation; One-way ANOVA test: ****—$p < 0.0001$.

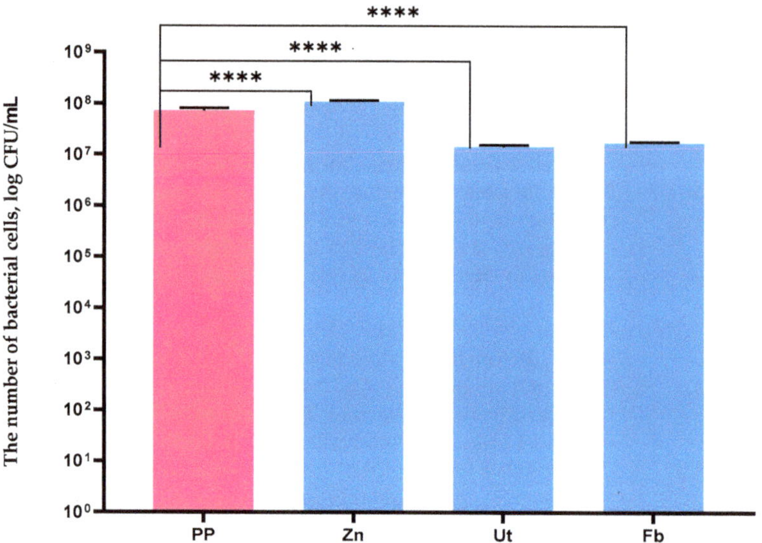

**Figure 2.** The influence of coatings on *P. syringae* growth. PP—PP film; Zn—PP film covered with the coating containing 0.06 g ZnO nanoparticles in 100 mL carrier; Ut—PP film covered with the coating containing 0.03 g of ZnO nanoparticles and *U. tomentosa* extract in 100 mL carrier; Fb—PP film covered with the coating containing 0.03 g of ZnO nanoparticles and *F. betulina* extract in 100 mL carrier; One-way ANOVA; ****—$p < 0.0001$.

## 3.2. SEM Analysis of the Coatings

It was shown through an SEM analysis that the PP film that had been used in the tests had a slightly rough surface of the uncovered film (Figure 3a. SEM images under magnification of 200× and 500×). As was emphasized in Figure 3b–d, active coatings (Zn, Ut, and Fb) had no effect on PP surface morphology. In addition, it was noted that PP film was homogenously and thoroughly coated with the active layers (covered foil samples micrographs reflected uncovered foil samples). Additionally, the Zn covering was not visible when compared to the Ut and Fb coatings. The covering of homogenous PP film with Zn, Ut, and Fb coatings can have an effect on the release of active agents on the whole layer surface of the food product. As was shown in the Figure 3d, convex, spherical particles were noted on the Fb layer. Additionally, as noted in Figure 3c, small pores were seen on the Ut coating.

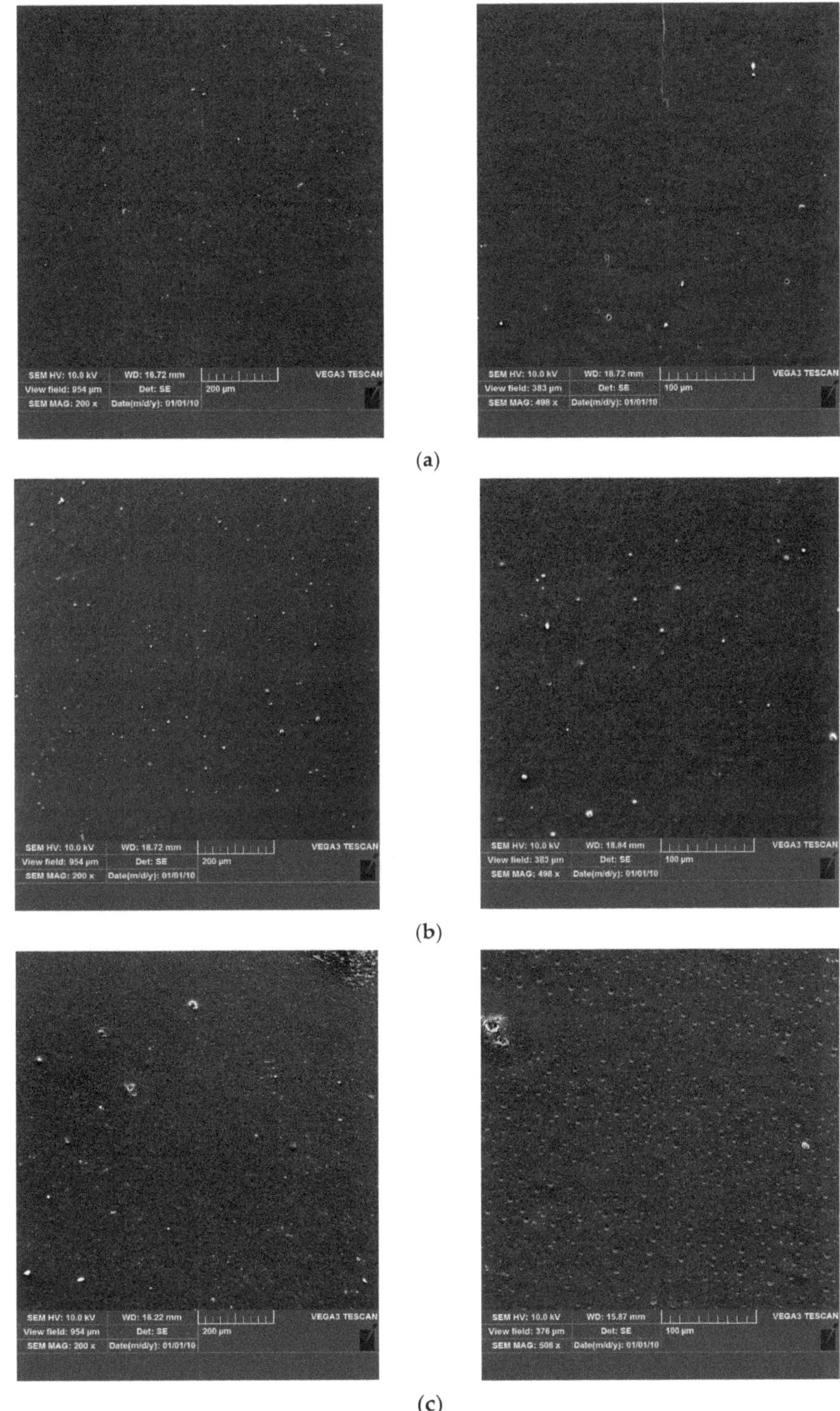

**Figure 3.** *Cont.*

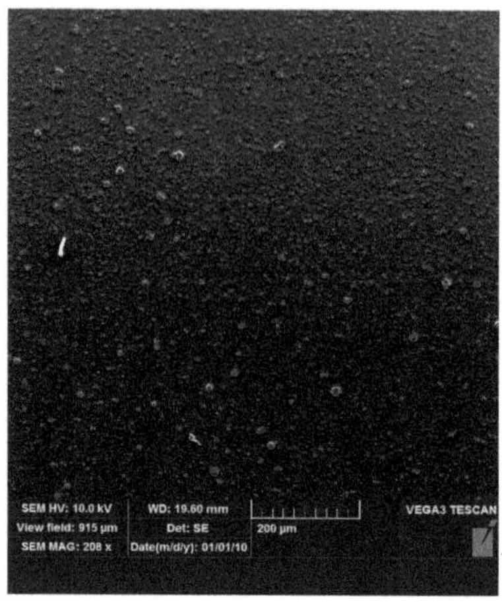 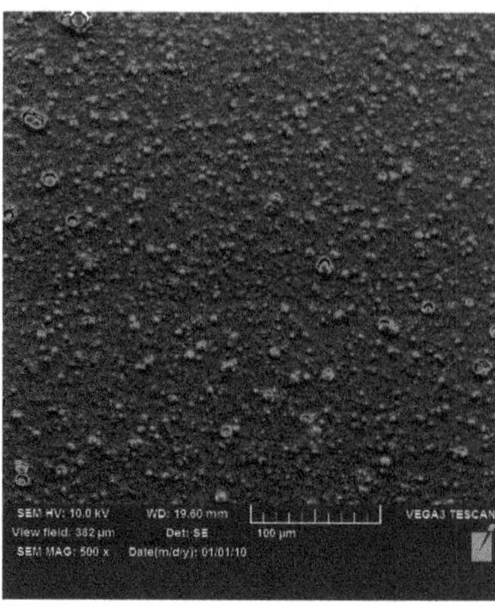

(d)

**Figure 3.** (**a**) The uncoated PP film. (**b**) The PP film covered with the Zn coating. (**c**) The PP film covered with the Ut coating. (**d**) The PP film covered with the Fb coating.

### 3.3. Microbial Purity Analysis

It was shown that the number of mesophilic bacteria found in the sliced cooked ham kept in the non-active coating bags (C—control sample) increased markedly after 48 h ($p < 0.001$) and 96 h ($p < 0.0001$) of storage at 5 °C in air conditions (than when placed in comparison to sample "0"—pre-storage). Figure 4 shows that a Ut coating containing 5% of the *U. tomentosa* extract and decreased amount of ZnO nanoparticles had an impact on the increase in mesophilic bacterial cells. It decreased the number of living cells after 48 h of storage compared to the samples, which were stored in bags devoid of coatings. Additionally, a higher than 1 log increase in bacteria number was seen (compared to sample "0"—pre-storage) for the slices stored 48 h in bags covered with the Ut coating. Unfortunately, the Fb coating did not decrease the number of microorganisms compared to the "0" sample and to the samples stored in bags without any coatings. It would be too easy to advance that only the coating with Ut extract was able to be active against mesophilic bacteria and that Ut active layer was significantly more effective than the Fb layer. As emphasized below (Figure 2), the PP films coated with the Ut layer had a significant influence on the total count compared to the PP bags covered with the Fb coatings and to the number of mesophilic microorganisms detected from the samples stored in the bags devoid of active coatings after storage of 96 h. There was clear evidence that the number of bacterial cells in the case of sliced ham samples stored in PP bags and in PP bags that were coated with the active layers was higher (than compared to the "0" sample—pre-storage) after 96 h storage. A clear statistical analysis showed that any variances between the number of microorganisms stored for 96 h were found to be significant ($p < 0.0001$). From the point of view of microbiological analysis, a limit that would be acceptable for a total viable count of products such as sliced meat is 6–7 $\log_{10}$ CFU/g [27]. Nevertheless, it is of note that according to Grzybowski et al. [28], a microbial load lower than 3 $\log_{10}$ CFU/g is a satisfactory/acceptable number of mesophilic bacteria for products made up of sliced meat, and the 4 $\log_{10}$ CFU/g is generally accepted as a maximum total count that would be seen as acceptable for sliced ready-to-eat meat products. A higher number of bacteria than 4 $\log_{10}$ CFU/g is seen as an unacceptable microbial load in the case of consumable sliced cooked ham. This means that all samples stored in PP bags that did not contain active layers or in active bags that were fully covered with the Ut or Fb active coatings were not suitable to be consumed after 48 and 96 h of storage (Figure 4). It has to be underlined that two bars of cooked ham had the same expiration date ($\pm 2$ days) on the day when they were purchased. Additionally, the ham was sliced by the shop assistant, who could contaminate the samples during slicing. It can be mentioned that the producer suggested the ham consumption within 2–3 days after the package opening. The results confirmed that the sliced ham was not acceptable for consumption after 48 h of their storage, but a lower increase in mesophilic microorganisms was noted for the samples stored in bags coated with the Ut layer. These findings suggest that the package covered with the more effective coating could decrease the number of bacteria to an acceptable level.

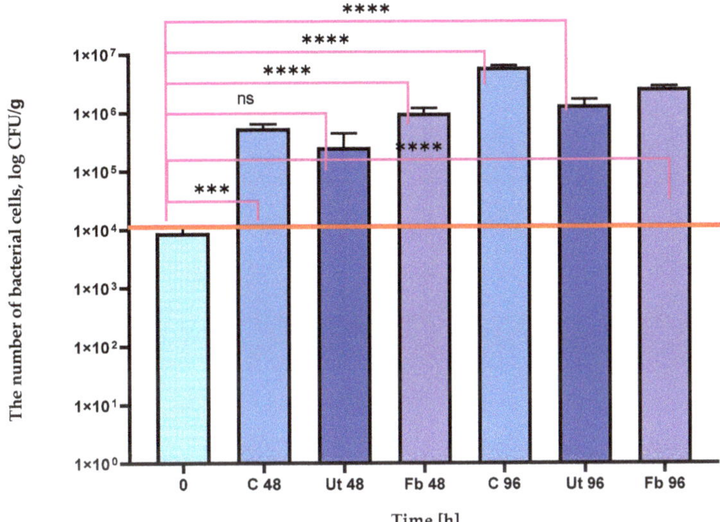


**Figure 4.** The total count of mesophilic bacteria detected from the sliced cooked ham stored for 48 and 96 h. Zero—sample "0"; (C) sliced pieces of ham, stored in PP bags, separated by uncoated PP spacers; (Ut) sliced pieces of ham, stored in PP bags covered with the Ut layer, separated by PP spacers coated with the Ut layer on each side; (Fb) sliced pieces of ham, stored in PP bags covered with the Fb layer, separated by PP spacers coated with the Fb layer on each side; One-way ANOVA ***—$p < 0.001$; ****—$p < -0.0001$; ns—$p > 0.05$.

The results of the study confirmed that Ut and Fb coatings inhibited the growth of *S. aureus*, *B. atrophaeus*, and *E. coli* strains (Table 1) and decreased the number of *P. syringae* cells (Figure 2). According to Grzybowski et al. [28], the number of coliform bacteria found in 10 mg of sliced cooked ham has to be zero. In contrast, the number of cells of *S. aureus* in 1 g of this product can be 100 to be suitable for consumption. Results from this study showed that neither the coliforms nor *S. aureus* cells were detected in the "0" sample and in all of the analyzed sliced cooked ham stored in uncovered PP bags and in PP bags coated with the Ut and Fb layers. With this in mind, all of the samples that were analyzed were suitable for consumption even after 96 h of storage. In accordance with Grzybowski et al. [28], the number of *Listeria monocytogenes* cells must not be found in 1 g of the ready to eat sliced meat products, and *Salmonella* sp. should not be found in 25 g of sliced ham. The activity of Ut and Fb coatings against *L. monocytogenes* and against *Salmonella* sp. was not determined in this research. It is why the presence of these microorganisms in sliced cooked ham before and after storage was not analyzed as well.

*3.4. The Textural Analysis*

The results of this research have shown that springiness in the sliced cooked ham increased after 48 h storage in the uncovered PP films. After 96 h, a decrease in this parameter was noted. Modifying the bags with the active layers containing the *F. betulina* and ZnO nanoparticles working as active agents led to a slightly lower enlargement of the springiness of the sliced cooked ham after 48 h of storage and a slightly greater decrease after 96 h of storage than when compared to bags that were uncoated. It was observed that 48 h storage of the cooked ham samples in bags covered with the Ut coating had not any influence on springiness, but it caused an uptick in springiness when it was compared to the uncoated bags (Figure 5). The differences between these parameter values were seen as insignificant, later confirmed through statistical analysis ($p > 0.05$).

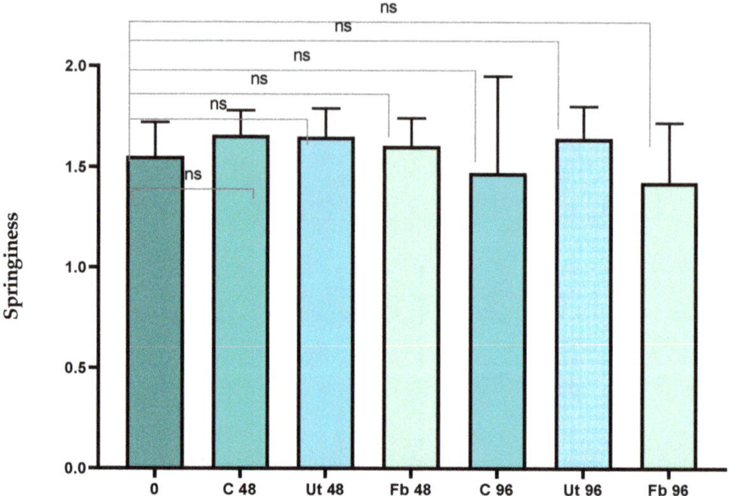


**Figure 5.** The springiness of the sliced cooked ham stored for 48 and 96 h. Zero—sample "0"; (C) sliced pieces of ham, stored in PP bags, separated by uncoated PP spacers; (Ut) sliced pieces of ham, stored in PP bags covered with the Ut layer, separated by PP spacers coated with the Ut layer on each side; (Fb) sliced pieces of ham, stored in PP bags covered with the Fb layer, separated by PP spacers coated with the Fb layer on each side; One-way ANOVA ns—$p > 0.05$.

As was observed below (Figure 6), the gumminess of the sliced cooked ham packed in bags devoid of active layers insignificantly rose after a period of 48 h of storage. After 96 h, the gumminess was noted to have risen compared to sample "0". In addition, this rise was seen to be higher when set in comparison to the sliced ham stored for more than 48 h. These differences were seen as insignificant and were confirmed by a detailed statistical analysis ($p > 0.05$). Similar results were obtained for packaging covered with the Fb coating. When analyzing the ham slices that were stored in packaging coated with the Ut active layer, it was observed that average gumminess decreased after 48 h of storage. After 96 h of storage, the gumminess of the sliced cooked ham kept in packaging with the Ut coating slightly increased when compared to that of the samples stored for 48 h and decreased compared to the "0" sample. The changes between the gumminess values were noted to be insignificant, again confirmed by statistical analysis ($p > 0.05$). Comparing the samples after 48 h of storage in coated PP packaging with a Ut layer with the samples stored for 96 h in these bags, the gumminess values of the sliced cooked ham were significantly lower. Contrary results were obtained for the sliced cooked ham after 48 h and 96 h of storage in bags coated with the Fb layers. The Fb coating did not influence gumminess values after 48 h of storage. Additionally, a slight, insignificant increase in this parameter in the case of samples kept for more than 96 h in packaging with an Fb layer was noted.

With regards to cohesiveness, it was seen that this particular value for the cooked sliced ham stored in uncoated PP bags increased after 48 h and after 96 h of storage. However, a rise in cohesiveness was seen to be less for samples stored for 96 h than for the slices stored for 48 h. These changes were insignificant and confirmed by statistical analysis (Figure 7). In the case of active packaging, a decrease in cohesiveness was noted in those samples stored for 48 h, but only in bags covered with Ut coating. After 96 h storage, an increase in the cohesiveness of the ham slices stored in these bags was observed. In addition, the differences in parameter values were found not to be significant. The average cohesiveness value, measured for the samples stored in bags with Fb layer, stored for 48 and for 96 h also increased, though this value was seen as insignificant.

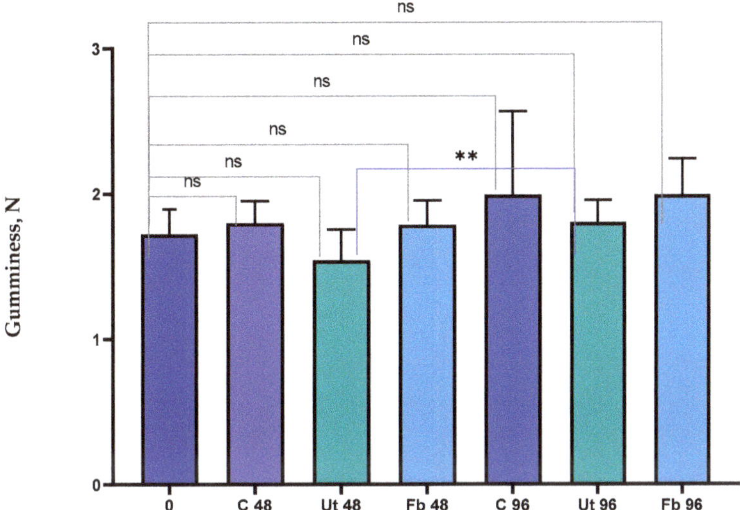

**Figure 6.** The gumminess of the sliced cooked ham stored for 48 and 96 h. Zero—sample "0"; (C) sliced pieces of ham, stored in PP bags, separated by uncoated PP spacers; (Ut) sliced pieces of ham, stored in PP bags covered with the Ut layer, separated by PP spacers coated with the Ut layer on each side; (Fb) sliced pieces of ham, stored in PP bags covered with the Fb layer, separated by PP spacers coated with the Fb layer on each side; One-way ANOVA **—$p < 0.01$; ns—$p > 0.05$.

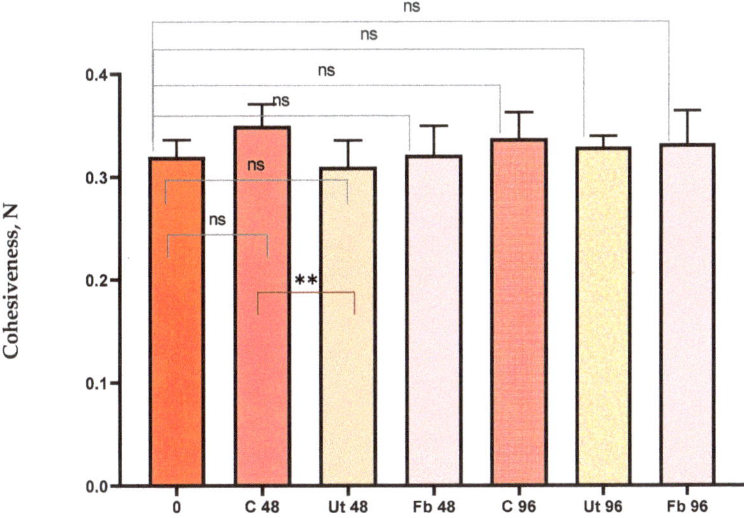

**Figure 7.** The cohesiveness of the sliced cooked ham stored for 48 and 96 h. Zero—sample "0"; (C) sliced pieces of ham, stored in PP bags, separated by uncoated PP spacers; (Ut) sliced pieces of ham, stored in PP bags covered with the Ut layer, separated by PP spacers coated with the Ut layer on each side; (Fb) sliced pieces of ham, stored in PP bags covered with the Fb layer, separated by PP spacers coated with the Fb layer on each side; One-way ANOVA **—$p < 0.01$; ns—$p > 0.05$.

Analyzing the decrease in the gumminess and cohesiveness values after short-term storage (48 h), it was demonstrated that a decrease in these two parameters was observed only for the samples stored in packaging coated with the Ut coating.

The results of the research also demonstrated that chewiness depended on the packaging material for 48 h of storage (Figure 8). Analyzing the chewiness of the sliced cooked ham slices, which were introduced into uncoated bags for 48 h, it was determined that these parameter values increased. Contrary results were obtained for the samples that were stored in bags covered with Ut and Fb coatings. The slight, insignificant increase in chewiness was seen in the slices kept in bags with a Fb layer. However, the Ut coating had no influence on the chewiness. After 96 h of being stored, a clear increase was noted as insignificant, this being confirmed by statistical analysis for sliced ham stored in all bags (coated or uncoated). However, a greater increase was noted for the samples that were introduced into the bags that were coated with Ut coating.

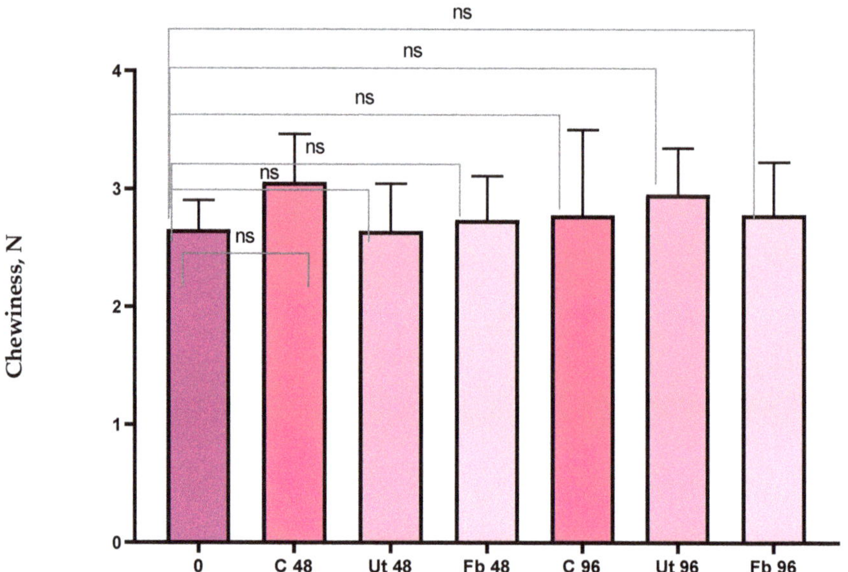


**Figure 8.** The chewiness of the sliced cooked ham stored for 48 and 96 h. Zero—sample "0"; (C) sliced pieces of ham, stored in PP bags, separated by uncoated PP spacers; (Ut) sliced pieces of ham, stored in PP bags covered with the Ut layer, separated by PP spacers coated with the Ut layer on each side; (Fb) sliced pieces of ham, stored in PP bags covered with the Fb layer, separated by PP spacers coated with the Fb layer on each side; One-way ANOVA; ns—$p > 0.05$.

### 3.5. Dry Mass Analysis

The results of this research have uncovered that the dry mass of cooked ham slices was 63.84%. Storing ham samples in PP bags led to a rise in the dry mass of the ham slices to 64.95% after 48 h of being stored and to 70.55% after 96 h. It was shown that the dry mass of sliced cooked ham stored in bags coated with Ut and Fb layers was greater than the dry mass of those ham slices stored in bags that remained uncoated after 48 h and 96 h of being stored. The Ut coating led to a slightly lower increase in the dry mass of the sliced ham after 48 h of storage than the Fb layer. Contrary results were uncovered in samples kept for 96 h (Table 2).

**Table 2.** The dry mass of sliced chicken sausage after 48 h and 96 h of storage.

| Time [h] | Dry Mass [%] | | |
|---|---|---|---|
| | **C** | **Ut** | **Fb** |
| 0 | $63.84 \pm 0.002$ | $63.84 \pm 0.002$ | $63.84 \pm 0.002$ |
| 48 | $64.95 \pm 0.094$ | $65.52 \pm 0.039$ | $65.92 \pm 0.015$ |
| 96 | $70.55 \pm 0.022$ | $72.23 \pm 0.012$ | $72.18 \pm 0.005$ |

(C) sliced pieces of ham, stored in PP bags, separated by uncoated PP spacers; (Ut) sliced pieces of ham, stored in PP bags covered with the Ut layer, separated by PP spacers coated with the Ut layer on each side; (Fb) sliced pieces of ham, stored in PP bags covered with the Fb layer, separated by PP spacers coated with the Fb layer on each side.

### 3.6. L*a*b* Analysis

It was demonstrated in this work that $\Delta E_{lab}$ had a great impact on the material used for packaging (Table 3). The $\Delta E_{lab}$ of sliced cooked ham, which was introduced into the uncovered bags for 48 h, was higher than the $\Delta E_{lab}$ of the samples that had been added to bags coated with Ut and Fb layers. This was contrary to PP films covered with a coating containing the *F. betulina* extract and ZnO nanoparticles; the $\Delta E_{lab}$ for ham slices kept in packaging that had been coated with a layer that contained *U. tomentosa* and nano ZnO was lower. Similar results were also found on 96 h storage. The highest $\Delta E_{lab}$ was noted in the case of samples of ham kept in bags that had not been coated. The lowest $\Delta E_{lab}$ was noted for active packaging. It was also observed that $\Delta E_{lab}$ was dependent on the material used for packaging. Tests on sliced cooked ham kept for 96 h showed that a lower $\Delta E_{lab}$ value was obtained for ham slices added to packaging with Ut covering than those only with the layer of Fb.

**Table 3.** The $\Delta E_{lab}$ of sliced chicken sausage after 72 h and 144 h of storage.

| | Time [h] | | C | Ut | Fb |
|---|---|---|---|---|---|
| $\Delta E_{lab}$ | | 0 | $2.31 \pm 0.19$ | $2.31 \pm 0.19$ | $2.31 \pm 0.19$ |
| $\Delta L$ | | | $2.25 \pm 0.37$ | $2.25 \pm 0.37$ | $2.25 \pm 0.37$ |
| $\Delta E_{lab}$ | | 48 | $3.40 \pm 0.61$ | $1.27 \pm 0.15$ | $2.19 \pm 0.73$ |
| $\Delta L$ | | | $7.60 \pm 1.06$ | $0.37 \pm 0.23$ | $0.76 \pm 0.50$ |
| $\Delta E_{lab}$ | | 96 | $12.03 \pm 0.46$ | $9.99 \pm 0.38$ | $11.74 \pm 0.21$ |
| $\Delta L$ | | | $11.99 \pm 0.44$ | $9.97 \pm 0.39$ | $11.62 \pm 0.26$ |

(C) sliced pieces of ham, stored in PP bags, separated by uncoated PP spacers; (Ut) sliced pieces of ham, stored in PP bags covered with the Ut layer, separated by PP spacers coated with the Ut layer on each side; (Fb) sliced pieces of ham, stored in PP bags covered with the Fb layer, separated by PP spacers coated with the Fb layer on each side.

### 3.7. SEM Analysis of the Sliced Ham

The microstructure of the sliced cooked ham before and after storage was analyzed using SEM. The cooked ham sample (Figure 9a) comprised a gel network including small and large fat particles entrapped in the protein network and unabsorbed protein structure. Although there were gullies on its surface, the overall surface of the sample was relatively smooth but not homogenous. The loose and porous structures with air bubbles (holes) were observed. Analyzing the influence of the storage on the microstructure of cooked ham, the samples were observed to have a gelatinous appearance (Figure 9b–g). After 48 h and 96 h of storage, a deterioration in myofibril led to an increase in water migration, which in turn led to a clearly visible irregular gully structure in the samples. The structure of the ham has become more compact (Figure 9b). However, a lower myofibril degradation was observed for the samples stored in active packaging than for the samples stored in uncoated bags. It was demonstrated that the surface of the slices of ham kept in bags of (active Ut and Fb) for a 96 h (Figure 9f,g) was less dense and compact than the microstructures of the samples stored in PP bags (Figure 9c), confirming water migration. In addition, small cracks were noted along the surface of the samples, as well as air bubbles (holes) visible on the sample surfaces of all items stored in the active packaging.

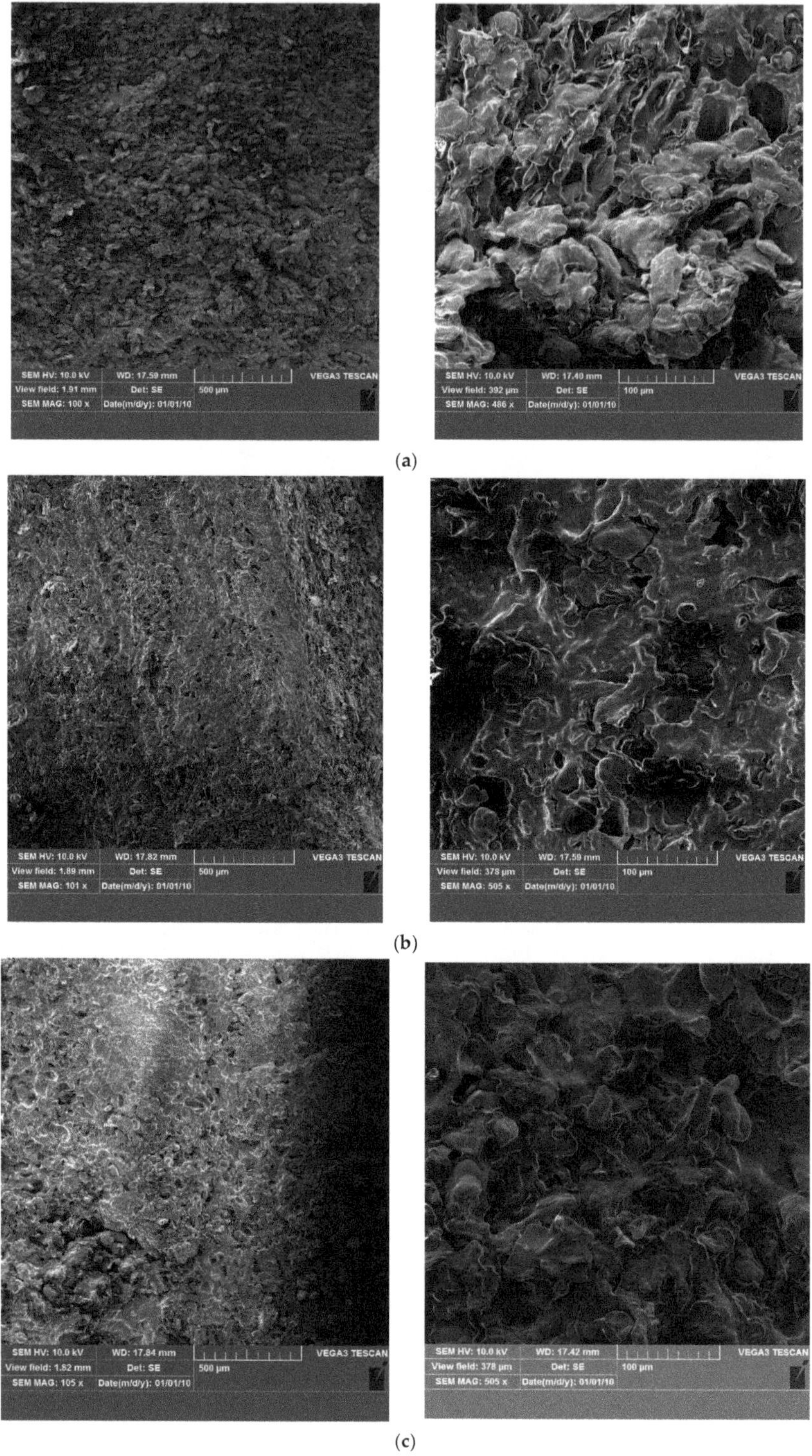

(a)

(b)

(c)

**Figure 9.** *Cont.*

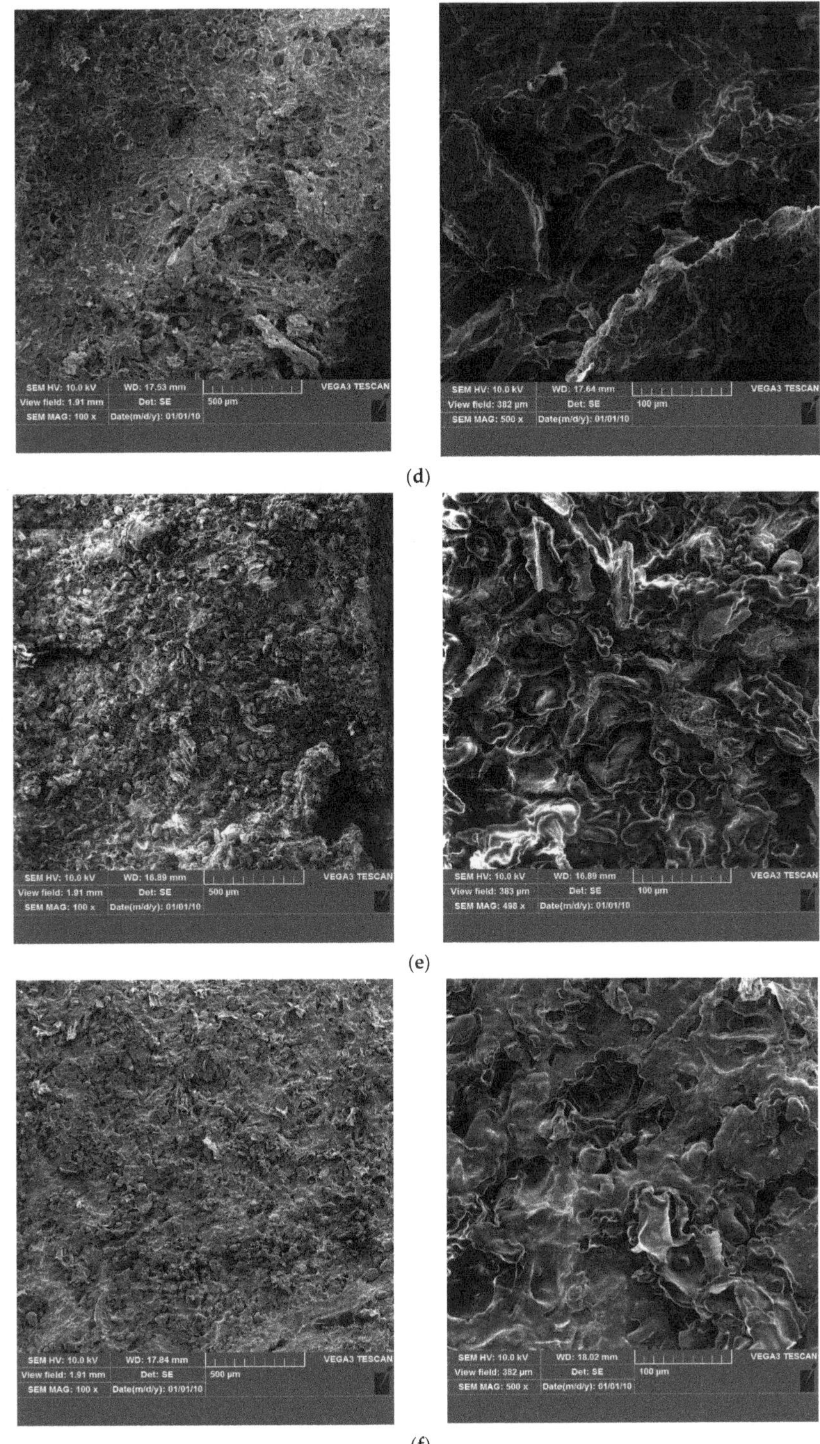

**Figure 9.** *Cont.*

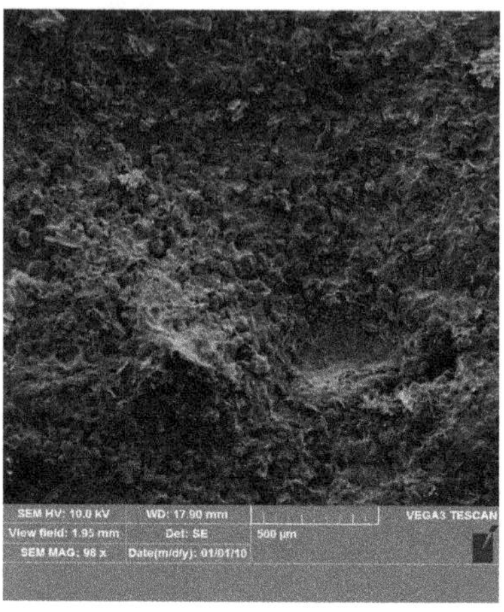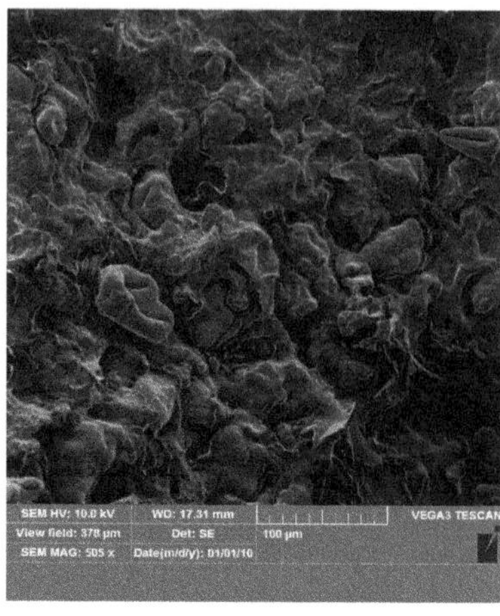

(**g**)

**Figure 9.** (**a**) Sliced cooked ham before storage (sample "0"). (**b**) Sliced pieces of ham stored for 48 h in uncoated PP bags, separated by uncovered PP spacers. (**c**) Sliced pieces of ham stored for 96 h in uncoated PP bags, separated by uncovered PP spacers. (**d**) Sliced pieces of ham, stored for 48 h in PP bags covered with the Ut layer, separated by PP spacers coated with the Ut layer on each side. (**e**) Sliced pieces of ham, stored for 96 h in PP bags covered with the Ut layer, separated by PP spacers coated with the Ut layer on each side. (**f**) Sliced pieces of ham, stored for 48 h in PP bags covered with the Fb layer, separated by PP spacers coated with the Fb layer on each side. (**g**) Sliced pieces of ham, stored for 96 h in PP bags covered with the Fb layer, separated by PP spacers coated with the Fb layer on each side.

## 4. Discussion

The results of the current study demonstrated that the amount of the ZnO nanoparticles should be increased (up to 0.06 g in 100 mL of the coating carrier) to maintain the bacteriolytic effect of the coating against selected Gram-negative and Gram-positive bacterial strains. The results of the previous work [12] demonstrated that a methyl-hydroxy-propyl-cellulose (MHPC) coating containing ZnO nanoparticles (0.082 g in 100 mL of the coating carrier) inhibited the growth of *S. aureus*, *E. coli*, *B. cereus*, and *P. aeruginosa*. It is worth mentioning that accelerated irradiation of this coating did not influence its antibacterial properties due to the shielding properties of ZnO nanoparticles [12–19]. The decreased amount of the nanoparticles in MHPC coating (0.041 g in 100 mL of the coating carrier) led to obtaining the active layer that inhibited the growth of *S. aureus* but only decreased the number of *E. coli* and *P. syringae* cells, confirming that its antibacterial activity was lower [11]. The previous work [11] also indicated that the coating containing a decreased amount of ZnO (0.041 g/100 mL of the coating carrier) and decreased amount of geraniol was more active against *S. aureus*, *E. coli*, and *P. syringae* strains than the coatings containing a decreased amount of geraniol or decreased amount of nano ZnO, confirming the synergistic effect of these active agents. The analysis of the antibacterial properties of *U. tomentosa* and *F. betulina* showed that the activity of these two extracts was very low [9]. A lower than 1-log reduction in the number of *S. aureus* cells was noticed for 5% of extracts of described plants. While analyzing the antimicrobial properties of *F. betulina* extracts on *B. subtilis* cells, it was demonstrated that they demonstrated lower than a 2-log reduction but higher than a 1-log reduction in the number of living cells. *U. tomentosa* extract was found to be even less active. Unfortunately, both extracts were not effective against *E. coli* and *P. syringae* strains. The findings of the current work indicated that the coatings containing 5% of *U. tomentosa* or *F. betulina* extracts with the addition of a decreased amount of the nanoparticles of ZnO (0.03 g/100 mL of the coating carrier) demonstrated high antibacterial activity. The synergistic effect between active agents caused a complete inhibition of the growth of *S. aureus*, *B. atropaheus*, and *E. coli* strains and a reduction in the number of *P. syringae* cells. Additionally, SEM analysis confirmed that the PP film was homogenously and thoroughly covered with the active Ut and Fb layers. However, small pores were observed on the Ut coating, which might have led to a faster release of the active substances from this coating than from the Fb coating or which might have been caused by the release of the active compounds from the Ut layer. Comparing the coatings Ut and Fb, it may be suggested that the Ut active layers were able to display a greater influence on the growth of microorganisms than Fb coating, but for a shorter time.

The results led to the assumption that packaging materials coated with Ut and Fb layers could be used to preserve cooked ham during short-term storage in air conditions.

The described findings demonstrated that coatings containing 5% of the *U. tomentosa* or 5% of the *F. betulina* extracts with the addition of a decreased amount of nano ZnO did not inhibit the growth of mesophilic bacterial strains for a sliced cooked ham that was stored for 96 h, but decreased their number when compared to the ham stored in uncovered bags. Comparing the influence of active layers on the total count after 48 h, it may be mentioned that the most effective packaging material was the bags covered with the Ut coating. Comparing SEM micrographs of the "0" samples and ham slices after storage, it was observed that the most similar microstructures were the surfaces of the ham before storage and the ham slices stored in bags with Ut coating, confirming its effectiveness. The results of this study determined that the number of mesophilic microbes from ready-prepared, sliced cooked ham samples purchased from a local market (sample "0") was almost $4 \log_{10}$ CFU/g, which is seen as an unsuitable number of microorganisms for the sliced ham to be consumed [28]. The expiration date ($\pm 2$ days) of the product could have been one of the reasons. Moreover, the producer suggested ham consumption within 2–3 days after package opening, meaning that a total count should be on the level acceptable for consumption after 48 h of ham storage. Unfortunately, according to Grzybowski et al. [28], ham samples were not acceptable for consumption after 2 days of storage. However, according to another author [29], an acceptable level of the total count of cooked ham products is generally set to $6 \log_{10}$ CFU/g. It means that ham slices stored in uncoated and in active packaging were still acceptable to be consumed after 48 h of storage. While after 96 h of storage, the only suitable samples for consumption were the ham slices stored in bags covered with the Ut coating. It should be added that the ham was purchased in a local market and sliced by the shop assistant after the package opening. The high level of mesophilic bacteria observed for the sample "0" proved that the ham must have been contaminated during slicing. It may be predicted and even suggested that the number of bacteria in sample "0" was too high to be decreased by the active substances which were released from the layers. Considering the presence of pathogenic microorganisms, it should be emphasized that neither the coliform bacteria nor *S. aureus* cells were detected in the cooked ham slices after 48 h and 96 h of storage in PP bags coated with the Ut and Fb layer. From this point of view, these ham slices were still suitable for consumption even after 96 h of storage. In summary, the influence of Ut coating on the quality of sliced, cooked ham was noted, confirming that packaging covered with this active layer can be used for short-term storage. It may be predicted that if ham slices are not contaminated during slicing, the samples will be stored for a longer time than the time suggested by producers. Similar results were obtained in the previous study [2]. The bags and spacers coated with the layer containing 50% of the mixture of *Glycyrrhiza* L. and *Scutelaria baicalensis* extracts were proved to be the best packaging material for ready-prepared chicken sausages which were purchased from a local butcher's shop and which were contaminated during slicing. This active packaging material preserved and maintained the quality of sausage slices after short-term storage (72 h) at 5 °C. Spampinato et al. [4] performed similar experiments. The authors purchased sliced cooked ham with an expiration date of $\pm 3$ days. According to the authors, the producer suggested ham consumption within 1–3 days after the package opening. It has to be underlined that the ham analyzed by the authors was not sliced by the shop assistant, but it was already sliced. The results of the authors' study showed that the total count ranged between 5.8 and 9.0 $\log_{10}$ CFU/g at 0 days. Furthermore, after 4, 8, and 12 days of fridge storage (without active packaging), total aerobic bacteria significantly increased. Additionally, the absence of viable *Salmonella* sp. in the product after 12 days of storage was satisfying for the authors; nevertheless, the presence of other opportunistic pathogens, such as *Klebsiella pneumoniae* and *E. coli*, was noted. The experiments of the authors revealed that, at the opening of packages, sliced ham presented a high microbial load and quite rich microbiota. It can be explained by other authors who suggested [30] that it is possible that post-cooking contamination of cooked ham occurred during the packaging phases of manufacturing, with the bacteria occurring in the environment of the facility, entering into contact and being packaged with virtually sterile sliced cooked ham.

The results of the study led to the assumption that bags and spacers coated with a Ut layer, even if though not recommended for the long-term safe keeping of sliced cooked ham, may be used for short-term storage. Many consumers buy ready-to-eat sliced ham that is sliced by shop assistants in a local market. It has to be underlined that ham is very often contaminated during the slicing process. Therefore, a packaging material, such as PP film, paper, or biopolymer foil coated with a Ut layer, could be used to preserve sliced cooked ham during transport and short-term storage (48–96 h). The findings of this work were similar to the results presented by Shiji et al. [31], who introduced chicken sausages into active packaging for short-term storage. The authors proved that after four days of storage at 4 °C, the active pouches containing silver nanoparticles as active agents were more effective than polyethylene pouches. Similar results were obtained by Shahrampour et al. [32], who confirmed that the use of active packaging for sausages storage at 4 °C led to detect a lower number of mesophilic microorganisms than the total count isolated from the samples which were stored in the uncoated polyethylene packaging. Rüegg et al. [6] prepared polyethylene terephthalate (PET) foil covered with active coatings containing thyme and rosemary essential oils as active agents. The authors used antimicrobial packaging to preserve a sliced cooked chicken breast during short-term storage (6 days). As was shown by the authors, the active layers reduced the number of mesophilic bacteria compared to the uncovered foil, though only slightly. Furthermore, it was confirmed that the MAP packaging system had a greater effect on microorganisms than the antimicrobial coverings, which is why MAP is recommended for long-term storage.

The texture is one of the most important parameters determining the freshness and quality of cooked ham. A product that is too soft or is not cohesive enough can create doubts in the consumers as to its quality and freshness. All the forces holding the food products together are represented by cohesiveness. Additionally, gummy, tough, or stringy food product is not suitable for consumers as it presents excessively strong resistance during mastication [2]. The results of this study determined that the cohesiveness of sliced cooked ham increased for active and uncoated packaging materials (when compared to the sample "0" after 96 h of storage. It is tempting to deduce that the storage of ham slices in uncovered PP bags and in the bags covered with the active layer had a positive effect on the texture of the ham. The contrary findings were noted by Patiño et al. [33], who proved that cohesiveness tended to decrease during storage which may be caused by the interaction of proteins and fats with the rest of the components, affecting the texture of the sausages. Furthermore, Araújo et al. [34] suggested that to increase the parameter values of the sausage samples during storage, an additive such as collagen powder should be added. The results of the current work indicated that the springiness of the ham samples increased for all analyzed packaging materials (when compared to the sample "0") after 48 h storage. Moreover, a decrease in this parameter was observed, except in the slices that were stored for 96 h in the bags coated with the Ut layer, confirming that the active coating containing *U. tomentosa* extract influenced the springiness improvement.

The present work indicated that, unfortunately, chewiness and gumminess increased in the samples stored in uncoated PP bags after 48 h and 96 h of storage, and this is a clear disadvantage. These findings were confirmed by SEM analysis, which showed a more compact microstructure of the cooked ham. Comparing the ham slices, which were stored in active packaging for 48 h, it has to be mentioned that the changes in chewiness values were not observed (sample "0"). In the case of gumminess, the decrease in this parameter was observed, but only for the samples stored in bags with a Ut layer. The bags covered with Fb coating did not influence the gumminess of the ham samples. This was backed up by a microbial purity analysis that showed that the bags coated with the Ut layer was the best packaging for the short-term (48 h) storage of the sliced cooked ham. It is also worth mentioning the disadvantage of the active bags with Ut and Fb layers; it was observed that water loss in sliced ham samples from these bags was higher than from uncoated bags. As was already suggested, the high water loss can release antibacterial substances from the active layers and thus could improve their activity. Analyzing the chewiness and gumminess for the ham slices stored in bags coated with Ut and Fb layers after 96 h of storage, the increase in these parameter values was noted. Simultaneously, the dry mass also increased. As was suggested by Zeraatpisheh et al. [35], when water loss increases, the meat proteins converge because of the formation of new crosslinks; consequently, the chewiness and gumminess of the samples are elevated. Gumminess is associated with food hardness; therefore, its increase is a clear disadvantage. Thus, in cooked ham stored in uncoated bags or in bags covered with a Fb layer with increased gumminess after storage, ham sample hardness may let to difficulties in swallowing [2]. While the ham stored in active packaging coated with Ut layer with decreased (after 48 h) or unchanged (after 96 h) gumminess, ham slices would still be acceptable for consumption.

The findings demonstrated in this work confirmed that the highest $\Delta E_{lab}$ values were noticed for cooked ham portions stored in the uncoated bags and into bags covered with Fb coating for 48 h and 96 h. It was also indicated that the highest $\Delta L$ values were noted for these portions. This indicates that the ham samples taken from this package were the lightest. Furthermore, it should be mentioned that these parameter values increased for samples stored in all analyzed packaging materials. The results were confirmed by Azlin-Hasim et al. [36], who indicated that L* values increased in meat samples stored for 6 and 12 days in packaging containing Ag nanoparticles as antimicrobial substances.

## 5. Conclusions

Microbial contaminants which may be present in cooked ham originates from incorrect cooking of meat product, inappropriate sanitization practices, and/or recontamination during packaging, following the increase in microorganisms during the primary shelf-life. It is worth mentioning that the package opening may be the cause of a sudden variation of the headspace gas composition and a loss of protection. Furthermore, if the ham is sliced by the shop assistant, it may be contaminated during the process. All the facts make that the secondary shelf-life is remarkably shorter than the primary shelf-life. The only possibility to maintain the quality of cooked ham is to decrease the microbial load on the surface of the slices using active packaging during transportation and secondary storage.

The total count of the ham samples before storage was high, proving that the cooked ham was contaminated. The spacers and bags coated with the antimicrobial layer containing *U. tomentosa* extract and the ZnO nanoparticles were observed to be the best bags for ready-prepared, sliced cooked ham that was purchased from a local market. It was confirmed by microbial purity, textural, and SEM analysis. These antimicrobial bags may be used to maintain the freshness and quality of ham sliced by the shop assistant during the secondary shelf-life (48 h) at 5 °C.

**Author Contributions:** M.M., M.O. (Magdalena Ordon), M.O. (Marcin Okręglicki) and J.S. conceived and designed the experiments; M.O. (Magdalena Ordon) wrote the paper; M.O. (Magdalena Ordon) and W.B. performed the microbiological tests; M.M. and M.O. (Magdalena Ordon) analyzed the data; M.O. (Magdalena Ordon) and W.B. performed mechanical tests; M.O. (Magdalena Ordon) and M.M. analyzed the data; M.O. (Magdalena Ordon) and W.B. performed Lab tests; M.M. performed SEM analysis; M.O. (Magdalena Ordon) analyzed the data; M.O. (Magdalena Ordon) performed dry mass tests and analyzed the data; M.O. (Magdalena Ordon), W.B., M.O. (Marcin Okręglicki) and J.S. contributed and prepared reagents/materials; M.O. (Marcin Okręglicki) and J.S. contributed analysis tools; M.O. (Magdalena Ordon) performed statistical analysis and analyzed the data. All authors have read and agreed to the published version of the manuscript.

**Funding:** This research work was funded by the project POIR.01.01.01-00-0740/21: "Innovative, ecological, active packaging made of plastics, biodegradable and compostable films and paper intended for packaging food products with special, protective and functional properties"; co-financed by the European Union and the National Center for Research and Development.

**Institutional Review Board Statement:** Not applicable.

**Informed Consent Statement:** Not applicable.

**Data Availability Statement:** Not applicable.

**Acknowledgments:** The authors acknowledge Alicja Tarnowiecka-Kuca for a support given during mechanical tests.

**Conflicts of Interest:** The authors declare no conflict of interest.

## References


1. Nikmaram, N.; Budaraju, S.; Barba, F.J.; Lorenzo, J.M.; Cox, R.B.; Mallikarjunan, K.; Roohinejad, S. Application of plant extracts to improve the shelf-life, nutritional and health-related properties of ready-to-eat meat products. *Meat Sci.* **2018**, *145*, 245–255. [CrossRef]
2. Ordon, M.; Burdajewicz, W.; Pitucha, J.; Tarnowiecka-Kuca, A.; Mizielińska, M. Influence of Active Packaging Covered with Coatings Containing *Mixtures of Glycyrrhiza L. and Scutellaria baicalensis* Extracts on the Microbial Purity and Texture of Sliced Chicken Sausages. *Coatings* **2023**, *13*, 795. [CrossRef]
3. Duthoo, E.; Rasschaert, G.; Leroy, F.; Weckx, S.; Heyndrickx, M.; De Reu, K. The Microbiota of Modified-Atmosphere-Packaged Cooked Charcuterie Products throughout Their Shelf-Life Period, as Revealed by a Complementary Combination of Culture-Dependent and Culture-Independent Analysis. *Microorganisms* **2021**, *9*, 1223. [CrossRef]
4. Spampinato, G.; Candeliere, F.; Amaretti, A.; Fabio, L.; Rossi, M.; Raimondi, S. Microbiota Survey of Sliced Cooked Ham During the Secondary Shelf Life. *Front. Microbiol.* **2022**, *13*, 842390. [CrossRef]
5. de Niederhäusern, S.; Bondi, M.; Camellini, S.; Sabia, C.; Messi, P.; Iseppi, R. Plant Extracts for the Control of Listeria monocytogenes in Meat Products. *Appl. Sci.* **2021**, *11*, 10820. [CrossRef]
6. Rüegg, N.; Teixeira, S.R.; Beck, B.M.; Monnard, F.W.; Menard, R.; Yildirim, S. Application of antimicrobial packaging based on modified calcium carbonate and EOs for RTE meat products. *Food Packag. Shelf Life* **2022**, *34*, 100982. [CrossRef]
7. Mizielińska, M.; Bartkowiak, A. Innovations in food packaging materials. In *Food Preservation and Packaging-Recent Process and Technological Advancements*; Tumuluru, Y.S., Ed.; IntechOpen: London, UK, 2022. [CrossRef]
8. Herrera, D.R.; Durand-Ramirez, J.E.; Falcão, A.; Nogueira da Silva, E.J.L.; dos Santos, E.B.; Figueiredo de Almeida Gomes, B.P. Antimicrobial activity and substantivity of *Uncaria tomentosa* in infected root, canal dentin. *Braz. Oral. Res.* **2016**, *30*, 61–66. [CrossRef]
9. Ordon, M.; Nawrotek, P.; Stachurska, X.; Schmidt, A.; Mizielińska, M. Mixtures of *Scutellaria baicalensis* and *Glycyrrhiza L.* Extracts as Antibacterial and Antiviral Agents in Active Coatings. *Coatings* **2021**, *11*, 1438. [CrossRef]
10. Sofrenić, I.; Anđelković, B.; Todorović, N.; Stanojković, T.; Vujisić, L.; Novaković, M.; Milosavljević, S.; Tešević, V. Cytotoxic triterpenoids and triterpene sugar esters from the medicinal mushroom *Fomitopsis betulina*. *Phytochemistry* **2021**, *181*, 112580. [CrossRef] [PubMed]
11. Mizielińska, M.; Nawrotek, P.; Stachurska, X.; Ordon, M.; Bartkowiak, A. Packaging Covered with Antiviral and Antibacterial Coatings Based on ZnO Nanoparticles Supplemented with Geraniol and Carvacrol. *Int. J. Mol. Sci.* **2021**, *22*, 1717. [CrossRef] [PubMed]
12. Mizielińska, M.; Łopusiewicz, Ł.; Mężyńska, M.; Bartkowiak, A. The Influence of Accelerated UV-A and Q-SUN Irradiation on the Antimicrobial Properties of Coatings Containing ZnO Nanoparticles. *Molecules* **2017**, *22*, 1556. [CrossRef] [PubMed]
13. Eskandari, F.; Mofidi, H.; Asheghi, B.; Mohammadi, F.; Gholami, A. Bringing resistance modulation to methicillin-resistant *Staphylococcus aureus* (MRSA) and vancomycin-resistant enterococci (VRE) strains using a quaternary ammonium compound coupled with zinc oxide nanoparticles. *World J. Microbiol. Biotechnol.* **2023**, *39*, 193. [CrossRef] [PubMed]
14. Karimi, M.; Sadeghi, E.; Khosravi Bigdeli, S.; Zahedifar, M. Optical properties, singlet oxygen, and free radical production ability with different UV irradiations and antimicrobial inhibitors against various bacterial species of ZnO: Eu nanoparticles. *Radiat. Phys. Chem.* **2023**, *212*, 111132. [CrossRef]
15. Wu, H.; Meng, Y.; Yu, M.; Yang, H. Modulating the antibacterial activity of ZnO/talc by balancing the monodispersity of ZnO nanoparticles. *Appl. Clay Sci.* **2023**, *242*, 107024. [CrossRef]
16. Sougandhi, P.R.; Jena, S.K.; Brahmanand, P.S.; Tej, M.B.; Kadiyala, N.K.; Rani, T.S.; Rao, M.V.B. Multifunctional Bio-Nanocomposite Films of Carboxymethyl Cellulose and Pectin with Incorporated Zinc Oxide Nanoparticles. *Lett. Appl. NanoBioSci.* **2023**, *12*, 85.
17. Nguyena, T.V.; Dao, P.H.; Khanh Linh Duong, K.L.; Duong, Q.H.; Vu, Q.T.; Nguyena, A.H.; Mac, V.P.; Lea, T.L. Effect of R-TiO₂ and ZnO nanoparticles on the UV-shielding efficiency of water-borne acrylic coating. *Prog. Org. Coat.* **2017**, *110*, 114–121. [CrossRef]

18. El-Feky, O.M.; Hassan, E.A.; Fadel, S.M.; Hassan, M.L. Use of ZnO nanoparticles for protecting oil paintings on paper support against dirt, fungal attack, and UV aging. *J. Cult. Herit.* **2014**, *15*, 165–172. [CrossRef]
19. Kairyte, K.; Kadys, A.; Luksiene, Z. Antibacterial and antifungal activity of photoactivated ZnO nanoparticles in suspension. *J. Photochem. Photobiol. Biol.* **2013**, *128*, 78–84. [CrossRef]
20. Kubra, I.R.; Kumar, D.; Rao, L.J.M. Effect of microwave-assisted extraction on the release of polyphenols from ginger (*Zingiber officinale*). *Int. J. Food Sci. Tech.* **2013**, *48*, 1828–1833. [CrossRef]
21. Mandal, V.; Mohan, Y.; Hemalatha, S. Microwave Assisted Extraction—An Innovative and Promising Extraction Tool for Medicinal Plant Research. *Pharmacogn. Rev.* **2007**, *1*, 7–18.
22. *ASTM E 2180-07*; Standard Test Method for Determining the Activity of Incorporated Antimicrobial Agent(s) in Polymeric or Hydrophobic Materials. ASTM: West Conshohocken, PA, USA, 2012.
23. PN-EN ISO 4833-2:2013-12. Available online: https://sklep.pkn.pl/pn-en-iso-4833-2-2013-12p.html (accessed on 1 September 2022).
24. PN-EN ISO 6888-1. Available online: https://sklep.pkn.pl/pn-en-iso-6888-1-2022-03e.html (accessed on 1 September 2022).
25. PN-ISO 4832:2007. Available online: https://sklep.pkn.pl/pn-iso-4832-2007p.html (accessed on 1 September 2022).
26. PN-ISO 11036:1999 Standard. Available online: http://sklep.pkn.pl/pn-iso-11036-1999p.html (accessed on 1 September 2022).
27. EC. Commission Regulation No 1441/2007 of 5 December 2007 Amending Regulation (EC) No 2073/2005 on Microbiological Criteria for Foodstuffs. 2007. Available online: https://www.google.com/url?sa=i&rct=j&q=&esrc=s&source=web&cd=&ved=0CAIQw7AJahcKEwiA27a4kLaAAxUAAAAAHQAAAAAQAg&url=https%3A%2F%2Feur-lex.europa.eu%2FLexUriServ%2FLexUriServ.do%3Furi%3DOJ%3AL%3A2007%3A322%3A0012%3A0029%3AEN%3APDF&psig=AOvVaw1DPdQ5p_OWuHv8R7u9rKch&ust=1690795927928650&opi=89978449 (accessed on 1 September 2022).
28. Grzybowski, J.; Reiss, J. *Praktyczna Bakteriologia Lekarska i Sanitarna*; Dom Wydawniczy Bellona: Warszawa, Polska, 2001; 346p.
29. Vasilopoulos, C.; De Vuyst, L.; Leroy, F. Shelf-life Reduction as an Emerging Problem in Cooked Hams Underlines the Need for Improved Preservation Strategies. *Crit. Rev. Food Sci. Nutr.* **2013**, *55*, 1425–1443. [CrossRef] [PubMed]
30. Raimondi, S.; Luciani, R.; Sirangelo, T.M.; Amaretti, A.; Leonardi, A.; Ulrici, A.; Foca, G.; D'Auria, G.; Moya, A.; Zuliani, V.; et al. Microbiota of sliced cooked ham packaged in modified atmosphere throughout the shelf life: Microbiota of sliced cooked ham in MAP. *Int. J. Food Microbiol.* **2019**, *289*, 200–208. [CrossRef] [PubMed]
31. Shiji, M.; Snigdha, S.; Jyothis, M.; Radhakrishnan, E.K. Biodegradable and active nanocomposite pouches reinforced with silver nanoparticles for improved packaging of chicken sausages. *Food Packag. Shelf Life* **2019**, *19*, 155–166. [CrossRef]
32. Shahrampour, D.; Razavi, S.; Sadeghi, A. Evaluation of green tea extract incorporated antimicrobial/antioxidant/biodegradable films based on polycaprolactone/polylactic acid and its application in cocktail sausage preservation. *Food Meas.* **2023**, *17*, 1058–1067. [CrossRef]
33. Patiño, J.H.; Henríquez, L.E.; Restrepo, D.A.; Lantero, M.I.; García, M.A. Influence of polyamide composite casings with silver–zinc crystals on the quality of beef and chicken sausages during their storage. *J. Food Sci. Technol.* **2022**, *59*, 75–85. [CrossRef]
34. Araújo, I.B.S.; Lima, D.A.S.; Pereira, S.F.; Paseto, R.P.; Madruga, M.S. Effect of storage time on the quality of chicken sausages produced with fat replacement by collagen gel extracted from chicken feet. *Poultry Sci.* **2021**, *100*, 1262–1272. [CrossRef]
35. Zeraatpisheh, F.; Tabatabaei, Y.F.; Shahidi, F. Investigation of effect of cold plasma on microbial load and physicochemical properties of ready-to-eat sliced chicken sausage. *J. Food Sci. Technol.* **2022**, *59*, 3928–3937. [CrossRef]
36. Azlin-Hasim, S.; Cruz-Romero, M.C.; Morris, M.A.; Cummins, E.; Kerry, J.P. Effects of a combination of antimicrobial silver low density polyethylene nanocomposite films and modified atmosphere packaging on the shelf life of chicken breast fillets. *Food Packag. Shelf Life* **2015**, *4*, 26–35. [CrossRef]

*Article*

# Effect of Lactic Acid Bacteria on Nutritional and Sensory Quality of Goat Organic Acid-Rennet Cheeses

Katarzyna Kajak-Siemaszko [1], Dorota Zielińska [1,*], Anna Łepecka [2], Danuta Jaworska [1], Anna Okoń [2], Katarzyna Neffe-Skocińska [1], Monika Trząskowska [1], Barbara Sionek [1], Piotr Szymański [2], Zbigniew J. Dolatowski [2] and Danuta Kołożyn-Krajewska [1]

[1] Department of Food Gastronomy and Food Hygiene, Institute of Human Nutrition Sciences, Warsaw University of Life Sciences (WULS–SGGW), Nowoursynowska 159c, 02-776 Warsaw, Poland
[2] Department of Meat and Fat Technology, Prof. Waclaw Dabrowski Institute of Agriculture and Food Biotechnology—State Research Institute, Rakowiecka 36 St., 02-532 Warsaw, Poland
* Correspondence: dorota_zielinska@sggw.edu.pl; Tel.: +48-22-59-37065

**Abstract:** The aim of this study was to evaluate the applicability of selected *Lactobacillus* strains, previously isolated from spontaneously fermented foods, as starter cultures in the production of organic dairy products—acid-rennet goat's cheeses under industrial conditions. The basic composition and the effect of starter cultures on the physicochemical, microbiological, sensory as well textural properties during the production and storage of goat's cheese were evaluated. Lactic acid bacteria count in cheese samples was at a high level of about 8 log CFU/g. The cheeses made with *Levilactobacillus brevis* B1 and *Lactiplantibacillus plantarum* Os2 bacterial cultures additions have showed more favorable Lipid Quality Indices than for the control one with the addition of acid whey. The time of ripening of the cheeses significantly ($p < 0.005$) changed their consistency—they became softer and more elastic and less moist. It is possible that the selected cultures of *L. brevis* B1 and *L. plantarum* Os2 isolated from traditional cheeses can be successfully applied to goat's milk cheese production. The strain *L. brevis* B1 is highly recommended as a starter culture for goat's milk cheese production, taking into account the good microbiological and sensory quality as well as the chemical composition.

**Keywords:** goat's cheese; *Lactobacillus* strains; sensory quality; fatty acid; lipolysis

Citation: Kajak-Siemaszko, K.; Zielińska, D.; Łepecka, A.; Jaworska, D.; Okoń, A.; Neffe-Skocińska, K.; Trząskowska, M.; Sionek, B.; Szymański, P.; Dolatowski, Z.J.; et al. Effect of Lactic Acid Bacteria on Nutritional and Sensory Quality of Goat Organic Acid-Rennet Cheeses. *Appl. Sci.* **2022**, *12*, 8855. https://doi.org/10.3390/app12178855

Academic Editors: Joanna Trafialek and Mariusz Szymczak

Received: 28 July 2022
Accepted: 31 August 2022
Published: 2 September 2022

**Publisher's Note:** MDPI stays neutral with regard to jurisdictional claims in published maps and institutional affiliations.

## 1. Introduction

In the last 60–80 years, agriculture has moved from traditional, natural plant cultivation and animal husbandry, used for thousands of years, to very intensive, often non-organic production [1–4]. Taking into account the latest findings on the effect of microbiota abundance and biodiversity on human health, it is claimed that the solution that may help avoid the plague of civilization diseases, which can be now observed, would be a harmonized diet based on organically produced, minimally processed, not sterilized food, coming from the areas where our ancestors lived for generations. Scientific research confirms that the microbiome is flexible enough to change as the environment changes. This phenomenon contributes to the ability of organisms to respond and adapt to environmental changes that can occur quicker than traditional body adaptation processes [5].

Developing food products with selected starter cultures is one way to improve the health status of consumers, as well as the texture and taste of the food, thereby improving its properties and reducing production costs. Moreover, the use of an appropriate starter culture can influence the nutritional composition, safety and overall quality of food. Some authors emphasize the health-promoting properties of specific environmental lactic acid bacteria and their effect on human health [6].

One of the promising raw materials of growing consumer interest, providing nutritional benefits, is goat's milk and its dairy products. Goat's milk, compared to cow's milk, contains higher amounts of calcium, magnesium and phosphorus [7,8] and has fat globules of a smaller size which facilitate digestion [9]. In addition, the oligosaccharides present in goat's milk are not only more concentrated, but also have a profile similar to human milk [10,11]. Goat's milk is often recommended for the prevention of cardiovascular diseases and allergies and is also used to stimulate the immune system. Goat products have a low allergenic potential, therefore, they are a good alternative for consumers who are sensitive or allergic to certain proteins present in cow's milk [12]. Goat's milk can be regarded as a natural functional food. The regular consumption of goat's milk and its products should be encouraged [9,13].

Goat's milk cheese is a highly valuable product, being a good source of protein, macronutrients, organic acids, vitamins and many other components of our diet. According to Santurino et al. [14], the consumption of goat's cheese naturally enriched with n-3 PUFA and CLA could play a potential role as a high nutritional health-enhancing food. It is also a potential source of bioactive peptides and a rich set of microorganisms, especially lactic acid bacteria, which have a positive effect on the microbiome and human health [15].

The role of starter cultures in the cheese production process is fairly well understood. The addition of bacteria as a starter plays a crucial role in cheese production technology. It aims to direct the fermentation processes during the processing of the cheese mass and inhibits the development of undesirable microbiota. During the ripening of the cheese mass, due to numerous factors, complicated biochemical processes take place, such as protein proteolysis, fat lipolysis, lactose glycolysis and the transformation of citrate. Protein hydrolysis is caused by the presence of enzymes (proteases) from the milk itself, coagulating preparations and bacteria. As a result of the processes, as the cheese matures, smaller protein fractions, such as peptides and free amino acids are formed. Fat lipolysis plays a role in developing the taste and aroma of the end product. During maturation, lipolytic enzymes hydrolyze the fat into short-chain and volatile fatty acids. The type of starting microbiota and the maturation time determine the level of lipid changes. As a result of lactic acid bacteria fermentation, the lactose contained in the milk is converted into lactic acid. The accumulation of lactic acid leads to a decreased pH value, protects against the development of undesirable microorganisms and helps with syneresis. The transformation of lactose and citrate contributes to the formation of compounds such as diacetyl, acetoin, ethanol and acetates [16–18].

Taking into account the above findings, the aim of this study was to evaluate the applicability of selected *Lactobacillus* strains, previously isolated from spontaneously fermented foods, as starter cultures in the production of organic dairy products—acid-rennet goat's cheeses—under industrial conditions and to examine the microbiological, nutritional and sensory quality of the obtained cheeses during two-month maturation storage. The idea arose from an emerging need for the restoration of the population's intestinal microbiota. Using the unique, selected environmental strains of lactic acid bacteria with potential probiotic properties for the production of organic goat's cheese can enrich the microbiota of consumers with valuable strains of environmental bacteria in the form of high-nutritional-value food products.

## 2. Materials and Methods

### 2.1. Bacteria Strains and Starter Cultures Preparation

Bacterial strains have been obtained from the internal collection of microorganisms of the Chair of Food Hygiene and Quality Management SGGW-WULS in Warsaw: *Levilactobacillus brevis* B1—isolated from "bundz"—and *Lactiplantibacillus plantarum* Os2—isolated from "oscypek", traditional cheeses made in Tatra mountain. The previous studies demonstrated their good technological as well as potentially probiotic properties [19–21]. These bacterial strains were selected at the preliminary stage of research, according to enzymatic activity (lipases, proteases and peptidases), the ability to ferment lactose into lactic acid, as well as sensory properties of fermented goat's milk.

The starter cultures were prepared as follows: bacteria were grown on de Man, Rogosa and Sharpe (MRS, LabM, Bury, UK) broth at 37 °C for 24 h to a concentration of

$10^9$ CFU/mL, then centrifuged ($6000\times g$, 10 min) and washed in sterile PBS (phosphate-buffered saline $1\times$, pH 7.4) in triplicate. Next, bacterial cells were resuspended in 1 L of goat's milk and again incubated at 37 °C for 24 h, obtaining an inoculum concentration of $10^8$ CFU/mL.

### 2.2. Cheese Samples

The research material consisted of acid-rennet cheeses made from goat's milk (Carpathian breed). The acid-rennet cheeses were produced in a small family's organic farm from the Podkarpackie Province in Poland. As a control sample, acid-rennet goat's cheese with the addition of acid whey (symbol AW) was used. The acid whey was obtained from organic cottage cheese production in the same organic farm and was used as a "natural" starter of bacteria cultures. The research samples were goat's acid-rennet cheeses with the addition of *L. brevis* B1 strain (symbol B1) and *L. plantarum* Os2 (symbol Os2).

To raw, unpasteurized goat's milk, acid whey (8% $w/v$) or starter culture of B1 or Os2 strains (2% $w/v$) were added. Next, the goat's milk was left at room temperature, left for a few hours, then heated in pots to 40 °C and microbial rennet was added. After 20–40 min, a clot (called a clag) formed on the surface, which was stirred with a ferula or by hand. After about 5 min, a small amount of boiling water was poured onto the surface, causing the curd to settle. In the meantime, appropriate forms were prepared, made of stainless steel and lined with cheese cloths. The appropriate amount of the curd was placed on the scarves, ironed and then the lids were put on and secured. In addition, the entire closed mold was poured over with boiling water in order to "close" the cheese. Pressing the cheese took about 4–6 h and then the cheese was brined in a saturated salt solution for 6–8 h. Two series of goat's cheese were made under industrial conditions.

The cheese samples (approx. 1 kg) were vacuum packed and stored at chilled temperature (4 °C) for 2 months of maturation. Once a month, the samples were taken and chemical, microbiological, textural and sensory analyses were made to determine the quality changes during storage.

### 2.3. Microbiological Analysis

The spread plate technique was used to determine the count of microorganisms. To determine the count of microorganisms of lactic acid bacteria (LAB) the following microbiological media and methodology tests were used: MRS agar (LabM, Heywood, UK) in combination with incubation under anaerobic conditions (AnaeroGen System, Oxoid, Hampshire, UK) was applied in accordance with the ISO15214:2002 [22], to determine the number of rods from the Enterobacteriaceae family (ENT), MacConkey agar No. 3 (LabM, Heywood, UK) in accordance with the ISO 21528-2 [23] and to determine the number of yeasts and molds (YM), YGC agar (Sabouraud Dextrose with chloramphenicol lab Agar, Biomaxima, Lublin, Poland) in accordance with the ISO 21527-1 [24] and ISO 21527-2 [25]. The number of microorganisms was expressed as the logarithm of the colony forming units per gram (log CFU/g).

The presence of selected pathogenic bacteria was determined using the enrichment culture method with the media indicated in the standards—XLD agar (Xylose Lysine Deoxycholate Agar, LabM, UK) and RAPID' *Salmonella* agar (Bio-Rad, Hecules, CA, USA)—to determine the presence of *Salmonella*, in accordance with the ISO 6579-1 [26], and ALOA agar (Listeria according to Ottaviani and Agosti Agar, Bio-Rad, Hecules, CA, USA) and PALCAM agar (LabM, Heywood, UK) to determine the presence of *Listeria monocytogenes*, in accordance with the ISO 11290-2 [27].

### 2.4. Chemical Composition

The selected parameters of water, protein, fat, lactose, NaCl and phosphorus content of the goat's cheeses were determined after production. The tests were performed in 4 repetitions.

- The water content [%] was determined by the weight method according to PN-ISO 1442:2000 [28]. The measured samples were dried at 103 °C for 30 min.
- The protein content was done using the Kjeldahl method according to PN-A-04018:1975/ Az3:2002 [29]. The method consists of determining the total nitrogen content, subsequently using the conversion factor of nitrogen content into protein content (for meat = 6.25).
- The content of free fat was made by the Weibull–Berntrop gravimetric method according to PN ISO 8262-3:2011 [30].
- Composition of fatty acids was determined by the gas chromatography (GC) method with flame ionization detection (HP 6890 II) according to PN-EN ISO 5508:1996 [31]. A BPX 70 high-polar capillary column (60 m $\times$ 0.25 mm, 25 μm) was used for ester separation. Analysis conditions: column temperature programmable in the range of 140–210 °C, injector temperature: 210 °C, detector temperature: 250 °C, carrier gas: helium. The tests were performed in 4 repetitions immediately after the production process, after 1 and 2 months of storage. The values are given in %.
- The Lipid Quality Indices were examined after the production process, after 1 and 2 months of storage. The indices were calculated using the following formulas [32–36]: The Index of Atherogenicity

$$AI = (C12{:}0 + (4 \times C14{:}0) + C16{:}0)/(\Sigma\ n\text{-}3\ PUFA + \Sigma\ n\text{-}6\ PUFA + \Sigma\ MUFA), \quad (1)$$

Index of Thrombogenicity

$$TI = (C14{:}0 + C16{:}0 + C18{:}0)/[(0.5 \times C18{:}1) + (0.5 \times other\ MUFA) + (0.5 \times \Sigma\ n-6\ PUFA) + (3 \times \Sigma\ n-3\ PUFA) + \Sigma\ n-3\ PUFA/\Sigma\ n-6\ PUFA)], \quad (2)$$

Hypocholesterolemic Fatty Acids

$$DFA = UFA + C18{:}0, \quad (3)$$

Hypercholesterolemic Fatty Acids

$$OFA = C12{:}0 + C14{:}0 + C16{:}0, \quad (4)$$

The ratio of hypocholesterolemic and hypercholesterolemic fatty acids

$$H/H = (C18{:}1\ n\text{-}9 + C18{:}2\ n\text{-}6 + C18{:}3\ n\text{-}3)/(C12{:}0 + C14{:}0 + C16{:}0), \quad (5)$$

Health-promoting index

$$HPI = \Sigma\ UFA/[C12{:}0 + (4 \times C14{:}0) + C16{:}0], \quad (6)$$

- Cholesterol content was made by the extraction of the lipid fraction from the cheese sample, esterification of fatty acids and derivatization of cholesterol in the presence of an internal standard. The sample was analyzed by GC method with flame ionization detection. The tests were performed in 4 repetitions immediately after the production process, after 1 and 2 months of storage. The cholesterol value is expressed in mg/100 g of product.
- Lactose content was performed by liquid chromatography (LC) using an RID refractive index detector. Lactose was separated and compared to the external standard method. The lactose value was expressed in mg/100 g.
- The phosphorus content was determined as total phosphorus content [%], expressed as $P_2O_5$, included the mineralization of the sample, precipitation of phosphorus in the form of choline phosphoromolybdate and weight determination of total phosphorus, according to PN-A-82060:1999 [37].

- The chloride content was determined using the potentiometric method according to PN-ISO 1841-2:2002 [38].

### 2.4.1. Water Activity

The water activity (aw) of cheese samples was determined by AQUALAB Pawkit Water Activity Meter (METER Group, Inc., Washington, DC, USA) according to the ISO 18787:2017 [39]. The samples were placed in special, closed vials, immediately after opening the package of cheese samples, and maintained at temperature of 25 °C for 1 h. The result is given as the value of the water activity with the standard deviation of three replications.

### 2.4.2. Determination of the Oxidation-Reduction Potential (ORP)

A 30 g sample of the previously ground product was transferred to a porcelain mortar. The cheese was ground thoroughly with a pestle, with addition of several small portions of distilled water (30 cm$^3$) at temperature of 40 °C in until a homogeneous emulsion was obtained. The resulting emulsion was adjusted to a temperature of 20 °C and ORP value was measured with a SevenCompactTM S220 (Mettler Toledo, Columbus, OH, USA) with an InLab Redox Pro electrode and expressed in mV. The tests were performed in 4 repetitions immediately after the production process, after 1 and 2 months of storage.

### 2.4.3. Acidity-pH Value

The measurement was made with the SevenCompactTM S220 device (Mettler Toledo, OH, USA) with an InLab electrode. The tests were performed in 4 repetitions immediately after the production process, after 1 and 2 months of storage. The results were presented with 0.01 accuracy with standard deviations.

### 2.4.4. Titratable Acidity

A 5 g sample of cheese of the previously ground product was transferred to a porcelain mortar. The cheese was thoroughly ground with a pestle, gradually adding 50 cm$^3$ of distilled water at temperature of 40 °C in small portions until obtaining a homogeneous emulsion. After that, 2 cm$^3$ of 2% alcoholic phenolphthalein solution was added and titrated with 0.25 M NaOH until a slightly pink color persisted for 30 s. The tests were performed in 4 repetitions immediately after the production process, after 1 and 2 months of storage. The values are given in °SH.

### 2.5. Sensory Evaluation

The sensory quality of the acid-rennet goat's cheese was determined using the scaling method [40]. The linear scale (100 mm) was converted to numerical values (0–10 c. u.). A set of cheese samples were presented to the assessors. Before proper evaluation, the distinguishing attributes were chosen, and in preliminary session their understanding was established and defined. The following set of attributes were established for the evaluation: 5 attributes of odor (milk's fermentation, goat's milk, fatty, sharp, mature cheese), 1 attribute of color (color intensity) and 3 texture attributes (softness, moisture, elasticity). Due to the severity of SARS-CoV-2 disease, it was necessary to eliminate the assessment of flavor attributes. It was assumed that flavor sensations in food are highly influenced by the aroma and texture compounds. It is well known from literature that the volatile compounds and their composition that occur in food products can affect the specific aroma of foods and the flavor of the resulting products [41]. The anchor marks of the tested attributes ranged from none (the left side) to very high intensity (the right side of the scale).

The cheese samples were portioned (average piece size of about 15 g) and put into disposable containers manufactured from food-safe plastic. The containers were then closed and individually coded with 3-digit codes and given in random order. The samples were assessed after production, after 1 month and after 2 months of storage.

The trained panel consisted of 8 members (7 women and 1 man, age 30–58), who were extensively and formally tested before being selected, according to the ISO standard

(ISO 8586:2012) [42]. The assessors had 4 to 18 years of practical experience with sensory procedures and sensory evaluation of different food products with different methods. The panelists' ability to differentiate product samples was verified by various concentrations of volatile and non-volatile stimuli.

The evaluators had the conditions for full concentration provided. The ambient temperature was 22–23 °C. The assessment and condition mode were established in accordance with Meilgaard et al. [43].

### 2.6. Instrumental Determination of Color

To determine the color, the CR-300 Chroma-Meter colorimeter (Konica Minolta, Tokyo, Japan) was used. The determinations were made under diffused illumination at 0° and a diaphragm diameter of 8 mm. For the determination, cut slices of cheese with a thickness of 20 mm were used. The apparatus was calibrated before use and calibration of the spectrophotometer was performed on a white standard. The color was measured in the CIE L*a*b* system. The tests were performed in 20 repetitions immediately after the production process, after 1 and 2 months of storage.

### 2.7. Instrumental Texture Evaluation

Texture profile analysis (TPA) was accomplished with CT3 Texture Analyzer (Brookfield Ametek, Middleborough, MA, USA) instruments. The speed of compression of cheese samples were 0.50 mm/s and the compression was made twice up to 50% of their original height with a cylindrical head with a diameter of 38.1 mm and a height of 20 mm with a 10,000 g maximum pressing force. The tested cheese had the shape of a cube with a side of 20 mm. The following textural parameters were determined: hardness 1 (force necessary to attain a given deformation; N), hardness 2 (force necessary to attain a given deformation; N), adhesiveness (work necessary to overcome the attractive forces between the surface of the food and the surface of the other materials with which the food comes into contact; mJ), cohesiveness (extent to which a material can be deformed before it ruptures; adimensional), springiness (rate at which a deformed material goes back to its undeformed condition after the deforming force is removed: mm), gumminess (energy required to disintegrate a semi-solid food to a state ready for swallowing: a product of a low degree of hardness and a high degree of cohesiveness; N) and chewiness (energy required to masticate a solid food to a state ready for swallowing: a product of hardness, cohesiveness and springiness; mJ) [44]. The measurements were carried out at room temperature, 20 min after the samples were taken out of the cooling chamber. The test was carried out in 6 replications.

### 2.8. Mathematical and Statistical Analysis of the Results

Two independent experimental series (batches) of the product were manufactured under industrial conditions. The measurements were repeated several times for each batch of the product. A one-way analysis of variance (ANOVA) at $p < 0.05$ was used, first to test the treatments effect (AW, B1, Os2) and then between the storage time (0, 1, 2 months). Means and standard deviations were calculated. The Tukey's test was conducted for a mean comparison ($p < 0.05$). The normality of distribution of all analyzed traits was checked by Shapiro–Wilk test. The Statistica 13.1 program (StatSoft Inc., Tulsa, OK, USA) was used to perform the calculations.

## 3. Results

### 3.1. Microbiological Quality

The microbiological quality of the cheese samples is presented in Table 1.

The lactic acid bacteria (LAB) count was at a high level of about 8 log CFU/g and was similar in all samples, however, the LAB count changed only in the B1 sample during storage ($p < 0.05$). The count of *Enterobacteriaceae* bacteria was in the range of 5–6 log CFU/g immediately after production and decreased during storage ($p < 0.05$) to approximately 4–5 log CFU/g in all samples. The initial yeast and mold contamination of the goat's cheeses was low, approx. 3 log CFU/g, and increased after storage ($p < 0.05$) in all of the cheese samples. No pathogenic bacteria (*L. monocytogenes* and *Salmonella*) were found in any of the tested cheese samples.

**Table 1.** Microbiological quality of the cheese samples during storage [log CFU/g].

| Count/Presence of Microorganisms [log CFU/g] | Time [Month] | | | Cheese Symbol |
| --- | --- | --- | --- | --- |
| | **0** | **1** | **2** | |
| LAB | 8.11 <sup>Aa</sup> ± 0.01 | 8.04 <sup>Aa</sup> ± 0.07 | 7.83 <sup>Aa</sup> ± 0.07 | |
| ENT | 5.90 <sup>Aa</sup> ± 0.01 | 4.39 <sup>Ba</sup> ± 0.16 | 4.37 <sup>Ba</sup> ± 0.01 | AW |
| YM | 3.68 <sup>Aa</sup> ± 0.01 | 4.69 <sup>Ba</sup> ± 0.21 | 4.65 <sup>Ba</sup> ± 0.07 | |
| SALM | nd | nd | Nd | |
| LIST | nd | nd | Nd | |
| LAB | 7.86 <sup>Ab</sup> ± 0.01 | 7.93 <sup>Ba</sup> ± 0.01 | 7.82 <sup>ABa</sup> ± 0.13 | |
| ENT | 5.57 <sup>Aa</sup> ± 0.09 | 5.16 <sup>ABa</sup> ± 0.18 | 4.36 <sup>Ba</sup> ± 0.01 | B1 |
| YM | 3.45 <sup>Aa</sup> ± 0.01 | 4.29 <sup>ABa</sup> ± 0.30 | 4.94 <sup>Ba</sup> ± 0.02 | |
| SALM | nd | nd | Nd | |
| LIST | nd | nd | Nd | |
| LAB | 8.08 <sup>Aa</sup> ± 0.01 | 8.09 <sup>Aa</sup> ± 0.07 | 7.82 <sup>Aa</sup> ± 0.16 | |
| ENT | 6.66 <sup>Ab</sup> ± 0.03 | 4.89 <sup>Ba</sup> ± 0.17 | 4.70 <sup>Bb</sup> ± 0.01 | Os2 |
| YM | 3.63 <sup>Aa</sup> ± 0.01 | 4.65 <sup>Ba</sup> ± 0.04 | 4.81 <sup>Ca</sup> ± 0.05 | |
| SALM | nd | nd | Nd | |
| LIST | nd | nd | Nd | |

Explanatory: AW—farmer's acid-rennet goat's cheese with acid whey, B1—farmer's acid-rennet goat's cheese with *L. brevis* B1, Os2—farmer's acid-rennet goat's cheese with *L. plantarum* Os2; LAB—the number of lactic acid bacteria; ENT—the number of Enterobacteriaceae; YM—the number of yeasts and molds; SALM—*Salmonella* sp.; LIST—*Listeria monocytogenes*; nd—not detected in 25 g of sample. The values were expressed as means ± SD; means in the same row followed by different uppercase letters within the same sample at different times are significantly different ($p < 0.05$); means in the same column followed by different lowercase letters within samples at the same time are significantly different ($p < 0.05$); (n = 3).

### 3.2. Chemical Composition

The basic composition, NaCl content and phosphorus content were different in the produced goat's cheeses (Table 2). The water content was in the range 46.60% to 48.10%. The protein content ranged from 23.85% for the Os2 cheese to 26.50% for the AW control cheese. In terms of water and protein content, all cheeses differed significantly ($p < 0.05$). The fat content in the AW and B1 cheeses was similar ($p > 0.05$) and amounted to 19.55–20.45%. The Os2 cheese (21.9%) had a significantly higher fat content. The NaCl content was similarly ranged from 1.31 to 1.36%, with the AW control cheese having a slightly higher content. The Os2 cheese had the lowest ($p < 0.05$) phosphorus content (1.21%). In all the tested samples, no lactose was found after cheese samples production (Table 2), which proves the high ability of the used LAB strains to ferment this sugar.

**Table 2.** The basic composition of acid-rennet goat's cheeses directly after production.

| | Cheese Symbol | | |
| --- | --- | --- | --- |
| **Composition** | **AW** | **B1** | **Os2** |
| Water [%] | 46.60 <sup>A</sup> ± 0.00 | 48.10 <sup>B</sup> ± 0.14 | 47.35 <sup>C</sup> ± 0.07 |
| Protein [%] | 26.50 <sup>C</sup> ± 0.28 | 25.60 <sup>B</sup> ± 0.00 | 23.85 <sup>A</sup> ± 0.07 |
| Fat [%] | 20.45 <sup>A</sup> ± 0.49 | 19.55 <sup>A</sup> ± 0.07 | 21.9 <sup>B</sup> ± 0.14 |
| NaCl [%] | 1.36 <sup>B</sup> ± 0.01 | 1.32 <sup>A</sup> ± 0.00 | 1.31 <sup>A</sup> ± 0.01 |
| Phosphorus [%] | 1.34 <sup>B</sup> ± 0.01 | 1.36 <sup>B</sup> ± 0.04 | 1.21 <sup>A</sup> ± 0.01 |
| Lactose [mg/100 g] | nd | nd | nd |

Explanatory: AW—goat's cheese with acid whey, B1—goat's cheese with *L. brevis* B1, Os2—goat's cheese with *L. plantarum* Os2; nd—not detected. The values were expressed as means ±SD; means in the same row followed by different uppercase letters represent significant differences ($p < 0.05$) (n = 4).

The most common fatty acids in goat's cheeses were presented in Table 3, while the full profile of fatty acids composition is available in the Table A1, Appendix A.

**Table 3.** Selected fatty acid composition, sums and cholesterol content in goat's acid-rennet cheeses after production, and after 1 and 2 months of storage.

| Parameter | Time [Month] | | | Cheese Symbol |
|---|---|---|---|---|
| | **0** | **1** | **2** | |
| Σ SFA [%] | 71.75 $^{Cb}$ ± 0.07 | 71.43 $^{Ca}$ ± 0.10 | 70.98 $^{Cc}$ ± 0.44 | AW |
| Σ MUFA [%] | 23.80 $^{Aa}$ ± 0.00 | 24.05 $^{Ab}$ ± 0.06 | 24.30 $^{Ac}$ ± 0.20 | |
| Σ PUFA [%] | 4.45 $^{Ab}$ ± 0.07 | 3.43 $^{Aa}$ ± 0.05 | 3.50 $^{Aa}$ ± 0.00 | |
| Trans [%] | 2.90 $^{Aa}$ ± 0.00 | 3.00 $^{Bb}$ ± 0.00 | 2.98 $^{Ab}$ ± 0.05 | |
| Omega-3 [%] | 1.20 $^{Aa}$ ± 0.00 | 1.20 $^{Aa}$ ± 0.00 | 1.20 $^{Aa}$ ± 0.00 | |
| Omega-6 [%] | 2.25 $^{Aa}$ ± 0.07 | 2.13 $^{Ab}$ ± 0.05 | 2.20 $^{Ab}$ ± 0.00 | |
| Omega-9 [%] | 17.30 $^{Aa}$ ± 0.00 | 17.25 $^{Aa}$ ± 0.06 | 17.15 $^{Aa}$ ± 0.10 | |
| Cholesterol [mg/100 g of product] | 51.10 $^{Ba}$ ± 1.56 | 49.08 $^{Ba}$ ± 1.09 | 59.85 $^{Ab}$ ± 0.00 | |
| Σ SFA [%] | 68.20 $^{bA}$ ± 0.00 | 66.73 $^{aA}$ ± 0.17 | 67.05 $^{aA}$ ± 0.42 | B1 |
| Σ MUFA [%] | 26.90 $^{aC}$ ± 0.00 | 27.20 $^{bC}$ ± 0.08 | 27.50 $^{bC}$ ± 0.45 | |
| Σ PUFA [%] | 4.90 $^{aC}$ ± 0.00 | 5.03 $^{bC}$ ± 0.10 | 5.10 $^{bC}$ ± 0.14 | |
| Trans [%] | 3.10 $^{aB}$ ± 0.00 | 3.20 $^{aC}$ ± 0.00 | 3.18 $^{aB}$ ± 0.13 | |
| Omega-3 [%] | 1.30 $^{aB}$ ± 0.00 | 1.40 $^{bC}$ ± 0.00 | 1.45 $^{bB}$ ± 0.10 | |
| Omega-6 [%] | 2.60 $^{Cb}$ ± 0.00 | 2.45 $^{Ca}$ ± 0.06 | 2.53 $^{Cb}$ ± 0.05 | |
| Omega-9 [%] | 20.10 $^{Ca}$ ± 0.00 | 20.05 $^{Ca}$ ± 0.06 | 20.18 $^{Ca}$ ± 0.33 | |
| Cholesterol [mg/100 g of product] | 48.30 $^{Aa}$ ± 1.27 | 47.55 $^{Aa}$ ± 3.63 | 62.98 $^{Ab}$ ± 2.60 | |
| Σ SFA [%] | 69.00 $^{Bc}$ ± 0.14 | 68.70 $^{Bb}$ ± 0.08 | 68.45 $^{Ba}$ ± 0.13 | Os2 |
| Σ MUFA [%] | 26.20 $^{Ba}$ ± 0.14 | 26.20 $^{Ba}$ ± 0.08 | 26.30 $^{Ba}$ ± 0.14 | |
| Σ PUFA [%] | 4.80 $^{Ba}$ ± 0.00 | 4.80 $^{Ba}$ ± 0.00 | 4.88 $^{Bb}$ ± 0.05 | |
| Trans [%] | 2.85 $^{Aa}$ ± 0.07 | 2.93 $^{Aa}$ ± 0.05 | 2.93 $^{Aa}$ ± 0.05 | |
| Omega-3 [%] | 1.30 $^{Ba}$ ± 0.00 | 1.30 $^{Ba}$ ± 0.00 | 1.43 $^{Bb}$ ± 0.05 | |
| Omega-6 [%] | 2.50 $^{Bb}$ ± 0.00 | 2.30 $^{Ba}$ ± 0.00 | 2.35 $^{Ba}$ ± 0.06 | |
| Omega-9 [%] | 19.65 $^{Bc}$ ± 0.07 | 19.38 $^{Bb}$ ± 0.05 | 19.18 $^{Ba}$ ± 0.10 | |
| Cholesterol [mg/100 g of product] | 54.45 $^{Ca}$ ± 1.77 | 53.75 $^{Ba}$ ± 0.37 | 74.88 $^{Bb}$ ± 2.63 | |

Explanatory: AW—goat's cheese with acid whey, B1—goat's cheese with *L. brevis* B1, Os2—goat's cheese with *L. plantarum* Os2; Σ SFA—all saturated fatty acids; Σ MUFA—all monounsaturated fatty acids; Σ PUFA—all polyunsaturated fatty acids; trans—all trans fatty acids. The values were expressed as means ± SD; means in the same row followed by different uppercase letters within the same sample at different times are significantly different ($p < 0.05$); means in the same column followed by different lowercase letters within samples at the same time are significantly different ($p < 0.05$) (n = 4).

The main fatty acids present in the tested acid-rennet cheeses were palmitic C16:0, oleic C18:1cis9, stearic C18:0 and myristic C14:0. Fatty acids capric C10:0, caprylic C8:0, caproic C6:0 and butyric C4:0 were observed in smaller amounts. The sum of the saturated fatty acids (SFA) of the fresh cheeses was 68.20–71.75%, the sum of the monounsaturated fatty acids (MUFA) was approx. 23–26% and of the polyunsaturated (PUFA) was approx. 4.4–4.9%. A statistically significant ($p < 0.05$) decrease in the SFA content was observed after 1 and 2 months of storage, with a simultaneous significant increase in the content of MUFA and PUFA in the AW and B1 cheeses. In the case of the B1 and Os2 cheeses, a significant increase in omega-3 after storage was noted ($p < 0.05$). In the fresh samples, the cholesterol content was 51.10, 48.30 and 54.45 mg/100 g of the cheese (AW, B1 and Os2, respectively). A significant ($p < 0.05$) increase in the cholesterol content was observed in all trials after 1 and 2 months of storage, with the highest increase observed in the case of the Os2 cheese.

Moreover, Table A2 in Appendix B shows the Lipid Quality Indices (LQI) for all types of cheeses immediately after production and after 1 and 2 months of refrigeration storage. Among the examined goat's cheeses, the highest AI index was found in AW (2.50) and it was lower in Os2 (2.18) and B1 (2.06). The examined cheeses showed a low TI index, ranging from 1.67 (B1) to 1.93 (AW). In the case of the B1 and Os2 goat's cheeses, a favorable share of DFA compared to OFA was observed. The AW goat's cheese was characterized by a higher OFA index. The AW goat's cheese was characterized by an H/H ratio of 0.49, while the B1 and Os2 cheeses were 0.63 and 0.60, respectively. In the present research, the B1 and Os2 cheeses were characterized with a higher HPI (0.50 and 0.47, respectively) than in the case of AW (0.41). After 1 and 2 months of storage, statistically favorable changes in the lipid indices were found ($p < 0.05$).

### 3.3. Physico-Chemical Properties of Cheese Samples

Acid-rennet cheeses made of goat's milk were characterized by the same water activity value of 0.960 immediately after production (Table 4). During storage, this value decreased significantly ($p < 0.05$) in the Os2, AW and B1 samples (to 0.932, 0.935 and 0.941, respectively), however, there were no statistical differences between them ($p > 0.05$).

**Table 4.** Mean values of water activity, pH value, oxidation-reduction potential and total acidity of acid-rennet goat's cheeses after production and after 1 and 2 months of storage.

| Parameter | Time [Month] | | | Cheese Symbol |
|---|---|---|---|---|
| | **0** | **1** | **2** | |
| Water activity ($a_w$) | 0.960 $^{Aa}$ ± 0.006 | 0.947 $^{Ba}$ ± 0.005 | 0.935 $^{Ba}$ ± 0.002 | |
| pH | 5.88 $^{Aa}$ ± 0.01 | 6.10 $^{Ba}$ ± 0.01 | 5.81 $^{Aa}$ ± 0.08 | AW |
| TA [°SH] | 30.00 $^{Ba}$ ± 1.41 | 29.50 $^{Ba}$ ± 1.12 | 40.25 $^{Bb}$ ± 4.82 | |
| ORP [mV] | 313.75 $^{Aa}$ ± 7.46 | 414.25 $^{Bb}$ ± 3.27 | 447.15 $^{Ca}$ ± 12.88 | |
| Water activity ($a_w$) | 0.960 $^{Aa}$ ± 0.002 | 0.950 $^{Ba}$ ± 0.001 | 0.941 $^{Ba}$ ± 0.004 | |
| pH | 6.16 $^{Bb}$ ± 0.00 | 6.27 $^{Ca}$ ± 0.01 | 6.02 $^{Ab}$ ± 0.05 | B1 |
| TA [°SH] | 22.00 $^{Aa}$ ± 1.41 | 24.75 $^{Aa}$ ± 0.83 | 29.25 $^{Ab}$ ± 1.30 | |
| ORP [mV] | 355.88 $^{Ab}$ ± 1.90 | 415.78 $^{Bb}$ ± 1.16 | 427.75 $^{Ba}$ ± 13.01 | |
| Water activity ($a_w$) | 0.960 $^{Aa}$ ± 0.006 | 0.940 $^{Ba}$ ± 0.001 | 0.932 $^{Ba}$ ± 0.003 | |
| pH | 5.86 $^{Ba}$ ± 0.02 | 6.01 $^{Ca}$ ± 0.02 | 5.72 $^{Aa}$ ± 0.02 | Os2 |
| TA [°SH] | 33.50 $^{Aa}$ ± 1.66 | 30.25 $^{Ba}$ ± 1.48 | 51.75 $^{Ca}$ ± 2.86 | |
| ORP [mV] | 428.53 $^{Bc}$ ± 3.79 | 401.73 $^{Aa}$ ± 8.39 | 448.15 $^{Ca}$ ± 3.75 | |

Explanatory: AW—farmer's acid-rennet goat's cheese with acid whey, B1—farmer's acid-rennet goat's cheese with *L. brevis* B1, Os2—farmer's acid-rennet goat's cheese with *L. plantarum* Os2; pH—pH value; ORP—oxidation-reduction potential; TA—titratable acidity. The values were expressed as means ± SD; means in the same row followed by different uppercase letters within the same sample at different times are significantly different ($p < 0.05$); means in the same column followed by different lowercase letters within samples at the same time are significantly different ($p < 0.05$) (n = 4).

The pH values of the cheeses after production were, respectively, for the control cheese AW, 5.88, for the B1 cheese, 6.16 and for the Os2 cheese, 5.86 (Table 4). In all the tests, after 1 month of storage, a significant ($p < 0.05$) increase in pH was observed and after 2 months of storage, the pH decreased again. Directly after production and after 2 months, the AW and Os2 cheeses did not differ from each other ($p > 0.05$).

Titratable acidity after production was 30.00°SH for the AW control sample, 22.00°SH for the B1 cheese and 33.50°SH for the Os2 cheese (Table 4). The B1 cheese showed a significant difference ($p < 0.05$). A decrease in the total acidity was observed in the case of the AW and Os2 cheeses after 1 month of storage, while in the case of the B1 cheese, there was an increase (not statistically significant, $p > 0.05$). On the other hand, after 2 months of storage, in the case of the AW and Os2 cheeses, a statistically significant ($p < 0.05$) increase in total acidity was observed.

In the tests of goat's cheeses, an increase in the oxidation-reduction potential during storage was observed in all samples (Table 4). However, this increase was lower in the case of goat's cheese samples with the addition of B1 and Os2 bacterial cultures in comparison to the control sample with the addition of whey (AW). A significant ($p < 0.05$) increase in ORP was observed in the control samples of AW and B1 after 1 month of storage. In the case of the Os2 cheese, a decrease in ORP was observed after 1 month of storage, followed by an increase after 2 months.

### 3.4. Sensory and Textural Evaluation of Acid-Rennet Goat's Cheese

The results of the sensory evaluation of goat's rennet cheese after production (month 0) and after 1 and 2 months of storage are presented in Table 5.

**Table 5.** Results of sensory evaluation of goat's acid-rennet cheeses after production and after 1 and 2 months of storage.

| Parameter | Time [Month] | | | Cheese Symbol |
|---|---|---|---|---|
| | **0** | **1** | **2** | |
| color | 5.16 $^{Aa}$ ± 1.90 | 5.02 $^{Aa}$ ± 1.24 | 5.46 $^{Aa}$ ± 1.54 | |
| milk's fermentation o. | 6.46 $^{Ab}$ ± 1.54 | 6.00 $^{Aa}$ ± 1.76 | 5.94 $^{Aa}$ ± 1.86 | |
| goat's milk o. | 6.11 $^{Aab}$ ± 1.65 | 6.00 $^{Aa}$ ± 1.87 | 5.99 $^{Aab}$ ± 1.67 | |
| fatty o. | 3.82 $^{ABa}$ ± 1.29 | 3.12 $^{Aa}$ ± 1.45 | 4.93 $^{Bb}$ ± 1.82 | AW |
| sharp o. | 4.22 $^{Ba}$ ± 1.83 | 4.11 $^{Ba}$ ± 1.73 | 3.67 $^{Aa}$ ± 1.92 | |
| mature cheese o. | 4.84 $^{Aa}$ ± 1.76 | 4.98 $^{Aa}$ ± 1.68 | 5.39 $^{Ba}$ ± 1.50 | |
| softness | 2.58 $^{Aa}$ ± 0.97 | 2.76 $^{Aa}$ ± 1.79 | 3.37 $^{Ba}$ ± 1.16 | |
| moisture | 7.53 $^{Ba}$ ± 1.25 | 7.55 $^{Ba}$ ± 1.95 | 5.09 $^{Aa}$ ± 1.93 | |
| elasticity | 7.18 $^{Ba}$ ± 1.61 | 7.32 $^{Ba}$ ± 1.44 | 5.48 $^{Aa}$ ± 1.52 | |
| color | 5.11 $^{Ba}$ ± 1.78 | 5.22 $^{Ba}$ ± 1.48 | 4.21 $^{Aa}$ ± 1.15 | |
| milk's fermentation o. | 5.72 $^{Aa}$ ± 1.70 | 5.97 $^{Aa}$ ± 1.97 | 6.01 $^{Aa}$ ± 1.53 | |
| goat's milk o. | 6.84 $^{Ab}$ ± 1.23 | 6.01 $^{Aa}$ ± 1.84 | 6.67 $^{Ab}$ ± 1.53 | B1 |
| fatty o. | 3.79 $^{Aa}$ ± 1.82 | 3.45 $^{Aa}$ ± 1.97 | 3.76 $^{Aa}$ ± 1.74 | |
| sharp o. | 4.46 $^{Ba}$ ± 0.98 | 4.66 $^{Ba}$ ± 1.09 | 3.53 $^{Aa}$ ± 1.65 | |
| mature cheese o. | 4.92 $^{Aa}$ ± 1.73 | 4.12 $^{Aa}$ ± 1.37 | 4.78 $^{Aa}$ ± 1.24 | |
| softness | 3.53 $^{Ab}$ ± 1.99 | 3.34 $^{Aa}$ ± 1.21 | 6.49 $^{Bc}$ ± 1.44 | |
| moisture | 7.53 $^{Aa}$ ± 1.47 | 7.01 $^{Aa}$ ± 1.76 | 7.11 $^{Ab}$ ± 1.47 | |
| elasticity | 7.38 $^{Aa}$ ± 1.14 | 7.77 $^{Aa}$ ± 1.93 | 7.49 $^{Ab}$ ± 1.35 | |
| color | 4.36 $^{Aa}$ ± 1.73 | 4.64 $^{Aa}$ ± 1.22 | 4.49 $^{Aa}$ ± 1.29 | |
| milk's fermentation o. | 5.37 $^{Aa}$ ± 1.76 | 5.32 $^{Aa}$ ± 1.63 | 5.14 $^{Aa}$ ± 1.82 | |
| goat's milk o. | 5.46 $^{Aa}$ ± 1.82 | 5.33 $^{Aa}$ ± 1.25 | 5.72 $^{Aa}$ ± 1.98 | |
| fatty o. | 3.66 $^{Aa}$ ± 1.76 | 3.23 $^{Aa}$ ± 1.89 | 5.39 $^{Bb}$ ± 1.71 | Os2 |
| sharp o. | 4.23 $^{Ba}$ ± 1.97 | 4.00 $^{ABa}$ ± 1.79 | 3.44 $^{Aa}$ ± 1.80 | |
| mature cheese o. | 4.70 $^{Aa}$ ± 2.63 | 4.97 $^{Aa}$ ± 1.57 | 4.97 $^{Aa}$ ± 1.49 | |
| softness | 3.49 $^{Ab}$ ± 1.87 | 3.02 $^{Aa}$ ± 1.26 | 5.28 $^{Bb}$ ± 1.82 | |
| moisture | 7.38 $^{Ba}$ ± 1.40 | 7.00 $^{Ba}$ ± 1.76 | 5.92 $^{Aa}$ ± 1.68 | |
| elasticity | 6.89 $^{Ba}$ ± 1.68 | 6.99 $^{Ba}$ ± 1.71 | 5.37 $^{Aa}$ ± 1.32 | |

Explanatory: AW—farmer's acid-rennet goat's cheese with acid whey, B1—farmer's acid-rennet goat's cheese with Levilactobacillus brevis B1, Os2—farmer's acid-rennet goat's cheese with Lactiplantibacillus plantarum Os2; o.—odor. The values were expressed as means ± SD; means in the same column followed by different lowercase letters within samples at the same time are significantly different ($p < 0.05$); means in the same row followed by different uppercase letters within the same sample at different times are significantly different ($p < 0.05$); (n = 16).

In the case of goat's acid-rennet cheese B1, i.e., the cheese obtained by adding a B1 strain isolated from "bundz" cheese, the color changed after two months of storage—the color of the cheese became lighter. A different phenomenon was observed in the case of the Os2 sample and the sample of AW, where after two months of storage, the cheese was more of a creamy yellow than after the production.

The odor of lactic acid fermentation after two months of storage became less noticeable both in the case of the Os2 and AW cheeses (5.14 and 5.94, respectively). The opposite is the case for the B1 cheese (6.01), i.e., the odor was more intense. The lowest value of the milk's fermentation odor corresponded to the Os2 cheese ($p > 0.05$). The intensity of the odor of goat's milk, in the case of the AW cheese as well as Os2, was at a similar level of intensity. On the other hand, differences were noticed in the case of the B1 cheese sample, where the odor of the goat's milk was most noticeable ($p < 0.05$) at the beginning, right after production (6.84). The intensity of the fatty odor in the Os2 cheese and AW cheese became more intense ($p < 0.05$) after two months of storage. Such a tendency was not observed in the case of sample B1. In contrast, for each sample, the pungent, irritating odor was less noticeable ($p < 0.05$) after two months of storage. On the other hand, the more intense odor of mature cheese was noticeable only in the case of the B1 sample. The time of ripening of the cheeses significantly ($p < 0.05$) changed their consistency to being softer, more elastic and less moist.

An instrumental examination of color is presented in Table 6. Based on the tests performed, it was found that the test samples (B1 and Os2) were significantly ($p > 0.05$) brighter and less yellow than the control sample AW (L* values after production AW 82.14, B1 84.60, Os2 84.50). After storage, a gradual, statistically significant lightening of the cheeses was observed, with different changes of yellow and red colors, depending on the storage time. Both the sensory analysis and instrumental determination of color showed that sample B1 turned lighter after 2 months of storage, while sample AW turned more yellow.

**Table 6.** Mean values of color parameters in cheese samples during storage.

| Parameter | Time [Month] | | | Cheese Symbol |
|---|---|---|---|---|
| | **0** | **1** | **2** | |
| L* | 82.14 [Aa] ± 2.48 | 79.83 [Ba] ± 1.55 | 79.41 [Ba] ± 2.33 | |
| a* | 1.39 [Aa] ± 0.33 | 1.37 [Aa] ± 0.30 | 1.53 [Aa] ± 0.22 | AW |
| b* | 8.71 [Ac] ± 0.58 | 9.40 [Bc] ± 0.36 | 8.49 [Aa] ± 0.38 | |
| L* | 84.60 [Ab] ± 2.41 | 81.60 [Ba] ± 1.62 | 80.67 [Ba] ± 0.64 | |
| a* | 1.41 [Aa] ± 0.23 | 1.72 [Bb] ± 0.16 | 1.93 [Cc] ± 0.08 | B1 |
| b* | 8.16 [Bb] ± 0.60 | 7.46 [Aa] ± 0.51 | 7.56 [Ab] ± 0.33 | |
| L* | 84.50 [Bb] ± 1.45 | 81.41 [Bb] ± 2.57 | 79.72 [Aa] ± 1.97 | |
| a* | 1.97 [Bb] ± 0.22 | 1.90 [Bc] ± 0.30 | 1.80 [Ab] ± 0.18 | Os2 |
| b* | 7.80 [Aa] ± 0.43 | 7.91 [Ab] ± 0.45 | 8.62 [Ba] ± 0.62 | |

Explanatory: AW—farmer's acid-rennet goat's cheese with acid whey, B1—farmer's acid-rennet goat's cheese with *L. brevis* B1, Os2—farmer's acid-rennet goat's cheese with *L. plantarum* Os2; color parameters—L* indicates lightness, a* is the red/green coordinate and b* is the yellow/blue coordinate. The values were expressed as means ± SD; means in the same row followed by different uppercase letters represent significant differences ($p < 0.05$); ($n = 20$).

The instrumental texture parameters of goat's acid-rennet cheeses after production and after 1 and 2 months of maturation were presented in Table 7.

Based on the results, changes in the texture of acid-rennet goat's cheese during the 2 months of storage can be noticed (Table 7). The values of hardness 1 and 2 were, significantly, the highest in the case of the AW cheese and statistically the lowest in the case of the Os2 cheese (93.50, 68.35, 51.54 and 35.78, respectively). During storage, hardness 1 and 2 decreased (in 1st month) and then increased (in 2nd month), regardless of the samples. The adhesiveness of samples B1 and Os2 were slightly lower during storage, however, there were not any statistical differences between the samples ($p < 0.05$). The cohesiveness of the acid-rennet cheeses during storage firstly increased in the 1st month and then decreased in the 2nd month, regardless of the samples. In the case of each sample in the 2nd month of storage, the springiness decreased significantly. Both the gumminess and chewiness, in the case of the Os2 and AW samples, decreased and were statistically lower after 2 months.

**Table 7.** Texture instrumental parameters of goat's acid-rennet cheeses after production and after 1 and 2 months of storage.

| Parameter | Time [Month] | | | Cheese Symbol |
|---|---|---|---|---|
| | **0** | **1** | **2** | |
| Hardness Cycle 1 [N] | 93.50 $^{Bb}$ ± 18.62 | 68.80 $^{Ab}$ ± 4.43 | 102.67 $^{Bc}$ ± 14.18 | |
| Adhesiveness [mJ] | 1.02 $^{Aa}$ ± 0.86 | 1.13 $^{Aa}$ ± 0.99 | 1.18 $^{Aa}$ ± 0.76 | |
| Hardness Cycle 2 [N] | 68.35 $^{Ab}$ ± 13.09 | 59.17 $^{Ac}$ ± 3.73 | 57.23 $^{Ab}$ ± 4.46 | |
| Cohesiveness | 0.61 $^{Ba}$ ± 0.05 | 0.74 $^{Cb}$ ± 0.01 | 0.38 $^{Aa}$ ± 0.06 | AW |
| Springiness [mm] | 8.87 $^{Ba}$ ± 0.07 | 8.67 $^{Ba}$ ± 0.14 | 8.11 $^{Aa}$ ± 0.56 | |
| Gumminess [N] | 57.20 $^{Bb}$ ± 11.77 | 51.07 $^{Bc}$ ± 2.87 | 39.45 $^{Ab}$ ± 9.45 | |
| Chewiness [mJ] | 506.83 $^{Bb}$ ± 101.56 | 442.37 $^{Bc}$ ± 20.15 | 324.06 $^{Ab}$ ± 91.61 | |
| Hardness Cycle 1 [N] | 64.21 $^{Aa}$ ± 15.75 | 50.00 $^{Aa}$ ± 8.03 | 72.50 $^{Ab}$ ± 19.50 | |
| Adhesiveness [mJ] | 1.77 $^{Aa}$ ± 0.15 | 1.28 $^{Aa}$ ± 0.68 | 1.00 $^{Aa}$ ± 0.77 | |
| Hardness Cycle 2 [N] | 45.30 $^{Aa}$ ± 12.07 | 43.30 $^{Ab}$ ± 6.79 | 56.42 $^{Ab}$ ± 14.54 | |
| Cohesiveness | 0.61 $^{Aa}$ ± 0.13 | 0.75 $^{Bb}$ ± 0.01 | 0.64 $^{ABb}$ ± 0.06 | B1 |
| Springiness [mm] | 8.82 $^{Ba}$ ± 0.14 | 8.67 $^{Ba}$ ± 0.12 | 8.46 $^{Aa}$ ± 0.13 | |
| Gumminess [N] | 38.72 $^{Aa}$ ± 10.61 | 37.45 $^{Ab}$ ± 5.94 | 46.26 $^{Ab}$ ± 12.32 | |
| Chewiness [mJ] | 341.27 $^{Aa}$ ± 93.96 | 325.15 $^{Ab}$ ± 54.34 | 391.35 $^{Ab}$ ± 104.33 | |
| Hardness Cycle 1 [N] | 51.54 $^{Ba}$ ± 4.93 | 42.94 $^{Aa}$ ± 4.63 | 46.38 $^{ABa}$ ± 5.77 | |
| Adhesiveness [mJ] | 1.27 $^{Aa}$ ± 0.68 | 1.07 $^{Aa}$ ± 0.89 | 0.95 $^{Aa}$ ± 1.10 | |
| Hardness Cycle 2 [N] | 35.78 $^{Aa}$ ± 4.27 | 35.85 $^{Aa}$ ± 3.10 | 31.49 $^{Aa}$ ± 2.53 | |
| Cohesiveness | 0.57 $^{Ba}$ ± 0.08 | 0.71 $^{Ca}$ ± 0.01 | 0.37 $^{Aa}$ ± 0.09 | Os2 |
| Springiness [mm] | 8.77 $^{Ba}$ ± 0.06 | 8.50 $^{Ba}$ ± 0.13 | 7.27 $^{Aa}$ ± 1.28 | |
| Gumminess [N] | 29.50 $^{Ba}$ ± 4.73 | 30.52 $^{Ba}$ ± 2.84 | 17.09 $^{Aa}$ ± 3.01 | |
| Chewiness [mJ] | 258.43 $^{Ba}$ ± 40.04 | 259.67 $^{Ba}$ ± 26.94 | 126.37 $^{Aa}$ ± 40.04 | |

Explanatory: AW—farmer's acid-rennet goat's cheese with acid whey, B1—farmer's acid-rennet goat's cheese with *L. brevis* B1, Os2—farmer's acid-rennet goat's cheese with *L. plantarum* Os2. The values were expressed as means ± SD; means in the same column followed by different lowercase letters within samples at the same time are significantly different ($p < 0.05$); means in the same row followed by different uppercase letters within the same sample at different times are significantly different ($p < 0.05$); (n = 6).

## 4. Discussion

Goat's milk is a very important food product in many regions of the world and the interest in the production of goat's milk products has increased over the past decade [45,46]. Goat's milk is also an excellent material for the development of such functional foods as cheeses [47,48]. In terms of suitability for processing, goat's milk differs significantly from cow's milk. Due to the lower content of casein and its smaller share in the total amount of proteins, the cheese yield is lower, and the curd obtained is more delicate and less firm [49]. Scientific interest in artisanal cheese is growing because it represents a source of environmental bacteria with specific, potential health benefits [50].

In the present work, the acid-rennet cheeses from organic goat's milk fermented with *L. brevis* B1 and *L. plantarum* Os2, or acid whey as a control, were obtained under industrial conditions and evaluated. According to FAO/WHO General Standard for Cheese [51], the produced cheeses can be classified as semi-hard or partially skimmed. The examples of semi-hard cheeses are Colby and Monterey (stirred-curd Cheddar-type cheeses), a number of British Territorial varieties (Caerphilly, Lancashire and Wensleydale) and cheeses such as Majorero (Spain) or Bryndza (Slovakia) and Bundz (Poland) [52]. The tested cheese samples were mostly similar to the Bundz type and prepared according to the producer recipe, replacing the acid whey with a dedicated starter culture B1 or Os2. It was found that lactic acid bacteria (LAB) were the predominant microbiota (about 8 log CFU/g) during cheese storage and the use of the B1 and Os2 culture and did not affect the microbiological quality of the cheeses. However, in case there was no confirmation of colonies of presumptive LAB on the medium (MRS agar), it was taken into consideration that other specific microorganisms could also grow, i.e., *Staphylococcus* sp. Essentially harmful are *S. aureus* enterotoxins producers and they should be controlled during the production process. When the detected value is $>10^5$ CFU/g, the cheese batch must be tested for staphylococcal enterotoxins [53]. It should also be emphasized that the cheese samples were made from unpasteurized, raw goat's milk, which was not free of native microbiota. However, even in the control sample, the acid whey was added to control the fermentation process. The endogenous microbiota can grow in the milk during fermentation, next to the starter cultures; however, carefully selected starter cultures of bacteria strains are able to dominate during the milk's fermentation [54]. Lactobacilli (predominantly *Lactobacillus brevis*, *Lactobacillus plantarum*, *Lactobacillus paracasei* and *Lactobacillus casei*) constitute the majority of nonstarter lactic acid bacteria in Cheddar, creamy, and some artisanal cheeses. Typical counts of *Lactobacillus* cells range from 10 to about $10^8$ CFU/g of the cheese samples [55–57]. In comparison, Mushtaq et al. [58] used *L. brevis* as a probiotic starter culture and found that counts of LAB during 30 days of storage period were approximately 9–6 log CFU/g. Although, in our study, we do not provide the identification of the LAB, the changes in the chemical and physical properties, as well as the texture and sensory characteristics, were observed in samples with *L. brevis* B1 and *L. plantarum* Os2, in comparison to the AW control. These observations gave an initial proof of domination of the selected starter cultures and could be a premise for further research.

We have found that the level of Enterobacteriaceae in the final product was very high (5.57–6.66 log CFU/g), which suggests the need to improve the hygienic condition of the process. The important observation is that the presence of the Enterobacteriaceae family was successfully reduced during storage. On the other hand, the samples were free of *Listeria monocytogenes* and *Salmonella* spp., which was in line with the UE regulation [53], as well as the studies by Dauber et al. [9], Andretta et al. [59], de Medeiros Carvalho et. al. [60] and De la Rosa-Alcaraz et al. [50]. Moreover, the selected bacterial strains *L. brevis* B1 and *L. plantarum* Os2 possess strong antibacterial properties [19,20]. The isolation of the LAB strains with anti-microbiological, mainly anti-listerial activity from artisanal cheese samples is quite common [16], which could be the reason of the phenomenon.

Microbial dynamics in cheese are affected by several factors, such as milk composition, salting, starter culture, rennet addition, temperature, relative humidity (85–97%), pH, redox potential, water activity and moisture [61]. Among the several factors that influence the physicochemical properties of cheese, the acidity is crucial as it directly affects the stability of casein micelles and milk minerals [62]. In our study, the pH value was in the range of 5–6, which is the typical value for most cheese varieties [52], however, it was lower than in study of Guimarães et al. [63]. The differences in the pH values could be caused by the different composition of the starter cultures and their interaction. After 1 month of storage, a significant ($p < 0.05$) increase in pH was observed, which may have been caused by the utilization of lactic acid, the formation of a non-acid decomposition product and the release of alkaline protein decomposition products [64,65]. Both the starter bacteria used, *L. plantarum* Os2 and *L. brevis* B1, are heterofermentative LABs that use the phosphoketolase pathway to produce a mixture of lactic acid, ethanol, acetic acid and $CO_2$ as products of hexose fermentation [55]. These bacterial strains represented a similar slight acidification model, however, the Os2 strain seems to be more effective as the TA value after 2 months of storage was the highest in the samples of cheeses fermented with the *L. plantarum* Os2 strain. In addition, the oxidation-reduction potential increased during storage in all samples and was higher than in the study by Abraham et al. [66] where cheeses such as Camembert, Cheddar and Comté reached an ORP between 100 and 350 mV. Some authors [67–69] claim that LAB strains show the antioxidant activity, which is why the high value of the oxidation-reduction potential can be explained by the high microbiological activity of the LAB strains and their enzymatic capacity. In the tested cheeses, no additional substances were used to influence the oxidation processes and the high values of the redox potential testify to the superiority of the oxidation reaction processes. The oxidation-reduction potential modifications can affect both the microbiological quality (biological activity of non-starter lactic acid bacteria) and the sensorial properties (synthesis and/or stability of aroma compounds) of cheeses [70].

The biochemical changes that occur during cheese maturation can be divided into primary metabolic changes such as glycolysis, lipolysis and proteolysis, followed by secondary biochemical changes such as fatty acid and amino acid metabolism, which play an important role in the production of secondary metabolites, including compounds responsible for the flavor bouquet [62]. The composition of cheese samples prepared in this study was typical and comparable to others [50,70–72].

Goat's milk contains slightly less lactose than cow's milk (goat's milk 4.1–4.5% of lactose, cow's milk 4.6–4.7% of lactose). Most of the lactose is found in whey, which is the liquid that is separated from the solid cheese curd during the cheese-making process. The longer the cheese has been aged, the less lactose will remain in the final product [73]. In our study, no lactose was found in the cheese samples. During the maturation process, protein and fat are continuously broken down. Casein is first hydrolyzed to long chain peptides which are then broken down into short chain peptides. Some of the casein eventually breaks down into amino acids and volatiles. When dispersed into small fat globules, the large fat globules are further broken down into ketones, aldehydes and lactones, which break down into volatile substances and free fatty acids [74]. In the present study, the main fatty acids present in the tested acid-rennet cheeses were palmitic C16:0, oleic C18:1cis9, stearic C18:0 and myristic C14:0, which were also observed in the studies by Popović-Vranješ et al. [75]. It should also be underlined that there was a significant increase in omega-3 and in the cholesterol content in all samples after storage was denoted. Similar results were obtained by Burgos et al. [76]. On the other hand, much higher values of the cholesterol content were obtained in the studies by Barłowska et al. [77], where 358–370 mg/100 g of fat were denoted. The results of the SFA, MUFA and PUFA content obtained in our work differ from the results presented by Paszczyk and Łuczyńska [32], where the sum of the saturated acids in the goat's cheeses was 58.08% on average, with the advantage for unsaturated acids MUFA and PUFA (23.66 and 3.49%, respectively). In turn, higher SFA values were observed in the studies by Schettino-Bermúdez et al. [78]. The high levels of SFA in milk naturally occur and dairy products are often correlated with the adverse effects of these foods and with the development of several civilization diseases, including cardiovascular disease, type 2 diabetes, obesity and cancer [79]. The differences in the chemical composition between the cited studies of the cheeses may have resulted from the different production and maturation conditions of the goat's cheeses, as well as the starter culture used for their production. Due to the observed changes in the fatty acid profile and in the microstructure of the goat's cheeses during the ripening period in the present study, it is recommended to mature the goat's cheese for at least 30 days before consumption or sale.

To assess the health properties of evaluated acid-rennet cheeses the Lipid Quality Indices were calculated. The Index of Atherogenicity (AI) shows the relationship between the sum of saturated fatty acids, which are considered pro-atherogenic and the sum of unsaturated fatty acids [33]. The AI index values in tested cheese samples were at an average level (2.06–2.50) in all the samples, which is a rather small value, due to the AI value ranging from 1.42 to 5.13 for dairy products [33]. The Index of Thrombogenicity characterizes the thrombogenic potential of fatty acids, indicating the tendency to form clots in blood vessels. The lower TI characterizes the healthier the product for the cardiac system [33].

Interestingly, in our study a favorable share of DFA compared to OFA in B1 and Os2 goat's cheeses was observed. On the other hand, control AW goat's cheese was characterized by a higher OFA index. In the studies by Paszczyk and Łuczyńska [32], higher OFA values were observed in goat's cheeses compared to DFA, as well as a lower AI value and a higher TI value. The hypocholesterolemic effect is mainly to reduce the absorption of cholesterol from the gastrointestinal tract [34]. The H/H ratio characterizes the relationship between the hypocholesterolemic fatty acids and hypercholesterolemic fatty acids and may more accurately reflect the effect of the fatty acids' composition on cardiovascular diseases. Additionally, the health-promoting index (HPI), which was proposed to assess the nutritional value of dietary fat and focuses on the effect of fatty acids composition on

CVD (Cardiovascular Disease) was found to be higher for B1 and Os2 goat's cheeses. High HPI dairy products are believed to be more beneficial to human health [33]. To sum up, the tested cheeses with the addition of *L. brevis* B1 and *L. plantarum* Os2 bacterial cultures were characterized by more favorable Lipid Quality Indices compared to the control with the addition of acid whey.

The sensory quality of foods is one of the main factors influencing the acceptance of a product on the market. For this reason, studying factors that influence the sensory quality are crucial in food product development. In the presented study, it was shown that some odor attributes, especially those with a negative sensory influence were of importance for the overall sensory characteristics. It is well known that even a slight increase in the intensity of negative attributes, mainly flavor or odor, are connected with a severe decrease of the overall sensory quality of food products [80–82]. In the present study, a negative odor intensity (sharp) was, on average, in the range of 3.44–4.66 c.u. and was not different between samples. Furthermore, the unpleasant odor typical for goat's milk was most noticeable at the beginning and decreased during the samples' storage. It could be due to the LAB metabolic activity, for example, the production of carboxylic acids in cheese due to their lipolytic activity. It was found that the typical, pleasant flavor of milky goat's cheese was related to the free fatty acids, especially free hexanoic acid, octanoic acid and *n*-decanoic acid [62]. Similarly, Calvo et al. [83] found that during the storage of ripened goat's cheese, as the external appearance features deteriorated and the intensity of the "goat" flavor decreased, it became less noticeable. Other attributes were not differentiated between the tested cheese samples. Due to the COVID-19 pandemic, oral evaluation was omitted, so the overall sensory quality was not assessed. However, it can be concluded that the odor of the tested cheese samples was similar in all samples.

Color and texture are important criteria for assessing the quality of a cheese, as these two parameters are important for consumers when making a decision to buy a product [84]. The color is one of the most important quality attributes, noticed at first glance, on the basis of which the consumer assesses the acceptability of the product [83]. The results of present study showed that the L* value decreases significantly, whereas the a* and b* values were found to remain unchanged in all the samples during storage, which is in agreement with other authors studies [58,85]. The texture characteristics become of significant importance when there are no flavor defects in the samples [44]. It must be considered by a production step that even the high quality of the texture attributes will not ensure the proper acceptance of the product. In the present study, we observed that in the 1st month of storage, the hardness decreased and then in the 2nd month, it increased regardless of the cheese samples. In turn, the cohesiveness of the acid-rennet cheeses during storage firstly increased and then decreased regardless of the samples. Similar trends were observed in the study of Zaravela et al. [73]. In the study of Jia et al. [62], the hardness of the semi-hard goat's cheese samples showed an overall upward trend with the increasing maturity and the cohesiveness decreased slightly as the maturation time progressed. This phenomenon is a consequence of protein and fat hydrolyzation during the maturation process.

During the 2 months of storage of the goat's cheese samples, the changes of the instrumentally evaluated texture were noted. The hardness increased regardless of the samples and was the highest in the case of the AW cheese. At the same time, the springiness, gumminess and chewiness in the case of samples Os2 and AW decreased. In the presented study, the sensory analysis of the cheeses' moisture was in agreement with the instrumental measurement of water activity and water content. Gámbaro et al. [71] showed that the moisture content of fresh cheese (48.7%) was higher than that of ripened cheese (42.7%), so the storage time influences the changes in cheese moisture. When comparing the sensory analysis of the texture and instrumental texture determination, there were some discrepancies. According to the evaluators, the cheese, after 2 month of storage, became softer and more elastic. In turn, TPA analysis shows that after 2 months of storage, the hardness increased and springiness decreased regardless of the sample, similar to the

García et al. [74] study. To sum up, sample B1 turned out to be more soft and elastic according to the sensory analysis, as well as being less hard and springy in the instrumental texture analysis.

## 5. Conclusions

When summarizing the obtained research results, it should be stated that the technology of acid-rennet curd production with the use of selected, environmental bacterial cultures is in line with the current trends related to the innovation of fermented food development with high nutritional quality. Scientific studies have shown that a daily diet enrichment with $10^6$–$10^9$ CFU/mL of beneficial lactic acid bacteria cells after just a few weeks may increase the number of natural killer cells in the blood serum, and increase the activity of macrophages and lymphocytes. Moreover, the immunomodulatory action of lactic acid bacteria may further reduce allergic reactions in humans.

According to our study it can be concluded that the selected cultures of *L. brevis* B1 and *L. plantarum* Os2 isolated from traditional cheeses from a Polish mountain can be applied to goat's milk cheese production. Bacteria cultures affected the microbiological, physical, chemical and sensory quality of the cheese samples. The lactic acid bacteria count in the cheese samples was at a high level of about 8 log CFU/g. Moreover, the cheeses with the addition of *L. brevis* B1 and *L. plantarum* Os2 bacterial cultures were characterized by more favorable Lipid Quality Indices than for the control with the addition of acid whey. The produced cheeses were at a similar sensory quality and after storage, the experimental cheeses (B1 and Os2) were more elastic and softer than the control sample (AW).

However, it should be underlined that after taking into account high number of the Enterobacteriaceae bacteria group, as well as high the pH and water activity values in the samples, the cheeses produced in this way cannot be recommended for consumption. Therefore, future studies should focus on increasing the microbiological quality and the deeper analysis of the cheese ripening processes, including the lipolysis and proteolysis rates. The next step could also be the identification of culture strains from food samples to provide proof of the potential probiotic properties of goat's cheeses. Such innovative products can be excellent carriers of environmental bacteria with potentially probiotic properties and can be widely used for the enrichment of the intestinal microbiota of humans, especially for the Polish population.

**Author Contributions:** Conceptualization, D.K.-K., Z.J.D. and P.S.; methodology, K.K.-S., A.O., A.Ł., D.J., M.T., K.N.-S., B.S. and D.Z.; software, K.K.-S., D.J., A.O. and A.Ł.; validation, A.O., A.Ł., K.K.-S. and B.S.; formal analysis, K.K.-S., D.Z. and D.J.; investigation, A.O., A.Ł., D.J., M.T., D.Z., B.S., K.K.-S. and K.N.-S.; resources, A.Ł., A.O., K.K.-S. and D.Z.; data curation, K.K.-S., D.Z. and D.J.; writing—original draft preparation, A.Ł.; K.K.-S., D.J. and D.Z.; writing—review and editing, D.Z., D.J., P.S. and D.K.-K.; visualization, K.K.-S. and D.Z.; supervision, D.K.-K. and Z.J.D.; project administration, P.S.; funding acquisition, Z.J.D. All authors have read and agreed to the published version of the manuscript.

**Funding:** This research was funded by the MINISTRY OF AGRICULTURE AND RURAL DEVELOPMENT, grant number JPR.re.027.7.2020. The APC was funded by Waclaw Dabrowski Institute of Agriculture and Food Biotechnology—State Research Institute.

**Institutional Review Board Statement:** Not applicable.

**Informed Consent Statement:** Not applicable.

**Conflicts of Interest:** The authors declare no conflict of interest.

## Appendix A

**Table A1.** Fatty acid composition and cholesterol content in goat's acid-rennet cheeses after production and after 1 and 2 months of storage (n = 4).

| Parameter | Time [Month] | | | | | | | | |
|---|---|---|---|---|---|---|---|---|---|
| | 0 | 1 | 2 | 0 | 1 | 2 | 0 | 1 | 2 |
| C4:0 [%] | 2.20 ± 0.00 | 2.13 ± 0.10 | 2.08 ± 0.10 | 2.10 ± 0.14 | 2.05 ± 0.06 | 2.13 ± 0.33 | 2.20 ± 0.00 | 2.08 ± 0.05 | 2.05 ± 0.06 |
| C6:0 [%] | 2.35 ± 0.07 | 2.28 ± 0.05 | 2.23 ± 0.05 | 2.20 ± 0.00 | 2.08 ± 0.05 | 2.05 ± 0.26 | 2.20 ± 0.00 | 2.13 ± 0.05 | 2.10 ± 0.00 |
| C8:0 [%] | 2.60 ± 0.00 | 2.53 ± 0.05 | 2.50 ± 0.00 | 2.25 ± 0.07 | 2.20 ± 0.00 | 2.08 ± 0.17 | 2.30 ± 0.00 | 2.30 ± 0.00 | 2.30 ± 0.00 |
| C10:0 [%] | 8.85 ± 0.07 | 8.88 ± 0.05 | 8.75 ± 0.06 | 7.40 ± 0.00 | 7.38 ± 0.05 | 6.95 ± 0.33 | 7.80 ± 0.00 | 7.83 ± 0.05 | 7.70 ± 0.08 |
| C10:1 [%] | 0.10 ± 0.00 | 0.10 ± 0.00 | 0.20 ± 0.00 | 0.10 ± 0.00 | 0.10 ± 0.00 | 0.10 ± 0.00 | 0.10 ± 0.00 | 0.10 ± 0.00 | 0.10 ± 0.00 |
| C12:0 [%] | 2.95 ± 0.07 | 3.00 ± 0.00 | 3.10 ± 0.00 | 2.50 ± 0.00 | 2.65 ± 0.06 | 2.55 ± 0.06 | 2.70 ± 0.00 | 2.80 ± 0.00 | 2.85 ± 0.06 |
| C13:0 [%] | 0.10 ± 0.00 | 0.10 ± 0.00 | 0.10 ± 0.00 | 0.10 ± 0.00 | 0.10 ± 0.00 | 0.10 ± 0.00 | 0.10 ± 0.00 | 0.10 ± 0.00 | 0.10 ± 0.00 |
| C14:0 [%] | 9.10 ± 0.00 | 9.03 ± 0.05 | 9.03 ± 0.05 | 8.75 ± 0.07 | 8.70 ± 0.00 | 8.50 ± 0.00 | 9.00 ± 0.00 | 9.00 ± 0.00 | 8.93 ± 0.05 |
| C14:1 [%] | 0.30 ± 0.00 | 0.30 ± 0.00 | 0.30 ± 0.00 | 0.30 ± 0.00 | 0.30 ± 0.00 | 0.30 ± 0.00 | 0.30 ± 0.00 | 0.30 ± 0.00 | 0.30 ± 0.00 |
| C15:0 br [%] | 0.40 ± 0.00 | 0.40 ± 0.00 | 0.40 ± 0.00 | 0.40 ± 0.00 | 0.40 ± 0.00 | 0.40 ± 0.00 | 0.40 ± 0.00 | 0.40 ± 0.00 | 0.40 ± 0.00 |
| C15:0 [%] | 0.80 ± 0.00 | 0.90 ± 0.00 | 0.90 ± 0.00 | 0.80 ± 0.00 | 0.85 ± 0.06 | 0.83 ± 0.05 | 0.80 ± 0.00 | 0.90 ± 0.00 | 0.90 ± 0.00 |
| C15:1 [%] | 0.20 ± 0.00 | 0.20 ± 0.00 | 0.20 ± 0.00 | 0.20 ± 0.00 | 0.20 ± 0.00 | 0.20 ± 0.00 | 0.20 ± 0.00 | 0.20 ± 0.00 | 0.20 ± 0.00 |
| C16:0 [%] | 28.85 ± 0.07 | 28.73 ± 0.15 | 28.35 ± 0.13 | 25.85 ± 0.07 | 25.50 ± 0.08 | 25.53 ± 0.22 | 26.65 ± 0.07 | 26.40 ± 0.00 | 26.20 ± 0.00 |
| C16:1 [%] | 0.80 ± 0.00 | 0.80 ± 0.00 | 1.08 ± 0.05 | 0.80 ± 0.00 | 0.80 ± 0.00 | 1.00 ± 0.08 | 0.85 ± 0.07 | 0.90 ± 0.00 | 1.10 ± 0.00 |
| C17:0 br | 0.80 ± 0.00 | 0.80 ± 0.00 | 0.80 ± 0.00 | 0.80 ± 0.00 | 0.83 ± 0.05 | 0.85 ± 0.06 | 0.80 ± 0.00 | 0.80 ± 0.00 | 0.90 ± 0.00 |
| C17:0 [%] | 0.60 ± 0.00 | 0.60 ± 0.00 | 0.60 ± 0.00 | 0.60 ± 0.00 | 0.60 ± 0.00 | 0.63 ± 0.05 | 0.60 ± 0.00 | 0.60 ± 0.00 | 0.60 ± 0.00 |
| C17:1 [%] | 0.20 ± 00 | 0.20 ± 0.00 | 0.20 ± 0.00 | 0.20 ± 0.00 | 0.20 ± 0.00 | 0.20 ± 0.00 | 0.20 ± 0.00 | 0.20 ± 0.00 | 0.20 ± 0.00 |
| C18:0 [%] | 11.85 ± 0.07 | 11.78 ± 0.05 | 11.85 ± 0.06 | 14.05 ± 0.07 | 13.93 ± 0.05 | 14.10 ± 0.41 | 13.15 ± 0.07 | 13.08 ± 0.05 | 13.03 ± 0.05 |
| C18:1 trans [%] | 2.90 ± 0.00 | 3.00 ± 0.00 | 2.98 ± 0.05 | 3.10 ± 0.00 | 3.20 ± 0.00 | 3.18 ± 0.13 | 2.85 ± 0.07 | 2.93 ± 0.05 | 2.93 ± 0.05 |
| C18:1 cis9 [%] | 17.20 ± 0.00 | 17.25 ± 0.06 | 17.15 ± 0.10 | 20.00 ± 0.00 | 20.05 ± 0.06 | 20.18 ± 0.33 | 19.55 ± 0.07 | 19.38 ± 0.05 | 19.18 ± 0.10 |
| C18:1 cis11 [%] | 0.70 ± 0.00 | 0.60 ± 0.00 | 0.60 ± 0.00 | 0.70 ± 0.00 | 0.70 ± 0.00 | 0.70 ± 0.00 | 0.70 ± 0.00 | 0.60 ± 0.00 | 0.60 ± 0.00 |
| C18:1 c other [%] | 1.30 ± 0.00 | 1.50 ± 0.00 | 1.50 ± 0.00 | 1.40 ± 0.00 | 1.55 ± 0.06 | 1.55 ± 0.06 | 1.35 ± 0.07 | 1.50 ± 0.00 | 1.60 ± 0.00 |
| C18:2 [%] | 1.95 ± 0.07 | 1.93 ± 0.05 | 2.00 ± 0.00 | 2.30 ± 0.00 | 2.25 ± 0.06 | 2.33 ± 0.05 | 2.20 ± 0.00 | 2.10 ± 0.00 | 2.25 ± 0.06 |
| C18:3 n3 [%] | 1.00 ± 0.00 | 1.00 ± 0.00 | 1.00 ± 0.00 | 1.10 ± 0.00 | 1.20 ± 0.00 | 1.18 ± 0.05 | 1.10 ± 0.00 | 1.10 ± 0.00 | 1.13 ± 0.05 |
| C18:2 c9 t11 [%] | 1.00 ± 0.00 | 1.00 ± 0.00 | 1.00 ± 0.00 | 1.00 ± 0.00 | 1.08 ± 0.05 | 1.03 ± 0.10 | 1.00 ± 0.00 | 1.10 ± 0.00 | 1.00 ± 0.00 |
| C20:0 [%] | 0.20 ± 0.00 | 0.20 ± 0.00 | 0.20 ± 0.00 | 0.20 ± 0.00 | 0.20 ± 0.00 | 0.28 ± 0.05 | 0.20 ± 0.00 | 0.20 ± 0.00 | 0.30 ± 0.00 |
| C20:1 [%] | 0.10 ± 0.00 | 0.10 ± 0.00 | 0.10 ± 0.00 | 0.10 ± 0.00 | 0.10 ± 0.00 | 0.10 ± 0.00 | 0.10 ± 0.00 | 0.10 ± 0.00 | 0.10 ± 0.00 |
| C20:2 [%] | 0.10 ± 0.00 | 0.10 ± 0.00 | 0.10 ± 0.00 | 0.10 ± 0.00 | 0.10 ± 0.00 | 0.10 ± 0.00 | 0.10 ± 0.00 | 0.10 ± 0.00 | 0.10 ± 0.00 |
| C20:3 n6 [%] | 0.10 ± 0.00 | 0.10 ± 0.00 | 0.10 ± 0.00 | 0.10 ± 0.00 | 0.10 ± 0.00 | 0.10 ± 0.00 | 0.10 ± 0.00 | 0.10 ± 0.00 | 0.00 ± 0.00 |
| C20:4 n6 [%] | 0.10 ± 0.00 | 0.10 ± 0.00 | 0.10 ± 0.00 | 0.10 ± 0.00 | 0.10 ± 0.00 | 0.10 ± 0.00 | 0.10 ± 0.00 | 0.10 ± 0.00 | 0.10 ± 0.00 |
| C20:5 EPA [%] | 0.10 ± 0.00 | 0.10 ± 0.00 | 0.10 ± 0.00 | 0.10 ± 0.00 | 0.10 ± 0.00 | 0.10 ± 0.00 | 0.10 ± 0.00 | 0.10 ± 0.00 | 0.10 ± 0.00 |
| C22:0 [%] | 0.10 ± 0.00 | 0.10 ± 0.00 | 0.10 ± 0.00 | 0.10 ± 0.00 | 0.10 ± 0.00 | 0.10 ± 0.00 | 0.10 ± 0.00 | 0.10 ± 0.00 | 0.10 ± 0.00 |
| C22:5 n3 [%] | 0.10 ± 0.00 | 0.10 ± 0.00 | 0.10 ± 0.00 | 0.10 ± 0.00 | 0.10 ± 0.00 | 0.18 ± 0.05 | 0.10 ± 0.00 | 0.10 ± 0.00 | 0.20 ± 0.00 |
| SFA [%] | 71.75 ± 0.07 | 71.43 ± 0.10 | 70.98 ± 0.44 | 68.20 ± 0.00 | 66.73 ± 0.17 | 67.05 ± 0.42 | 69.00 ± 0.14 | 68.70 ± 0.08 | 68.45 ± 0.13 |
| MUFA [%] | 23.80 ± 0.00 | 24.05 ± 0.06 | 24.30 ± 0.20 | 26.90 ± 0.00 | 27.20 ± 0.08 | 27.50 ± 0.45 | 26.20 ± 0.14 | 26.20 ± 0.08 | 26.30 ± 0.14 |
| PUFA [%] | 4.45 ± 0.07 | 3.43 ± 0.05 | 3.50 ± 0.00 | 4.90 ± 0.00 | 5.03 ± 0.10 | 5.10 ± 0.14 | 4.80 ± 0.00 | 4.80 ± 0.00 | 4.88 ± 0.05 |
| Trans [%] | 2.90 ± 0.00 | 3.00 ± 0.00 | 2.98 ± 0.05 | 3.10 ± 0.00 | 3.20 ± 0.00 | 3.18 ± 0.13 | 2.85 ± 0.07 | 2.93 ± 0.05 | 2.93 ± 0.05 |
| Omega-3 [%] | 1.20 ± 0.00 | 1.20 ± 0.00 | 1.20 ± 0.00 | 1.30 ± 0.00 | 1.40 ± 0.00 | 1.45 ± 0.10 | 1.30 ± 0.00 | 1.30 ± 0.00 | 1.43 ± 0.05 |
| Omega-6 [%] | 2.25 ± 0.07 | 2.13 ± 0.05 | 2.20 ± 0.00 | 2.60 ± 0.00 | 2.45 ± 0.06 | 2.53 ± 0.05 | 2.50 ± 0.00 | 2.30 ± 0.00 | 2.35 ± 0.06 |
| Omega-9 [%] | 17.30 ± 0.00 | 17.25 ± 0.06 | 17.15 ± 0.10 | 20.10 ± 0.00 | 20.05 ± 0.06 | 20.18 ± 0.33 | 19.65 ± 0.07 | 19.38 ± 0.05 | 19.18 ± 0.10 |
| Cholesterol [mg/100 g of products] | 51.10 ± 1.56 | 49.08 ± 1.09 | 59.85 ± 0.00 | 48.30 ± 1.27 | 47.55 ± 3.63 | 62.98 ± 2.60 | 54.45 ± 1.77 | 53.75 ± 0.37 | 74.88 ± 2.63 |
| Cheese symbol | AW | | | B1 | | | Os2 | | |

## Appendix B

**Table A2.** The Lipid Quality Indices of goat's cheeses after production and after 1 and 2 months of storage.

| LQI | Time [Month] | | | Cheese Symbol |
|---|---|---|---|---|
| | 0 | 1 | 2 | |
| AI | 2.50 $^{Bc}$ ± 0.01 | 2.48 $^{Bc}$ ± 0.00 | 2.44 $^{Ac}$ ± 0.15 | |
| TI | 1.93 $^{Ac}$ ± 0.00 | 1.91 $^{Ab}$ ± 0.01 | 1.90 $^{Ab}$ ± 0.02 | |
| DFA | 40.10 $^{Aa}$ ± 0.00 | 39.26 $^{Aa}$ ± 0.11 | 39.65 $^{Aa}$ ± 0.06 | AW |
| OFA | 40.90 $^{Cc}$ ± 0.15 | 40.76 $^{Bc}$ ± 0.25 | 40.48 $^{Ac}$ ± 0.12 | |
| H/H | 0.49 $^{Aa}$ ± 0.00 | 0.50 $^{Aa}$ ± 0.01 | 0.50 $^{Aa}$ ± 0.06 | |
| HPI | 0.41 $^{Aa}$ ± 0.00 | 0.41 $^{Aa}$ ± 0.10 | 0.41 $^{Aa}$ ± 0.08 | |
| AI | 2.06 $^{Ba}$ ± 0.02 | 2.03 $^{Ba}$ ± 0.00 | 1.97 $^{Aa}$ ± 0.17 | |
| TI | 1.67 $^{Aa}$ ± 0.01 | 1.62 $^{Aa}$ ± 0.12 | 1.60 $^{Aa}$ ± 0.05 | |
| DFA | 45.85 $^{Ac}$ ± 0.15 | 46.16 $^{Bc}$ ± 0.02 | 46.70 $^{Cc}$ ± 0.05 | B1 |
| OFA | 37.10 $^{Ca}$ ± 0.00 | 36.85 $^{Ba}$ ± 0.05 | 36.58 $^{Aa}$ ± 0.12 | |
| H/H | 0.63 $^{Ab}$ ± 0.02 | 0.64 $^{Ac}$ ± 0.01 | 0.65 $^{Ab}$ ± 0.05 | |
| HPI | 0.50 $^{Ab}$ ± 0.01 | 0.51 $^{Ab}$ ± 0.06 | 0.53 $^{Ac}$ ± 0.00 | |

**Table A2.** *Cont.*

| LQI | Time [Month] | | | Cheese Symbol |
|---|---|---|---|---|
| | **0** | **1** | **2** | |
| AI | 2.18 [Ab] ± 0.02 | 2.19 [Ab] ± 0.01 | 2.15 [Ab] ± 0.00 | |
| TI | 1.71 [Ab] ± 0.01 | 2.59 [Cc] ± 0.01 | 2.51 [Bc] ± 0.09 | |
| DFA | 44.15 [Bb] ± 0.05 | 44.08 [Ab] ± 0.06 | 44.21 [Bb] ± 0.23 | Os2 |
| OFA | 38.35 [Bb] ± 0.05 | 38.20 [Ab] ± 0.00 | 37.98 [Cb] ± 0.12 | |
| H/H | 0.60 [Ab] ± 0.01 | 0.60 [Ab] ± 0.01 | 0.60 [Ab] ± 0.05 | |
| HPI | 0.47 [Ab] ± 0.13 | 0.48 [Ab] ± 0.02 | 0.48 [Ab] ± 0.00 | |

Explanatory: AW—goat's cheese with acid whey, B1—goat's cheese with *L. brevis* B1, Os2—goat's cheese with *L. plantarum* Os2; LQI—Lipid Quality Indices; AI—Index of atherogenicity, TI—Index of thrombogenicity, DFA—Hypocholesterolemic fatty acids, OFA—Hypercholesterolemic fatty acids, H/H—The ratio of hypocholesterolemic and hypercholesterolemic fatty acids; HPI—Health-promoting index. The values were expressed as means ± SD; means in the same row followed by different uppercase letters within the same sample at different times are significantly different ($p < 0.05$); means in the same column followed by different lowercase letters within samples at the same time are significantly different ($p < 0.05$) (n = 4).

# References


1. Cabral, C.F.S.; Veiga, L.B.E.; Araújo, M.G.; de Souza, S.L.Q. Environmental Life Cycle Assessment of goat cheese production in Brazil: A path towards sustainability. *LWT* **2020**, *129*, 109550. [CrossRef]
2. Bonczar, G.; Filipczak-Fiutak, M.; Pluta-Kubica, A.; Duda, I.; Walczycka, M.; Staruch, L. The range of protein hydrolysis and biogenic amines content in selected acid-and rennet-curd cheeses. *Chem. Pap.* **2018**, *72*, 2599–2606. [CrossRef] [PubMed]
3. Pombo, A.F.W. Cream cheese: Historical, manufacturing, and physico-chemical aspects. *Int. Dairy J.* **2020**, *117*, 104948. [CrossRef]
4. Rozos, G.; Voidarou, C.; Stavropoulou, E.; Skoufos, I.; Tzora, A.; Alexopoulos, A.; Bezirtzoglou, E. Biodiversity and microbial resistance of lactobacilli isolated from the traditional Greek cheese Kopanisti. *Front. Microbiol.* **2018**, *9*, 517. [CrossRef]
5. Voolstra, C.R.; Ziegler, M. Adapting with microbial help: Microbiome flexibility facilitates rapid responses to environmental change. *BioEssays* **2020**, *42*, 2000004. [CrossRef] [PubMed]
6. Zielińska, D.; Kołożyn-Krajewska, D. Food-origin lactic acid bacteria may exhibit probiotic properties: Review. *BioMed Res. Int.* **2018**, *2018*, 5063185. [CrossRef]
7. Raynal-Ljutovac, K.; Lagriffoul, G.; Paccard, P.; Guillet, I.; Chilliard, Y. Composition of goat and sheep milk products: An update. *Small Rumin. Res.* **2008**, *79*, 57–72. [CrossRef]
8. Yadav, A.K.; Singh, J.; Yadav, S.K. Composition, nutritional and therapeutic values of goat milk: A review. *Asian J. Dairy Food Res.* **2016**, *35*, 96–102. [CrossRef]
9. Dauber, C.; Carreras, T.; Britos, A.; Carro, S.; Cajarville, C.; Gámbaro, A.; Jorcin, S.; López, T.; Vieitez, I. Elaboration of goat cheese with increased content of conjugated linoleic acid and transvaccenic acid: Fat, sensory and textural profile. *Small Ruminant Res.* **2021**, *199*, 106379. [CrossRef]
10. Meyrand, M.; Dallas, D.C.; Caillat, H.; Bouvier, F.; Martin, P.; Barile, D. Comparison of milk oligosaccharides between goats with and without the genetic ability to synthesize αs1-casein. *Small Rumin. Res.* **2013**, *113*, 411–420. [CrossRef]
11. Sousa, Y.R.F.; Medeiros, L.B.; Pintado, M.M.E.; Queiroga, R.C.R.E. Goat milk oligosaccharides: Composition, analytical methods and bioactive and nutritional properties. *Trends Food Sci. Technol.* **2019**, *92*, 152–161. [CrossRef]
12. Li, X.Z.; Yan, C.G.; Lee, H.G.; Choi, C.W.; Song, M.K. Influence of dietary plant oils on mammary lipogenic enzymes and the conjugated linoleic acid content of plasma and milk fat of lactating goats. *Anim. Feed Sci. Technol.* **2012**, *174*, 26–35. [CrossRef]
13. Mituniewicz–Małek, A.; Zielińska, D.; Ziarno, M. Probiotic monocultures in fermented goat milk beverages–sensory quality of final product. *Int. J. Dairy Technol.* **2019**, *72*, 240–247. [CrossRef]
14. Santurino, C.; López-Plaza, B.; Fontecha, J.; Calvo, M.V.; Bermejo, L.M.; Gómez-Andrés, D.; Gómez-Candela, C. Consumption of goat cheese naturally rich in omega-3 and conjugated linoleic acid improves the cardiovascular and inflammatory biomarkers of overweight and obese subjects: A randomized controlled trial. *Nutrients* **2020**, *12*, 1315. [CrossRef]
15. Reps, A.; Wiśniewska, K.; Bohdziewicz, K. Sery podpuszczkowe, twarogowe i topione. In *Produkty Mleczne. Technologia i Rola w Żywieniu Człowieka*; Gawęcki, J., Pikul, J., Eds.; Uniwersytetu Przyrodniczego w Poznaniu: Poznań, Poland, 2018; pp. 135–142.
16. Campagnollo, F.B.; Margalho, L.P.; Kamimura, B.A.; Feliciano, M.D.; Freire, L.; Lopes, L.S.; Alvarenga, V.O.; Cadavez, V.A.P.; Gonzales-Barron, U.; Schaffner, D.W.; et al. Selection of indigenous lactic acid bacteria presenting anti-listerial activity, and their role in reducing the maturation period and assuring the safety of traditional Brazilian cheeses. *Food Microbiol.* **2018**, *73*, 288–297. [CrossRef]
17. Laranjo, M.; Potes, M.E.; Elias, M. Role of starter cultures on the safety of fermented meat products. *Front. Microbiol.* **2019**, *10*, 853. [CrossRef]
18. Campos, G.Z.; Lacorte, G.A.; Jurkiewicz, C.; Hoffmann, C.; Landgraf, M.; de Melo Franco, B.D.G.; Pinto, U.M. Microbiological characteristics of Canastra cheese during manufacturing and ripening. *Food Control* **2021**, *121*, 107598. [CrossRef]

19. Ołdak, A.; Zielińska, D.; Rzepkowska, A.; Kołożyn-Krajewska, D. Comparison of antibacterial activity of *Lactobacillus plantarum* strains isolated from two different kinds of regional cheeses from Poland: Oscypek and Korycinski cheese. *BioMed Res Int.* **2017**, *2017*, 6820369. [CrossRef]

20. Ołdak, A.; Zielińska, D.; Łepecka, A.; Długosz, E.; Kołożyn-Krajewska, D. *Lactobacillus plantarum* strains isolated from Polish regional cheeses exhibit anti-Staphylococcal activity and selected probiotic properties. *Probiotics Antimicrob. Proteins* **2020**, *12*, 1025–1038. [CrossRef]

21. Polish Patent No. p-426002. Available online: https://ewyszukiwarka.pue.uprp.gov.pl/search/pwp-details/P.426002?lng=pl (accessed on 25 July 2022).

22. *ISO 15214:2002*; Microbiology of Food and Animal Feeding Stuffs—Horizontal Method for the Enumeration of Mesophilic Lactic Acid Bacteria—Colony-Count Technique at 30 °C. ISO: Geneva, Switzerland, 2012.

23. *ISO 21528-2:2017*; Microbiology of Food and Feeding Stuffs—Horizontal Method for the Detection and Enumeration of Enterobacteriaceae—Part 2: Colony Count Method. ISO: Geneva, Switzerland, 2017.

24. *ISO 21527-1:2008*; Microbiology of Food and Animal Feeding Stuffs—Horizontal Method for the Enumeration of Yeasts and Moulds—Part 1: Colony Count Technique in Products with Water Activity Greater than 0.95. ISO: Geneva, Switzerland, 2008.

25. *ISO 21527-2:2008*; Microbiology of Food and Animal Feeding Stuffs—Horizontal Method for the Enumeration of Yeasts and Moulds—Part 2: Colony Count Technique in Products with Water Activity Less than or Equal to 0.95. ISO: Geneva, Switzerland, 2008.

26. *ISO 6579-1:2017*; Microbiology of the Food Chain—Horizontal Method for the Detection, Enumeration and Serotyping of Salmonella—Part 1: Detection of Salmonella spp. ISO: Geneva, Switzerland, 2017.

27. *ISO 11290-1:2017*; Microbiology of Food and Animal Feeding Stuffs—Horizontal Method for the Detection and Enumeration of Listeria Monocytogenes and Others Listeria spp.—Part 1: Detection Method. ISO: Geneva, Switzerland, 2017.

28. *PN-ISO 1442:2000*; Mięso i Przetwory Mięsne—Oznaczanie Zawartości Wody (Metoda Odwoławcza). ISO: Geneva, Switzerland, 2000.

29. *PN-A-04018:1975/Az3:2002*; Produkty Rolniczo-Żywnościowe—Oznaczanie Azotu Metodą Kjeldahla i Przeliczanie na Białko. ISO: Geneva, Switzerland, 2002.

30. *PN-ISO 8262-3:2011*; Przetwory mleczne i Żywność na Bazie Mleka—Oznaczanie Zawartości Tłuszczu Metodą Grawimetryczną Weibulla-Berntropa (Metoda Odniesienia)—Część 3: Przypadki Szczególne. ISO: Geneva, Switzerland, 2011.

31. *PN-EN ISO 5508:1996*; Oleje i Tłuszcze Roślinne Oraz Zwierzęce—Analiza Estrów Metylowych Kwasów Tłuszczowych Metodą Chromatografii Gazowej. ISO: Geneva, Switzerland, 1996.

32. Paszczyk, B.; Łuczyńska, J. The Comparison of Fatty Acid Composition and Lipid Quality Indices in Hard Cow, Sheep, and Goat Cheeses. *Foods* **2020**, *9*, 1667. [CrossRef]

33. Chen, J.; Liu, H. Nutritional indices for assessing fatty acids: A mini-review. *Int. J. Molecular Sci.* **2020**, *21*, 5695. [CrossRef] [PubMed]

34. Osmari, E.K.; Cecato, U.; Macedo, F.A.F.; Souza, N.E. Nutritional quality indices of milk fat from goats on diets supplemented with different roughages. *Small Rumin. Res.* **2011**, *98*, 128–132. [CrossRef]

35. Ulbricht, T.L.V.; Southgate, D.A.T. Coronary heart disease: Seven dietary factors. *Lancet* **1991**, *338*, 985–992. [CrossRef]

36. Ivanova, A.; Hadzhinikolova, L. Evaluation of nutritional quality of common carp (*Cyprinus carpio* L.) lipids through fatty acid ratios and lipid indices. *Bulg. J. Agric. Sci.* **2015**, *21*, 180–185.

37. *PN-A-82060:1999*; Mięso i przetwory Mięsne—Oznaczanie Zawartości Fosforu. ISO: Geneva, Switzerland, 1999.

38. *PN-ISO 1841-2:2002*; Mięso i Przetwory MIĘSNE—Oznaczanie Zawartości Chlorków—Część 2: Metoda Potencjometryczna. ISO: Geneva, Switzerland, 2002.

39. *ISO 18787:2017*; Food Stuffs—Determination of Water Activity. ISO: Geneva, Switzerland, 2017.

40. *ISO 4121:2003*; Sensory Analysis—Guidelines for the Use of Quantitative Response Scales. ISO: Geneva, Switzerland, 2003.

41. Calín-Sánchez, Á.; Carbonell-Barrachina, Á.A. Flavor and Aroma Analysis as a Tool for Quality Control of Foods. *Foods* **2021**, *10*, 224. [CrossRef]

42. *ISO 8586:2012*; Sensory Analysis. General Guidelines for the Selection, Training and Monitoring of Selected Assessors and Expert Sensory Assessors. ISO: Geneva, Switzerland, 2012.

43. Meilgaard, M.; Civille, G.V.; Carr, B.T. *Sensory Evaluation Techniques*, 4th ed.; CRC Press: Abingdon, UK, 2006.

44. Surmacka Szczesniak, A. Texture is a sensory property. *Food Qual. Pref.* **2002**, *13*, 215–225. [CrossRef]

45. Gómez-Cortés, P.; Cívico, A.; de la Fuente, M.A.; Sánchez, N.N.; Blanco, F.P.; Marín, A.L.M. Short term evolution of nutritionally relevant milk fatty acids of goats fed a cereal-based concentrate enriched with linseed oil. *Innov. Food Sci. Emerg. Technol.* **2019**, *51*, 107–113. [CrossRef]

46. Lu, C.D.; Miller, B.A. Current status, challenges and prospects for dairy goat production in the Americas. *Asian-Australas. J. Anim. Sci.* **2019**, *32*, 1244. [CrossRef]

47. Kocak, A.; Sanli, T.; Anlib, E.A.; Hayaloglu, A.A. Role of using adjunct cultures in release of bioactive peptides in white brined goat-milk cheese. *LWT—Food Sci. Technol.* **2020**, *123*, 109127. [CrossRef]

48. Del Toro-Gipson, R.S.; Rizzo, P.V.; Hanson, D.J.; Drake, M.A. Consumer perception of smoked Cheddar cheese. *J. Dairy Sci.* **2021**, *104*, 1560–1575. [CrossRef]

49. Vacca, G.M.; Stocco, G.; Dettori, M.L.; Bittante, G.; Pazzola, M. Goat cheese yield and recovery of fat, protein, and total solids in curd are affected by milk coagulation properties. *J. Dairy Sci.* **2020**, *103*, 1352–1365. [CrossRef] [PubMed]
50. De la Rosa-Alcaraz, M.D.L.Á.; Ortiz-Estrada, Á.M.; Heredia-Castro, P.Y.; Hernández-Mendoza, A.; Reyes-Díaz, R.; Vallejo-Cordoba, B.; González-Córdova, A.F. Poro de Tabasco cheese: Chemical composition and microbiological quality during its artisanal manufacturing process. *J. Dairy Sci.* **2020**, *103*, 3025–3037. [CrossRef] [PubMed]
51. General Standard for Cheese. Available online: https://www.fao.org/fao-who-codexalimentarius/sh-proxy/es/?lnk=1&url=https%253A%252F%252Fworkspace.fao.org%252Fsites%252Fcodex%252FStandards%252FCXS%2B283-1978%252FCXS_283e.pdf (accessed on 29 August 2022).
52. Fox, P.F.; Guinee, T.P.; Cogan, T.M.; McSweeney, P.L. Principal families of cheese. In *Fundamentals of Cheese Science*; Springer: Boston, MA, USA, 2017; pp. 27–69.
53. Commission Regulation (EC). No 2073/2005 of 15 November 2005 on Microbiological Criteria for Foodstuffs. OJEU, n° L338. 22 December 2005, pp. 1–26. Available online: https://eur-lex.europa.eu/legal-content/EN/TXT/PDF/?uri=CELEX:32005R2073&from=ES (accessed on 25 July 2022).
54. Salazar, J.K.; Gonsalves, L.J.; Fay, M.; Ramachandran, P.; Schill, K.M.; Tortorello, M.L. Metataxonomic Profiling of Native and Starter Microbiota During Ripening of Gouda Cheese Made with Listeria monocytogenes—Contaminated Unpasteurized Milk. *Front. Microbiol.* **2021**, *12*, 642789. [CrossRef] [PubMed]
55. Peterson, S.D.; Marshall, R.T. Nonstarter lactobacilli in Cheddar cheese: A review. *J. Dairy Sci.* **1990**, *73*, 1395–1410. [CrossRef]
56. Williams, A.G.; Choi, S.C.; Banks, J.M. Variability of the species and strain phenotype composition of the non-starter lactic acid bacterial population of Cheddar cheese manufactured in a commercial creamery. *Food Res. Int.* **2002**, *35*, 483–493. [CrossRef]
57. Gantzias, C.; Lappa, I.K.; Aerts, M.; Georgalaki, M.; Manolopoulou, E.; Papadimitriou, K.; De Brandt, E.; Tsakalidou, E.; Vandamme, P. MALDI-TOF MS profiling of non-starter lactic acid bacteria from artisanal cheeses of the Greek island of Naxos. *Int. J. Food Microbiol.* **2020**, *323*, 108586. [CrossRef]
58. Mushtaq, M.; Gani, A.; Masoodi, F.A.; Ahmad, M. Himalayan cheese (Kalari/Kradi)–Effect of different probiotic strains on oxidative stability, microbiological, sensory and nutraceutical properties during storage. *LWT-Food Sci. Technol.* **2016**, *67*, 74–81. [CrossRef]
59. Andretta, M.; Almeida, T.T.; Ferreira, L.R.; Carvalho, A.F.; Yamatogi, R.S.; Nero, L.A. Microbial safety status of Serro artisanal cheese produced in Brazil. *J. Dairy Sci.* **2019**, *102*, 10790–10798. [CrossRef]
60. de Medeiros Carvalho, M.; de Fariña, L.O.; Strongin, D.; Ferreira, C.L.L.F.; Lindner, J.D.D. Traditional Colonial-type cheese from the south of Brazil: A case to support the new Brazilian laws for artisanal cheese production from raw milk. *J. Dairy Sci.* **2019**, *102*, 9711–9720. [CrossRef]
61. Moreira, R.V.; Costa, M.P.; Frasao, B.S.; Sobral, V.S.; Cabral, C.C.; Rodrigues, B.L.; Mano, S.B.; Conte-Junior, C.A. Effect of ripening time on bacteriological and physicochemical goat milk cheese characteristics. *Food Sci. Biotechnol.* **2020**, *29*, 459–467. [CrossRef]
62. Jia, R.; Zhang, F.; Song, Y.; Lou, Y.; Zhao, A.; Liu, Y.; Peng, H.; Hui, Y.; Ren, R.; Wang, B. Physicochemical and textural characteristics and volatile compounds of semihard goat cheese as affected by starter cultures. *J. Dairy Sci.* **2021**, *104*, 270–280. [CrossRef]
63. Guimarães, D.H.P.; Barros, D.; de Moraes Gomes Rosa, M.T. Sensorial and Rheological Parameters of Cured Cheese Produced with Goat Milk. In *Proceedings of the 4th Brazilian Technology Symposium (BTSym'18), Smart Innovation, Systems and Technologies*; Iano, Y., Arthur, R., Saotome, O., Vieira Estrela, V., Loschi, H., Eds.; Springer: Cham, Switzerland, 2019; Volume 140. [CrossRef]
64. Farahnaky, A.; Mousavi, S.H.; Nasiri, M. Role of salt in Iranian ultrafiltered feta cheese: Some textural and physicochemical changes during ripening. *Int. J. Dairy Technol.* **2013**, *66*, 359–365. [CrossRef]
65. Fresno, M.; Álvarez, S. Chemical, textural and sensorial changes during the ripening of Majorero goat cheese. *Int. J. Dairy Technol.* **2012**, *65*, 393–400. [CrossRef]
66. Abraham, S.; Cachon, R.; Colas, B.; Feron, G.; De Coninck, J. Eh and pH gradients in Camembert cheese during ripening: Measurements using microelectrodes and correlations with texture. *Int. Dairy J.* **2007**, *17*, 954–960. [CrossRef]
67. Hansen, E.B. Redox reactions in food fermentations. *Curr. Opin. Food Sci.* **2018**, *19*, 98–103. [CrossRef]
68. Lin, X.; Xia, Y.; Yang, Y.; Wang, G.; Zhou, W.; Ai, L. Probiotic characteristics of Lactobacillus plantarum AR113 and its molecular mechanism of antioxidant. *LWT* **2020**, *126*, 109278. [CrossRef]
69. Shi, Y.; Cui, X.; Gu, S.; Yan, X.; Li, R.; Xia, S.; Chen, H.; Ge, J. Antioxidative and probiotic activities of lactic acid bacteria isolated from traditional artisanal milk cheese from Northeast China. *Probiotics Antimicrob. Proteins.* **2019**, *11*, 1086–1099. [CrossRef]
70. Cachon, R.; Jeanson, S.; Aldarf, M.; Divies, C. Characterisation of lactic starters based on acidification and reduction activities. *Le Lait* **2002**, *82*, 281–288. [CrossRef]
71. Gámbaro, A.; Gonzalez, V.; Jimenez, S.; Arechavaleta, A.; Irigaray, B.; Callejas, N.; Grompone, M.; Vieitez, I. Chemical and sensory profiles of commercial goat cheeses. *Int. Dairy J.* **2017**, *69*, 1–8. [CrossRef]
72. do Oriente, S.F.; Barreto, F.; Tomaszewski, C.A.; Barnet, L.S.; Souza, N.C.; Lisboa Oliveira, H.M.; de Bittencourt Pasquali, M.A. Retention of vitamin A after goat milk processing into cheese: A nutritional strategy. *J. Food Sci. Technol.* **2020**, *57*, 4364–4370. [CrossRef]
73. Zaravela, A.; Kontakos, S.; Badeka, A.V.; Kontominas, M.G. Effect of adjunct starter culture on the quality of reduced fat, white, brined goat cheese: Part I. Assessment of chemical composition, proteolysis, lipolysis, texture and sensory attributes. *Eur. Food Res. Technol.* **2021**, *247*, 2211–2225. [CrossRef]

74. García, V.; Rovira, S.; Teruel, R.; Boutoial, K.; Rodríguez, J.; Roa, I.; López, M.B. Effect of vegetable coagulant, microbial coagulant and calf rennet on physicochemical, proteolysis, sensory and texture profiles of fresh goats cheese. *Dairy Sci. Technol.* **2012**, *92*, 691–707. [CrossRef]

75. Popović-Vranješ, A.; Pihler, I.; Paskaš, S.; Krstović, S.; Jurakić, Ž.; Strugar, K. Production of hard goat cheese and goat whey from organic goat's milk. *Mljekarstvo Časopis Unaprjeđenje Proizvodnje Prerade Mlijeka* **2017**, *67*, 177–187. [CrossRef]

76. Burgos, L.S.; Pece, N.; Maldonado, S. Fatty acid composition and microstructure of ripened goat cheese. *Lat. Am. Appl. Res. -Int. J.* **2021**, *51*, 43–48. [CrossRef]

77. Barłowska, J.; Pastuszka, R.; Rysiak, A.; Król, J.; Brodziak, A.; Kędzierska-Matysek, M.; Wolanciuk, A.; Litwińczuk, Z. Physico-chemical and sensory properties of goat cheeses and their fatty acid profile in relation to the geographic region of production. *Int. J. Dairy Technol.* **2018**, *71*, 699–708. [CrossRef]

78. Schettino-Bermúdez, B.; y León, S.V.; Gutierrez-Tolentino, R.; Perez-Gonzalez, J.J.; Escobar, A.; Gonzalez-Ronquillo, M.; Vargas-Bello-Perez, E. Effect of dietary inclusion of chia seed (*Salvia hispanica* L.) on goat cheese fatty acid profile and conjugated linoleic acid isomers. *Int. Dairy J.* **2020**, *105*, 104664. [CrossRef]

79. Lordan, R.; Zabetakis, I. Invited review: The anti-inflammatory properties of dairy lipids. *J. Dairy Sci.* **2017**, *100*, 4197–4212. [CrossRef]

80. Mehanna, N.M.; Swelam, S.; Ragab, W.A.; Dawoud, M.A. Composition and some Properties of Processed Cheese Spread Made from Blends Containing different Quantities of The Same Main Ingredients. *J. Food Dairy Sci.* **2020**, *11*, 285–288. [CrossRef]

81. Papetti, P.; Carelli, A. Composition and Sensory Analysis for Quality Evaluation of a Typical Italian Cheese: Influence of Ripening Period. *Czech J. Food Sci.* **2013**, *31*, 438–444. [CrossRef]

82. Jaworska, D.; Kolanowski, W.; Waszkiewicz-Robak, B.; Świderski, F. Relative importance of texture properties in sensory quality and acceptance of natural yoghurts. *Int. J. Dairy Technol.* **2005**, *58*, 39–46. [CrossRef]

83. Calvo, M.V.; Castialo, I.; Diaz-Barcos, V.; Requena, T.; Fontecha, J. Effect of a hygienized rennet paste and a defined strain starter on proteolysis, texture and sensory properties of semi-hard goat cheese. *Food Chem.* **2007**, *102*, 917–924. [CrossRef]

84. Pinho, O.; Mendes, E.; Alves, M.; Ferreira, I. Chemical, Physical, and Sensorial Characteristics of "Terrincho" Ewe Cheese: Changes during Ripening and Intravarietal Comparison. *J. Dairy Sci.* **2004**, *87*, 249–257. [CrossRef]

85. Verruck, S.; Prudêncio, E.S.; Müller, C.M.O.; Fritzen-Freire, C.B.; Amboni, R.D.D.M.C. Influence of Bifidobacterium Bb-12 on the physicochemical and rheological properties of buffalo Minas Frescal cheese during cold storage. *J. Food Eng.* **2015**, *151*, 34–42. [CrossRef]

 *applied sciences*

*Article*

# Effect of Marinating Temperature of Atlantic Herring on Meat Ripening, Peptide Fractions Proportion, and Antioxidant Activity of Meat and Brine

Mariusz Szymczak [1,*], Patryk Kamiński [1], Marta Turło [2], Justyna Bucholska [2], Damir Mogut [2], Piotr Minkiewicz [2], Małgorzata Mizielińska [3] and Magdalena Stobińska [3]

[1] Department of Toxicology, Dairy Technology and Food Storage, Faculty of Food Sciences and Fisheries, West Pomeranian University of Technology in Szczecin, Papieza Pawla VI 3, 71-459 Szczecin, Poland; patryk-kaminski@zut.edu.pl

[2] Department of Food Biochemistry, Faculty of Food Science, University of Warmia and Mazury in Olsztyn, Pl. Cieszyński 1, 10-726 Olsztyn, Poland; marta.turlo@uwm.edu.pl (M.T.); justyna.michalska@uwm.edu.pl (J.B.); damir.mogut@uwm.edu.pl (D.M.); minkiew@uwm.edu.pl (P.M.)

[3] Center of Bioimmobilisation and Innovative Packaging Materials, Faculty of Food Sciences and Fisheries, West Pomeranian University of Technology Szczecin, Janickiego 35, 71-270 Szczecin, Poland; mmizielinska@zut.edu.pl (M.M.); mstobinska@zut.edu.pl (M.S.)

* Correspondence: mariusz.szymczak@zut.edu.pl or mszymczak@zut.edu.pl; Tel.: +48-91-4496503

Citation: Szymczak, M.; Kamiński, P.; Turło, M.; Bucholska, J.; Mogut, D.; Minkiewicz, P.; Mizielińska, M.; Stobińska, M. Effect of Marinating Temperature of Atlantic Herring on Meat Ripening, Peptide Fractions Proportion, and Antioxidant Activity of Meat and Brine. *Appl. Sci.* **2023**, *13*, 7225. https://doi.org/10.3390/app13127225

Academic Editor: Anabela Raymundo

Received: 29 May 2023
Revised: 12 June 2023
Accepted: 13 June 2023
Published: 16 June 2023



**Abstract:** The temperature has a significant effect on cathepsin activity, but the effect of temperature on the ripening of marinades, and the formation of protein hydrolysis products, is less studied than other technological factors. The results of this study showed that herring marinated at 2 °C showed a higher mass yield, but lower non-protein nitrogen (NPN), peptides, and free amino acid fraction content, than after marinating at 7 and 12 °C. The higher temperature increased the free amino acid content the most, and decreased the hardness, as measured via sensory assessment, of the marinated meat. This was confirmed by the hardness measurement in the texture profile analysis. The highest activity of cathepsins D and B in the meat was found at 7 °C, while cathepsin L was found at 2 °C. Increasing the temperature by 10 °C increased the diffusion/loss of nitrogenous substances from the meat to the brine by 36%. The meat and brine showed high antioxidant activity, which depended on the marinating temperature, and originated mainly from the 5–10 or <5 kDa fraction. The meat had a higher ABTS (2,2′-azinobis-3-ethylbenzothiazoline-6-sulfonate) activity than the brine, opposite to the DPPH (2,2-diphenyl-1-picrylhydrazyl radical) activity, while the FRAP (Ferric Reducing Antioxidant Power) capacity was similar for meat and brine. The fractionation of the meat and brine extracts increased the antioxidant potential of FRAP and ABTS only for the brine. The most hydrophobic peptides were released during marinating at 7 °C. The meat and brine were dominated by 2–4 kDa peptides, followed by 4–6 and 0.5–2 kDa. The higher temperature favored a higher proportion of <4 kDa than >4 kDa peptides in the brine.

**Keywords:** herring; marinating; temperature; cathepsins; antioxidant activity; peptides; RP-HPLC

## 1. Introduction

The marinating of herring has been known of since ancient times, and the process has not changed significantly to this day. The industry most often uses the classic method, which involves a single stage of marination in a solution of acetic acid and table salt [1]. Cold fish marinades are a unique food product, as they are made from wild-caught herring, which contains fatty acids with positive effects on human health. The lipids in cold marinades are not heat-treated, so they undergo minor hydrolysis and oxidation processes [2]. In addition, marinades contain complete protein and protein hydrolysis products (PHP), which are released through the activity of muscle proteases called cathepsins. In general, the quality of fish products depends on many factors, and the production process plays a special role [3]. Probably the greatest influence on the ripening of marinades is the activity of cathepsins D, B and L, which are activated in an acid-salt environment [4,5] The released PHP (peptides and amino acids), and the products of their interaction with other components of the muscle tissue, give the marinades their characteristic taste and odor. During marination, PHPs diffuse into the surrounding marinating brine, resulting in nutrient loss [6]. The PHP content of the brine is often higher than that of the marinated meat [7]. The peptides present in the brine show antioxidant activity, which has been used to develop a new method of marinating herring [8], the marinating of not herring-fish [9], and as an ingredient in the glaze during the frozen storage of fatty fish, or as an additive for minced herring tissue [10]. At present, the spent brine left over after marinating herring is a huge problem for the industry and the public, as it ends up in wastewater treatment plants, causing economic and environmental problems [11]. An increased knowledge of the brine's composition, especially of the biologically active substances, could contribute to its faster valorization in the food industry.

In addition to their antioxidant activity, peptides also possess antibacterial, antifungal, antiviral, immunomodulatory, antiproliferative, anticoagulant, and antihypertensive functions, as angiotensin-converting enzyme inhibitors (ACE-inhibitors), and exhibit hemolytic, and opioid- and calcium-binding activity. Certain peptides, usually 2–20 amino acid residues in length, can exhibit a range of bioactivities, the activity being dependent on the sequence of amino acids [12,13].

Currently, in the industry, the ripening process of marinated herring is regulated by several basic technological factors, such as the concentration of acetic acid and salt, the marinating time, and the ratio of fish to brine, but the least studied is still the influence of the temperature [4,5,7,14]. The classical marinating of herring takes place at 7 °C, but a two-stage method is known, in which 12 °C is used for the first 2–4 days of marinating, and then the process continues at 7 °C. Nowadays, the industry uses much lower temperatures of 0–4 °C, to obtain the highest possible mass yield of marinated herring, and to reduce the proliferation of microorganisms. The temperature also affects the activity of muscle cathepsins, which in fish are highly active at refrigeration temperatures [4,5,15]. Fish cathepsins show comparable activity at 7 °C to mammalian cathepsins at 37 °C. Despite this, there is a lack of research on the effect of the temperature on the marinating process, and on the cathepsin activity, composition, and antioxidant properties of the released peptides. Until now, the effect of the temperature has only been studied as a factor in regulating the microbiological stability of marinated fish [16].

In the case of salted whole herring, in which digestive proteases with a different specificity to cathepsins are active, the content of only a few peptides in the brines is known [17]. These brines have demonstrated significant iron(II)-chelating activity, reducing power, and radical-scavenging activity; however, in addition to the peptides, the brines contain phenolic compounds from spices, which also exhibit strong antioxidant activity.

Therefore, the aim of this research was to determine the effect of different temperatures of the marinating process on the main indicators of the marinade's quality, on the cathepsin activity, and on the profile and antioxidant activity of the peptides present in both the meat and the brine.

## 2. Materials and Methods

### 2.1. Raw Materials and Marinating Process

Atlantic herring was purchased as frozen skinless fillets (4/8 size) in 12 kg blocks packed in plastic bags stored at $-18$ °C. Prior to marinating, the frozen fillets in a bag were thawed in continuous circulation of water at $10 \pm 2$ °C. One kg of fillets was placed in 2 L glass jars, and then filled with the solution of marinating brine, including 4.5% acetic acid and 6% NaCl. The fish to brine ratio was 1.5:1.0 ($w/w$). The process of marinating was conducted over 7 days, at $2 \pm 0.1$ °C, $7 \pm 0.1$ °C, and $12 \pm 0.1$ °C. Non-iodinated rock salt and 10% spirit vinegar were used to prepare the marinating brine.

### 2.2. Mass, Salt, Total Acidity, pH, Moisture, and Lipid Content Analysis

Prior to analysis, the marinated fillets and brine were transferred to a large sieve for 3 min, until the liquid and solid fractions were fully separated, and then weighed. Six fillets were randomly chosen for sensory and texture analyses, and the remaining marinated fillets and raw material were minced. The pH value was determined using a digital pH meter (F20, Mettler Toledo, Columbus, OH, USA), in water extract (1:5, *w:v*). The moisture (no. 950.46B) and total lipids (no. 960.39) in the meat, the total acidity (no. 935.57), and the salt (no. 937.09) content in the meat and in the brine were determined using standard AOAC analytical techniques.

### 2.3. Total Nitrogen and Non-Protein Nitrogen Fractions

Total nitrogen was assayed using the Kjeldahl method. In 5% TCA extracts from the meat and brine, we determined: (i) the non-protein nitrogen using the Kjeldahl method (AOAC no. 940.25); and (ii) the protein hydrolysis products of peptides fraction (PHP(R)), and the free amino acids fraction (PHP(A)), using the Lowry method, with modification by Kołakowski [18].

### 2.4. Cathepsins Activity

The activity of aspartyl cathepsin (D + E) and cysteine cathepsins B and L against Mca-GKPILFFRLK(Dnp)-r-NH2, with and without pepstatin-A, Z-RR-MCA, and Z-FR-MCA, respectively, was measured according to Szymczak [4]. One unit of cathepsin activity (U) was defined as 1 fluorescence unit change per minute at 37 °C. All chemicals were purchased from Peptide Institute Inc. (Osaka, Japan).

### 2.5. Sensory Assessment

The marinated herring fillets were analyzed by sensory profiling, performed by a sensory panel of seven people trained according to ISO 8586:2023, using a five-point unstructured scale, with 0.1 point accuracy anchored at their extremes, with minimum and maximum degrees of acceptance (ISO 11035). A higher note signifies better texture attributes (0 points for the worst/extremely disliked, 5 points for the best/extremely liked). Briefly, three skinned fillets from each sample were served on porcelain trays. Each assessor was cut into three pieces (2–3 cm width each), one from each fillet to test. The tested area of the fillets ranged from 2/10 to 6/10 fillet, with the length measured from the head side. The evaluations were performed in separate cubicles, under daylight and at ambient temperature. The assessors used water and flat bread to cleanse their palate between samples.

### 2.6. Texture Profile Analyses (TPA)

The TPA hardness was determined in three fillets from each sample, using a TAXTplus Texture Analyzer (Stable Micro Systems, Godalming, UK). The test included two-fold compression using a cylindrical sonde P10 (10 mm diameter), with sample deformation up to 50% of height at the speed of 5 mm·s$^{-1}$. The course of the test was recorded as curves representing changes of force in time. The tests were conducted for each fillet separately (in 3–4 repetitions each). The tested area of the fillets ranged from 2/10 to 6/10 fillet, with the length measured from the head side.

### 2.7. ISP Extracts of Meat and Brine

The meat and marinating brine were subjected to protein removal using isoelectric solubility and precipitation (ISP). Briefly, minced fillets were mixed with water (1:5, *w:v*) for 15 min at pH 12.0, and centrifuged (10 min, 9000× *g*, 4 °C), the precipitate was discarded, and the supernatant was adjusted to pH 5.3 for 15 min; centrifuged again to discard the precipitate, and the supernatant was adjusted to pH 7.0 to obtain ISP-meat-extract. In turn, the marinating brine was filtered through 20 μm, adjusted to pH 2.0 and, after 15 min, was centrifuged to remove the precipitate; the resulting supernatant was adjusted to pH 11.0 for 15 min, centrifuged and the supernatant was adjusted to pH 7.0 to obtain ISP-brine-extract.

### 2.8. Antioxidant Activity of Protein Hydrolysis Products

Extracts of meat and brine obtained using the ISP method were subjected to ultrafiltration (UF) into fractions >10, 5–10 and <5 kDa using 10 and 5 kDa centrifuge filters (Amicon-Ultra, Merck Millipore, Ireland). A ferric reducing antioxidant power (FRAP) assay was measured as the reduction of ferric tripyridyltriazine complex ($Fe^{3+}$-TPTZ). The 5 µL of unfractionated ISP extract and its UF fractions were mixed with 150 µL of TPTZ and, after 30 min, the absorbance was measured at 593 nm. We used $FeSO_4$ as a positive control, and to obtain a standard curve. One unit [1 U] of FRAP was expressed in mM $FeSO_4$ reduced by 100 g of meat or 100 mL brine in 30 min. The radical-scavenging activity (RSA) was determined against ABTS radical cation ($ABTS^+$), according to Ginger at al. [17]. Ten µL of sample was mixed with 190 µL ABTS working solution and read after 5 min at room temperature at 734 nm. A DPPH free-radical-scavenging-capacity assay was measured by mixing 50 µL of DPPH solution (0.1 mM in methanol) with 150 µL of sample solution and, after 30 min at room temperature, the absorbance was measured at 517 nm. Trolox (2 mM, in methanol) was used as a positive control and a standard curve in the RSA and ABTS assays. One unit [1 U] of RSA or DPPH was expressed in µM Trolox, reduced by 100 g of meat or 100 mL of brine in 5 min. All assays were measured on a microplate spectrophotometer (Synergy 2 Multi-Mode Microplate Reader, BioTek® Instruments, Inc., Winooski, VT, USA). As the control, distilled water was used instead of the sample.

### 2.9. Lyophylization Process

The ISP extracts of meat and brine were freeze-dried in a freeze-dryer (CHRIST BETA, Martin Christ, Germany) without any protective medium. The frozen ISP extracts were placed in a tray and lyophilized. As the first step of the process, the samples were loaded onto a shelf at 0 °C for 2 h and cooled to −35 °C (7 h). The primary drying was carried out in the temperature range −35 °C to 0 °C (24 h). Secondary drying was performed at 0 °C to 20 °C/11 h. The samples were unloaded at 20 °C and stored 1 week in Falcon tubes in vacuum bags before the analyses.

### 2.10. Chromatography and Mass Spectrometry of ISP Extracts

The brine and meat extract samples were prepared using solid-phase extraction (SPE), according to Nagai et al. [19]. The Supelclean LC-18 SPE Tube (45 µm, 60 Å, 3 mL) columns by Supelco (Bellefonte, PA, USA) was used. Samples were dissolved in 0.1% ($v/v$) of trifluoroacetic acid (TFA) in HPLC-suitable water (Sigma Aldrich, St. Louis, MO, USA, cat. no. 270733). The columns were initially conditioned using 6 mL 0.1% ($v/v$) TFA in HPLC-suitable acetonitrile (ACN) gradient grade (Sigma Aldrich, cat. no. 34851) and equilibrated using 6 mL 0.1% ($v/v$) TFA in water. Samples with a flow rate of 1–2 drops per second were loaded on columns, and then washed five times using 3 mL portions of 0.1% ($v/v$) of (TFA) in water. Elution was performed using 5 mL of 0.1% TFA in acetonitrile. Eluates were evaporated using a vacuum concentrator (Martin Christ, Osterode am Harz, Germany). The remaining pellets were dissolved in water, lyophilized, and stored at −86 °C before further analysis.

Samples were dissolved in 300 µL of buffer with pH 6.6, containing 0.1 M BIS-TRIS and 4 M urea. Samples were reduced by the addition of 20 µL of 2-mercaptoethanol, followed by shaking for 1 h at room temperature, $1000 \times min^{-1}$ (ThermoMixer®, Eppendorf, Hamburg, Germany). The reaction was stopped by the addition of 680 µL of 6 M urea in water (pH 2.2 adjusted using TFA), and samples were centrifuged (10 min, 4 °C, $10,000 \times g$). The final concentration was $1 mg \times mL^{-1}$ for brine samples, and $3 mg \times mL^{-1}$ for meat samples [20].

RP-HPLC analyses were carried out using the Shimadzu (Tokyo, Japan) assembly-containing controller CBM-20A, the DGU-20A5 degasser, the SIL-20AC HT autosampler, two LC-20AD pumps, the CTO-10AS thermostat, and the SPD-M20A photodiode array detector. The assembly was equipped with the Jupiter Proteo (Phenomenex, Torrance, CA, USA) C18 column, 250 × 2 mm, with a particle size of 5 µm, and pore size of 300 Å. The mobile phase consisted of solvents A and B—0.075% ($v/v$) TFA in water, and 0.1% ($v/v$) TFA in ACN, respectively, to minimize baseline drift [21]. The following gradient of solvent B was applied: analysis: 5–20% from 0.00 to 30.00 min; 20–70% from 30.01 to 60.00 min; column washing: from 70–100% from 60.01 to 61.00 min; 100% from 61.01 to 75.00 min; column equilibration: 100–5% from 75.00 to 76.00 min; 5% from 76.01 to 90.00 min. The injection volume was 50 μL, flow rate: 0.2 mL $\times$ min$^{-1}$, and column temperature: 30 °C. Chromatograms were acquired at wavelength 220 nm. Data were collected and processed using Lab Solution (LC Solution) software (Shimadzu). RP-HPLC analyses were replicated twice.

Samples were reduced using the same procedure as previously. Before analysis, samples were purified using C18 ZipTips (Merck Millipore, Burlington, VT, USA) with bed capacity of 0.6 μL, according to provider instruction. The resin was washed twice using 10 μL ACN (wetting solution), and twice using 10 μL of 0.1% (*v*/*v*) TFA in water (equilibration solution). The samples were loaded by passing them ten times through ZipTip. Then, the ZipTip was washed five times with equilibration solution. Finally, 1.5 μL of 50% acetonitrile in water with 0.1% TFA addition (*v*/*v*) in clear tubes was aspirated and dispensed through the ZipTip for sample elution.

Matrix-Assisted Laser Desorption Ionization–Time of Flight (MALDI-ToF) mass spectrometry was carried out using MALDI-7090 (Kratos, Shimadzu Group, Manchester, UK), according to provider instruction. Samples were loaded onto target (Kratos) using the matrix pre-coated targets method according to the provider instruction, as follows: 0.5 μL of matrix solution containing 5 mg/mL of sinapinic acid (Sigma-Aldrich) in ACN containing 0.1% TFA for proteins analysis, was spotted and left for a few seconds. The surplus was removed using a pipette and left to dry. Next, 0.5 μL of desalted sample was spotted and, immediately, 0.5 μL of matrix solution was added. The spotted target plate was left for several minutes at room temperature, until samples were dried. Next, the target was inserted into the instrument. The TOF-mix kit (LaserBio Labs, Valbonne, France) was used for internal calibration. The samples were analyzed using positive linear mode within the range of 100–300 000 Da (proteins). The 20 laser shots per sample with a frequency of 1000 Hz were used to generate each mass spectrum. The number of profiles was 109. Laser power of 140 arbitrary units for proteins analysis was used. The spectra were processed using the MALDI Solutions v. 2.6 software (Kratos-Shimadzu).

### 2.11. SEM Analysis

Lyophilized ISP extracts of meat and brine were analyzed using a scanning electron microscope (SEM). The microscopic analysis was performed using a Vega 3 LMU microscope (Tescan, Brno-Kohoutovice, Czech Republic). The experiments were necessary to compare the morphology of the lyophilizate particles of meat and brine. An analysis was performed at room temperature with a tungsten filament, and an accelerating voltage of 10 kV was used to capture SEM images. All specimens were viewed from above.

### 2.12. Statistical Analysis

If not mentioned, all analyses were performed in three replications. Results were analyzed statistically (Statistica 13.1, Tulsa, OK, USA) using one-way analysis of variance (ANOVA), $p$ value was set at 0.05, and the differences between treatments were examined using the post hoc Tukey's honestly significant difference test ($p < 0.05$) [22].

## 3. Results and Discussion

### 3.1. Mass Yield, Moisture, Salt, and pH of Meat

Atlantic herring, the most used in marinade production, was used in the study, and contained $17.5 \pm 0.5\%$ lipids, and an average activity of cathepsins D, B, and L (Table 1). The fillets were from fresh fish, as they had a pH of 6.52, and a low content of PHB(R) and PHB(A) fractions. The herring were probably caught during the feeding season, before spawning. A higher marinating temperature had a significant effect on the lower mass yield of the marinades, which decreased according to an exponential function by 0.5–1.1 percentage points for each 1 °C (Table 1). The herring marinated at 2 °C (89.5%) had the highest mass yield, followed by the herring marinated at 7 °C (87.2%), and the lowest using 12 °C (81.5%). The lower mass of marinades was the result of several phenomena simultaneously. The higher temperature increased the acid and salt content of the marinades, so the higher moisture may have been due to the higher pH of the meat (Table 1), but the differences were not so great as to significantly reduce the yield of the fatty herring marinades [14]. In addition, the small differences (about 0.1%) in the acid and salt content of the marinades were not related to their content in the brine (Table 2). Therefore, the higher acidity of the brine at a higher temperature may be due to a greater diffusion of acidic substances (free amino acids, free fatty acids) from the meat into the brine, or a greater proliferation of lactic acid bacteria. Shenderyuk and Bykowski [1] found an effect of higher temperature (3 °C vs. 13 °C) on faster salt and acid penetration, mainly in the first days of marinating, and lower marinating yield, mainly using a higher (2:1 vs. 1.5:1) proportion of fish to brine. In turn, Rodger et al. [23] found that an increase in marinating temperature for herring from 2.3 °C to 5.0 °C accelerated acid penetration (from 75 h to 51 h) more than salt penetration (from 110 h to 94 h), while an increase from 5.0 °C to 10.3 °C accelerated salt penetration (from 94 h to 45 h) more than acid penetration (from 51 h to 48 h).

**Table 1.** Basic composition, nitrogen fractions content, and cathepsin activity in the raw meat and marinated meat; TN—total nitrogen, NPN—non-protein nitrogen, PHP(R)—peptides fraction, PHP(A)—free amino acids fraction; [abcd] means within the same row with different lowercase letters differing significantly.

| Analyses | Fresh Herring | Marinating Temperature | | |
| --- | --- | --- | --- | --- |
| | | 2 °C | 7 °C | 12 °C |
| Mass yield (%) | 100 | 89.5 | 87.2 | 81.5 |
| pH | 6.52 ± 0.01 [a] | 4.15 ± 0.01 [b] | 4.16 ± 0.01 [b] | 4.17 ± 0.01 [b] |
| NaCl (%) | 0.30 ± 0.01 [d] | 2.41 ± 0.01 [b] | 2.43 ± 0.01 [b] | 2.54 ± 0.01 [a] |
| Total acidity (%) | 0.22 ± 0.01 [d] | 1.56 ± 0.02 [c] | 1.62 ± 0.01 [b] | 1.67 ± 0.01 [a] |
| Moisture (%) | 66.6 ± 0.29 [a] | 59.9 ± 0.6 [d] | 62.0 ± 0.3 [b] | 61.3 ± 0.3 [c] |
| TN (g/100 g) | 2.30 ± 0.00 [b] | 2.45 ± 0.01 [a] | 2.43 ± 0.03 [a] | 2.44 ± 0.02 [a] |
| NPN (mg/100 g) | 233.6 ± 3.16 [b] | 218.8 ± 1.6 [c] | 237.9 ± 1.6 [b] | 280.5 ± 0.0 [a] |
| PHP(R) (mg/100 g) | 6.4 ± 1.8 [d] | 208.6 ± 3.5 [c] | 228.3 ± 2.3 [b] | 252.4 ± 6.8 [a] |
| PHP(A) (mg/100 g) | 12.9 ± 0.2 [d] | 35.2 ± 0.9 [c] | 42.3 ± 0.2 [b] | 55.2 ± 0.4 [a] |
| Cathepsin D (U) | 3259 ± 32 [d] | 6077 ± 22 [c] | 6315 ± 45 [a] | 6118 ± 22 [b] |
| Cathepsin B (U) | 214.7 ± 1.2 [a] | 136 ± 1.5 [d] | 190 ± 4.8 [b] | 168 ± 2.3 [c] |
| Cathepsin L (U) | 390 ± 3.2 [a] | 188 ± 1.6 [b] | 168 ± 0.5 [c] | 131.1 ± 2.8 [d] |

**Table 2.** Basic composition, nitrogen fractions content, and cathepsin activity in the marinating brine; TN—total nitrogen, NPN—non-protein nitrogen, PHB(R)—peptides fraction, PHB(A)—free amino acids fraction; [abc] means within the same row with different lowercase letters differing significantly.

| Analyses | Marinating Temperature | | |
| --- | --- | --- | --- |
| | 2 °C | 7 °C | 12 °C |
| Mass yield (%) | 111.3 | 114.6 | 125.2 |
| pH | 4.28 ± 0.00 [a] | 4.27 ± 0.00 [b] | 4.26 ± 0.00 [c] |
| NaCl (%) | 1.73 ± 0.00 [a] | 1.73 ± 0.00 [a] | 1.73 ± 0.00 [a] |
| Total acidity (%) | 1.98 ± 0.00 [b] | 1.98 ± 0.00 [b] | 2.04 ± 0.00 [a] |
| TN (mg/100 mL) | 355.4 ± 4.18 [c] | 396.3 ± 1.36 [b] | 444.9 ± 4.7 [a] |
| NPN (mg/100 mL) | 326.4 ± 0.8 [c] | 362.2 ± 0.0 [b] | 429.9 ± 0.8 [a] |
| PHB(R) (mg/100 mL) | 194.0 ± 3.5 [b] | 200.9 ± 4.2 [ab] | 200.9 ± 3.4 [a] |
| PHB(A) (mg/100 mL) | 33.1 ± 0.2 [c] | 41.4 ± 0.0 [b] | 61.7 ± 0.9 [a] |
| Cathepsin D (U) | 1817 ± 52 [a] | 1196 ± 32 [b] | 609 ± 41 [c] |
| Cathepsin B (U) | 1194 ± 30 [c] | 1363 ± 21 [b] | 1463 ± 58 [a] |
| Cathepsin L (U) | 1411 ± 48 [b] | 1689 ± 35 [a] | 1374 ± 22 [b] |

### 3.2. Nitrogen Indicators of Ripening

The higher marinating temperature did not determine the total protein content, but significantly increased the NPN (326 vs. 430 mg/100 g), peptide fraction (194 vs. 201 mg/100 g), and amino acid content (33 vs. 62 mg/100 g) in marinated meat (Table 1). Zamojski [24] found that the ripening of herring at 18 °C was 35% faster than at 10 °C, obtaining 500 and 440 mg of NPN in 100 g of marinated meat, respectively. In our study, increasing the temperature by 10 °C increased the content of the amino acid fraction 2–3 times more than that of the peptide fraction, whereby the peptide content increased by the same amount in both the 2–7 °C and 7–12 °C ranges, while the amino acid content increased twice as much in the 7–12 °C range as in the 2–7 °C range. On the one hand, this may indicate that exopeptidases are more sensitive to temperature change than endopeptidases, and therefore an increase in temperature is more conducive to the release of amino acids. On the other hand, a higher exopeptidase activity depends on more peptides being released by higher endopeptidase activity. A more accurate assessment of protein proteolysis in marinades requires analysis of both the meat and the brine, into which large amounts of nitrogenous substances diffuse [6]. The use of a higher temperature significantly increased the NPN content of the brine by 10.4 mg for every 1 °C, while the peptide and amino acid fractions increased by 1.4 mg and 2.9 mg, respectively (Table 2). Thus, using a higher marinating temperature, the NPN and amino acid content increased more in the brine than in the meat, while the opposite was the case for peptides (Tables 1 and 2). The sum of the nitrogen fractions in the meat and the brine showed that the higher temperature reduced the proportion of amino acids in the meat to the brine from 1.59 to 1.34, while for peptides, the proportion increased from 1.61 to 1.88. This means that the higher temperature was more conducive to amino acid loss than peptide loss, which may be related to the described greater increase in the amino acid content of the meat. The results may also indicate that an increase in temperature does not have a major effect on endopeptidase activity, in contrast to an increase in exopeptidase activity, confirming Zamojski's [24] results that high temperature caused an increase in mainly free amino acids. Both the quantitative and qualitative composition of the amino acids and peptides released during cathepsin activity have important effects on the sensory quality of marinades [7,14].

### 3.3. Cathepsins Activity

Increasing the temperature of the marinating process from 2 to 7–12 °C increased cathepsin D activity by only 1–4% (Table 1). In the case of cathepsin B, activity increased by 23–40% (most at 7 °C), while the highest increase in cathepsin L activity of 28–43% occurred in marinades ripened at the lowest temperature. Refrigeration temperature strongly decreases protease activity, but fish cathepsin activity remains at around 20% [25]. Even the application of negative temperatures (superchilling) does not inhibit cathepsin activity. Cooling close to 0 °C was found to contribute to an increase in cathepsin L activity in the sarcoplasm, but also a decrease in activity in the cellular structure, because of the tight tangle of more muscle fibers [26]. Szymczak [4] noted that a decrease in temperature from 12 to 5 °C reduced the total proteolytic activity (GPA) in marinades, and the proportion of cathepsin D in GPA, particularly up to day 7. The cathepsins tested are endopeptidases, with cathepsin B (B1) also showing dipeptidase activity, which may be related to its higher activity in meat and brine under higher marinating temperatures. In optimal conditions, cathepsin D releases very short tetrapeptides and, together with cathepsin B, acts on myofibrillar proteins, while cathepsin L and, to a lesser extent, cathepsin B are specific to collagen, which has a greater effect on meat texture than myofibrillar proteins.

In contrast, the effect of temperature on the activity of cathepsins D and L in the brine was different from that in the meat. At 12 °C, cathepsin D activity was 67% lower compared to 2 °C (Table 2). Cathepsin B activity increased from 1194 to 1463 U when the temperature was increased by 10 °C, while cathepsin L had the highest activity at 7 °C. Cathepsins are present in the brine because of their diffusion from the muscle tissue, which increases using frozen-thawed raw material and a high concentration of salt [5].

### 3.4. Sensory Assesment of Marinades

The sensory quality of the marinades ripened at higher temperatures was evidently the best, but the increase in the overall sensory acceptability score from 4.45 to 4.84 points was not statistically significant (Table 3). The fillets marinated at 12 °C reached consumer ripeness, while the other samples only reached technological ripeness, and included fragments of less-ripened tissue, which significantly reduced the texture score from 4.91 to 4.37–4.42 points. In addition, the TPA hardness analysis showed a significant increase in hardness by 2–3 N (20–30%) for marinades obtained at the lowest temperature (Table 3), indicating a reduced activity of cathepsins D and B (Table 1). The marinades obtained at 12 °C had the highest rating for taste and odor, which were harmonized, and characteristic of ripened marinades, in contrast to the other marinades, which were dominated by the taste of acetic acid and salt. This indicates that the higher marinating temperature favored the activity of alanine aminopeptidase and cathepsins A and C, which release sweet amino acids. The lower temperature and slower ripening only had a beneficial effect on the appearance of the marinades (Table 3), mainly for less dissection of the myomeres throughout the length of the fillets, and their bright color.

**Table 3.** Sensory assessment and TPA hardness of marinated meat; [ab] means within the same row with different lowercase letters differing significantly.

| Analyses | Marinating Temperature | | |
| --- | --- | --- | --- |
|  | **2 °C** | **7 °C** | **12 °C** |
| Appearance | 4.95 ± 0.09 [a] | 4.91 ± 0.10 [a] | 4.79 ± 0.23 [a] |
| Odor | 4.60 ± 0.51 [a] | 4.65 ± 0.41 [a] | 4.79 ± 0.25 [a] |
| Flavor | 4.30 ± 0.42 [a] | 4.39 ± 0.53 [a] | 4.78 ± 0.23 [a] |
| Texture | 4.37 ± 0.35 [b] | 4.42 ± 0.23 [b] | 4.91 ± 0.07 [a] |
| Overall sensory evaluation | 4.45 ± 0.40 [a] | 4.51 ± 0.37 [a] | 4.84 ± 0.17 [a] |
| TPA hardness | 10.49 ± 0.95 [a] | 8.03 ± 0.57 [b] | 7.40 ± 1.14 [b] |

### 3.5. Chromatography and Mass Spectrometry of ISP Extracts

The RP-HPLC chromatograms of samples of brine and meat ISP extracts are presented in Figure 1 in the main text, and in Figures S1–S6 in the Supplementary Materials. Figure 1 presents chromatograms within the time interval up to 60 min. Chromatograms presented in this figure show peaks with retention times between 15 and 60 min. Entire chromatograms are presented in the Supplementary Materials (Figures S1–S6). Our resignation from the interpretation of the injection peaks may appear controversial, due to the fact that short and hydrophilic peptides may be not retained at the column, but on the other hand, absorbance at the injection peak apex is c.a. 4, and is too high to provide predictable dependence between the composition of the solution and the output from the detector. The peaks at the start of the chromatogram contain many low-molecular compounds belonging to various classes. It is impossible to estimate the contribution of peptides in this fraction. In our previous works [27–29], interpretation of the area of injection peaks did not provide valuable information, in contrast to fractions with retention times exceeding 15 min. A time interval exceeding 60 min contains so-called system peaks, artifacts that are useless from the point of view of information about the sample [30]. The sample concentration of the meat extracts was three times higher than the concentration of the brine samples. The retention times of peptides depend on their hydrophobicity and length. More hydrophobic and longer chains have longer retention times than those that are shorter and more hydrophilic. Reversed-phase HPLC does not provide a simple correlation between retention time and molecular mass [31,32], but experiments using proteins from herring [33] and other sources [27,28,34–37] may serve as examples, illustrating the fact that retention times of proteolysis products are usually shorter than those of entire proteins. Thus, RP-HPLC may serve as a tool for the comparison of the proteolysis degree if it is not determined directly.

Chromatograms of samples with an unknown composition may be subjected to simple interpretation by comparison of the relative area of peaks within various retention time intervals [27–29]. The relative areas of peaks within the intervals 15–34.99 min (shorter) and 35–60 min (longer) are presented in Table 4. It is possible to discriminate the chromatographic patterns of the brine and meat extracts. The brine extracts contain more material within a shorter retention time interval, compared with the meat extracts. This suggests the predominant extraction of hydrophilic or short peptides during marinating. The extracts from meat, obtained after marinating, contained a relatively higher content of hydrophobic or long peptides than brine. The brine, after marinating at a temperature of 12 °C, contained larger amounts of compounds forming peaks within a shorter retention time segment, than other brines did. This may indicate a larger extent of hydrolysis of herring proteins at 12 °C than in lower temperatures. This agrees with previous results revealing that proteolysis progress causes an increase in the relative area of peaks within shorter retention time intervals [27,28].

**Table 4.** Relative areas of peaks in the chromatograms of brine and meat ISP extracts, measured using RP-HPLC.

| Sample | Relative Area of Chromatographic Peaks (%) [1] | |
| --- | --- | --- |
| | Retention Time Interval 15–34.99 min | Retention Time Interval 35–60 min |
| Meat, 2 °C | 52.32 | 47.68 |
| Meat, 7 °C | 60.24 | 39.76 |
| Meat, 12 °C | 51.41 | 48.59 |
| Brine, 2 °C | 69.72 | 30.28 |
| Brine, 7 °C | 71.22 | 28.78 |
| Brine, 12 °C | 93.36 | 6.63 |
| All meat extracts [2,3] | 54.66 ± 4.86 [a] | 45.34 ± 4.86 [a] |
| All brines [2,3] | 78.10 ± 13.23 [b] | 21.90 ± 13.24 [b] |

[1] Area of peaks within the retention time interval between 15 and 60 min was considered as 100%; [2] Mean ± SD; [3] [a,b] means within the same row with different lowercase letters differing significantly.

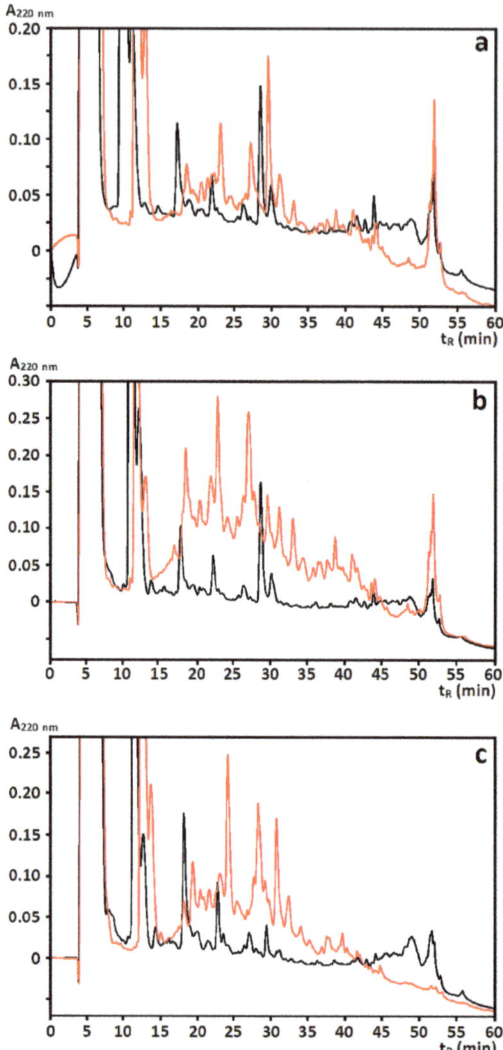

**Figure 1.** RP-HPLC chromatograms of (black line) meat ISP extracts, and (red line) brine ISP-extracts from marinades ripened at (**a**) 2 °C, (**b**) 7 °C, and (**c**) 12 °C.

The MALDI-ToF mass spectra of brine and meat ISP extracts within the $m/z$ (mass to charge ratio) range up to 10 kDa are presented in Figure 2. Entire MALDI-ToF mass spectra are presented in Figures S7–S12 in the Supplementary Materials. MALDI-ToF serves mainly as a tool for the qualitative analysis of peptides, proteins, and their mixtures [38,39]. Quantitation of individual proteins or peptides, although possible, is difficult, and requires many repeats to obtain a significant correlation between the amount of compound and the peak intensity [38]. On the other hand, mass spectrometry measurements are not affected by hydrophobic or any other interactions (in contrast to chromatography) and provide a precise mass to charge ratio. The last value measured using MALDI-ToF mass spectrometry should also be interpreted carefully. Proteins form not only singly protonated ions $(M + H)^+$, but also may be doubly protonated or appear as aggregates containing two or three analyte molecules with one proton [39–41].

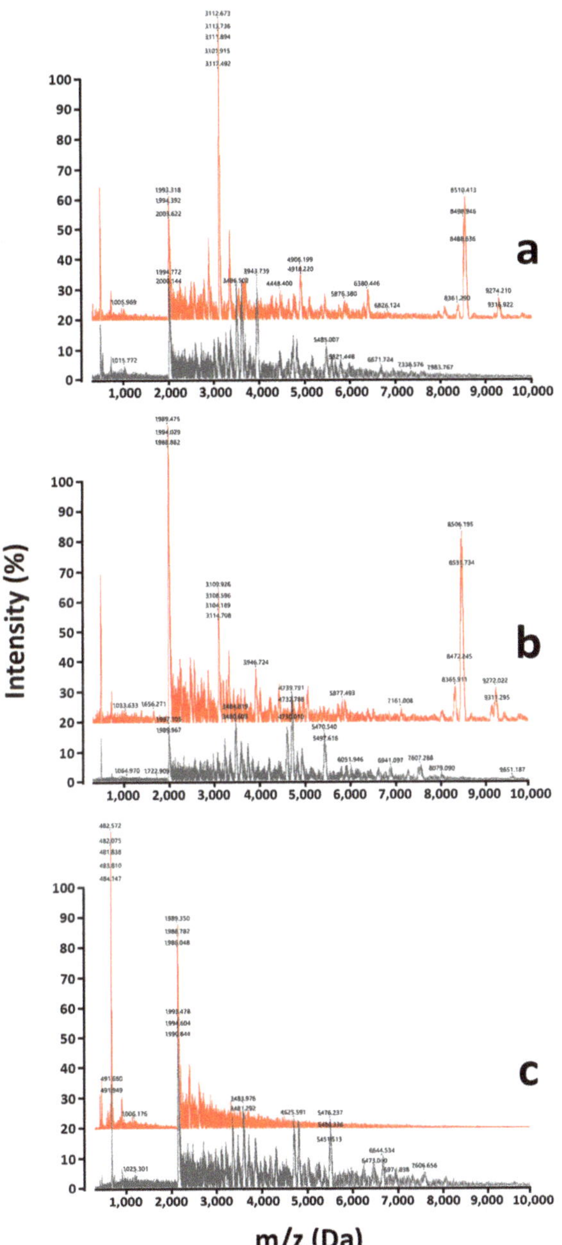

**Figure 2.** MALDI-ToF mass spectra of (black line) meat ISP extracts and (red line) brine ISP extracts from marinades ripened at (**a**) 2 °C, (**b**) 7 °C, and (**c**) 12 °C.

Taking the above into account, we decided to interpret the MALDI-ToF mass spectra in a similar way to chromatograms, i.e., to compare relative areas of peaks within particular $m/z$ segments. For the purpose of comparison, a few samples were obtained and analyzed in the same way we could use relative peak areas within defined $m/z$ intervals. We did not utilize $m/z$ lower than 500 Da. The matrix (in this experiment, sinapinic acid) clusters predominantly possess masses below 500 Da, although they may also reveal higher masses [42,43]. Relative areas of peaks within various $m/z$ segments are presented in Table 5.

**Table 5.** Relative areas of peaks on the MALDI-ToF MS spectra of meat and brine ISP extracts.

| Sample | Relative Area of Peaks (%) [1] | | | | | |
|---|---|---|---|---|---|---|
| | $m/z$ Range 0.5–2 kDa | $m/z$ Range 2–4 kDa | $m/z$ Range 4–6 kDa | $m/z$ Range 6–8 kDa | $m/z$ Range 8–10 kDa | $m/z$ Range >10 kDa |
| Meat, 2 °C | 14.99 | 47.31 | 22.29 | 9.68 | 3.47 | 2.26 |
| Meat, 7 °C | 11.95 | 43.71 | 26.21 | 12.03 | 3.74 | 2.36 |
| Meat, 12 °C | 14.08 | 45.40 | 23.01 | 11.27 | 3.59 | 2.64 |
| Brine, 2 °C | 14.95 | 44.64 | 21.11 | 9.05 | 5.89 | 4.36 |
| Brine, 7 °C | 14.57 | 45.83 | 18.04 | 8.28 | 5.92 | 7.36 |
| Brine, 12 °C | 18.26 | 60.88 | 14.49 | 4.48 | 1.31 | 0.59 |
| All meat extracts [2,3] | 13.68 ± 1.56 [a] | 45.47 ± 1.80 [a] | 23.83 ± 2.09 [a] | 10.99 ± 1.20 [a] | 3.60 ± 0.14 [a] | 2.42 ± 0.20 [a] |
| All brines [2,3] | 15.93 ± 2.03 [a] | 50.45 ± 9.05 [a] | 17.88 ± 3.31 [b] | 7.27 ± 2.45 [b] | 4.37 ± 2.66 [a] | 4.10 ± 3.39 [a] |
| Ratio meat to brine, 2 °C | 1 | 1.06 | 1.06 | 1.07 | 0.59 | 0.52 |
| Ratio meat to brine, 7 °C | 0.82 | 0.95 | 1.45 | 1.45 | 0.63 | 0.32 |
| Ratio meat to brine, 12 °C | 0.77 | 0.75 | 1.59 | 2.52 | 2.74 | 4.47 |

[1] Area of all peaks with mass to charge ($m/z$) ratio exceeding 500 Da are considered as 100%; [2] Mean ± SD; [3] Difference between values denoted by the same letter in column is statistically insignificant at $p < 0.05$ as judged by Student $t$-test; [ab] means within the same row with different lowercase letters differing significantly.

Some results of the MALDI-ToF MS confirmed our interpretation of the RP-HPLC chromatograms. The brine after marinating at 12 °C contained the most peptides with $m/z$ below 4000 Da, and the least compounds with $m/z$ above 4000 Da, characteristic of polypeptides and intact proteins. This result is consistent with the RP-HPLC, revealing a lower amount of material with long retention times, compared to other samples. On the one hand, this suggests higher activity of endoproteases and exoproteases at the higher temperature of 12 °C, as compared with lower temperatures. On the other hand, temperature changes may affect not only proteolysis, but also the extraction of oligo- and polypeptides from meat to brine during marination. Changes in the brine composition cannot be thus considered as results of proteolysis only. The 2–4 kDa fraction dominated the meat and the brine, followed by 4–6 kDa, and 0.5–2 kDa (Table 5). An increase in the marinating temperature had no effect on the proportion of fractions > 8 kDa in meat, while a temperature of 7 °C favored a higher proportion of 4–6 and 6–8 kDa fractions, at the expense of 2–4 and 0.5–2 kDa fractions, than using 2 and 12 °C. All brines contained fewer polypeptides with a mass to charge ratio within the range 4–8 kDa than meat extracts did. There was no significant difference between the brine and meat extracts in their content of material within other segments of $m/z$, annotated in Table 5. In contrast, the meat to brine ratio for the fractions <4 kDa decreased with an increasing marinating temperature from 1.0–1.06 to 0.75–0.77, in contrast to fractions >4 kDa, where the meat to brine ratio changed from 0.52–1.06 to 1.59–4.47. This means that the higher marinating temperature promotes both higher protease activity, and a greater diffusion of peptides from the meat into the brine. Søtoft et al. [44], using the SDS-PAGE method, found that the brine was dominated by the 10–20 kDa protein fraction, but after membrane fractionation of the brine, a higher total amino acid content was found in the nanofiltration fraction than in the ultrafiltration fraction.

All samples contained only a slight amount of material with a high molecular mass ($m/z$ exceeding 10 kDa). This may suggest that the main compounds extracted to the brine during marinating, as well as during post-marinating extraction, were proteolysis products. RP-HPLC and MALDI-ToF mass spectrometry may be recommended as tools for monitoring proteolysis, as alternatives to polyacrylamide gel electrophoresis (SDS-PAGE), which is more commonly used for this purpose [45]. Both HPLC and MS may serve for simultaneous detection compounds in a broad range of molecular masses, from oligopeptides to intact proteins. SDS-PAGE serves as a method for the analysis of polypeptides and proteins, but not oligopeptides.

### 3.6. Antioxidant Activity

To further characterize the ripening process, three in vitro assays were used to determine the potential for antioxidant activity of protein hydrolysis products in meat and brine (Table 6). Marinades obtained at 2 and 12 °C showed the highest FRAP reduction capacity, 10% higher than using marinades at 7 °C. FRAP activity in meat was mostly from the 5–10 kDa or <5 kDa fraction for herring marinated at 2 or 12 °C, respectively. The ability to inhibit lipid peroxidation in unfractionated brine ISP extracts was 26.6–31.2 mM $FeSO_4$/100 mL, which was as high as in marinated meat 30.4–34.9 mM $FeSO_4$/100 g. FRAP activity in the brine was derived almost entirely equally from the two fractions 5–10 and <5 kDa. The fractionation of ISP extracts did not significantly affect the sum of FRAP-reducing capacities, indicating that fractionation does not alter the synergistic action of peptides of marinades and/or the proportion of lysine and sulfur-containing amino acids in peptides of different molecular weights [46].

**Table 6.** Antioxidant activity of ISP extracts from meat and brine before (unfractionated) and after fractionation (>10, 5–10, and <5 KDa); [a–prs] means within the same column with different lowercase letters differing significantly.

| Sample | Fraction | FRAP [U] | TEAC [U] | DPPH [U] |
|---|---|---|---|---|
| Meat, 2 °C | Unfractionated | 34.9 ± 3.9 [ab] | 135.7 ± 1.8 [a] | 281.1 ± 2.2 [f] |
|  | >10 kDa | 4.0 ± 0.0 [k] | 8.6 ± 0.2 [o] | 38.3 ± 1.7 [m] |
|  | 5–10 kDa | 20.3 ± 1.4 [d] | 78.5 ± 3.7 [d] | 70.7 ± 3.9 [k] |
|  | <5 kDa | 10.9 ± 0.9 [i] | 64.9 ± 5.1 [ef] | 168.5 ± 9.7 [h] |
| Meat, 7 °C | Unfractionated | 30.4 ± 0.8 [bc] | 106.9 ± 5.3 [c] | 312.7 ± 18.1 [e] |
|  | >10 kDa | 4.6 ± 0.5 [j] | 13 ± 0.8 [n] | 22.6 ± 1.4 [P] |
|  | 5–10 kDa | 26.4 ± 3.2 [c] | 61.9 ± 4.7 [fg] | 35.2 ± 0.5 [o] |
|  | <5 kDa | 10.5 ± 3.0 [i] | 55.7 ± 3.8 [g] | 113.7 ± 12.3 [i] |
| Meat, 12 °C | Unfractionated | 33.7 ± 2.0 [a] | 124.2 ± 3.8 [b] | 251.5 ± 14.3 [g] |
|  | >10 kDa | 4.3 ± 0.1 [j] | 6.9 ± 0.1 [P] | 36.2 ± 0.1 [n] |
|  | 5–10 kDa | 13.3 ± 1.5 [ghi] | 60.5 ± 3.2 [fg] | 41 ± 3.9 [m] |
|  | <5 kDa | 16.1 ± 0.9 [ef] | 71.1 ± 4.1 [de] | 138.9 ± 17.4 [ij] |
| Brine, 2 °C | Unfractionated | 31.2 ± 1.3 [ab] | 35.7 ± 0.1 [i] | 615.5 ± 38.9 b |
|  | >10 kDa | 1.8 ± 0.0 [l] | 1.7 ± 0.0 [r] | 35.2 ± 1.2 [no] |
|  | 5–10 kDa | 15.7 ± 0.9 [fg] | 29.3 ± 0.8 [k] | 128.6 ± 7.8 [ij] |
|  | <5 kDa | 14.1 ± 0.4 [h] | 24.0 ± 0.9 [m] | 325.3 ± 22.4 [e] |
| Brine, 7 °C | Unfractionated | 26.6 ± 1.5 [c] | 34.3 ± 0.3 [j] | 638.4 ± 19.3 [b] |
|  | >10 kDa | 1.6 ± 0.1 [l] | 1.5 ± 0.0 [s] | 57.3 ± 5.4 [l] |
|  | 5–10 kDa | 15.5 ± 0.6 [f] | 27.4 ± 0.9 [l] | 121.3 ± 1.3 [j] |
|  | <5 kDa | 17.6 ± 1.0 [e] | 27.5 ± 0.9 [l] | 410.6 ± 15.5 [c] |
| Brine, 12 °C | Unfractionated | 29.1 ± 1.0 [bc] | 36.6 ± 0.1 [h] | 679.9 ± 15.3 [a] |
|  | >10 kDa | 1.8 ± 0.1 [l] | 1.5 ± 0.0 [s] | 12.8 ± 1.2 [r] |
|  | 5–10 kDa | 20.3 ± 0.7 [d] | 34.9 ± 0.9 [ij] | 372.5 ± 9.3 [d] |
|  | <5 kDa | 18.3 ± 1.3 [de] | 27.8 ± 1.5 [kl] | 392 ± 8.8 [c] |

The antioxidant activity against the ABTS (the TEAC method) ranged from 107 to 136 μM TE/100 g of the marinades; the highest was in marinades ripened at 2 °C, and the lowest at 7 °C (Table 6). Using the TEAC method, activity was mainly formed by the 5–10 kDa fraction for marinades at 2–7 °C, and the <5 kDa fraction for marinades at 12 °C, similar to the FRAP method, but the differences in activity between these fractions were much smaller. In the brine, the antioxidant activity was 34.3–36.6 μM TE/100 mL; however, the sum of the activities of the three fractions was almost two times higher than the activity in the brine before fractionation. This may be due to (i) the release of substances with antioxidant properties, similar to the phenomenon of cathepsin release from lysosomes [5]; (ii) the removal of enzymes with oxidative activity [17], which remained in the >10 kDa fraction and reduced its activity almost completely (Table 6); and/or (iii) an increase in the synergistic effect of individual peptides after purification/fractionation. A synergistic effect, and a 68% increase in antioxidant activity, was also described for the cod hydrolysate and its RP-HPLC fractions [46].

The free-radical-scavenging capacity of the unfractionated ISP extract and its UF fractions were also tested by their ability to scavenge the stable DPPH radical. The DPPH radical scavenging activity in the meat was 251.5–312.7 µM TE/100 g, while in the brine, it was 615.5–679.9 µM TE/100 ml. (Table 6). The highest DPPH activity was using marination at 7 °C for the meat, and 12 °C for the brine. The DPPH activity in the meat consisted mainly of the <5 kDa fraction, while in the brine, in addition to the <5 kDa fraction, the 5–10 kDa fraction contributed significantly, especially as the marinating temperature became higher. Increasing the marinating temperature increased the DPPH activity in the unfractionated ISP extract of the meat more than the activity of the sum of the three UF fractions, while the opposite was the case for the brine.

Despite the methodological differences, similar relationships in activity between methods were found when herring was marinating in reused brine [8]. There was no statistically significant effect of non-protein nitrogen content, peptide and amino acid fractions, or cathepsins activity on antioxidant activity. The >10 kDa fraction contained mainly proteins and had the lowest antioxidant activity in the three in vitro analyses. This confirms that antioxidant activity is mainly possessed by peptides and free amino acids. Moreover, in the brine after the salting of the herring, higher antioxidant activity was shown in the <10 kDa fraction than in the >10 kDa fraction [17]. Similar relationships have been shown for the hydrolysis products of proteins of milk and plants. Girgih et al. [46] showed that the DPPH activity depended on the elution time (25–37 > 14–25 > 0–14 = 37–48 min), and the hydrophobic amino acid content, as more hydrophobic peptides interact with hydrophobic radicals. Similarly, in our study, the highest positive correlation ($r^2$ = 0.883) was found between the DPPH activity and the peak area with elution times of 15–35 min. In turn, the ABTS activity in the meat correlated ($-0.995$, $p$ = 0.02) with the proportion of the peptides fraction 4–10 kDa in the MALDI-Tof assay.

### 3.7. SEM Analyses

To investigate the effect of the marinating temperature of Atlantic herring on the microstructure of freeze-dried ISP extracts, SEM (200×) micrographs were determined. No cryoprotectants, which might modify the growth of the ice crystals during the freezing process, thus forming a large or small number of small ice crystals, were added to the ISP extracts before the freeze-drying. All the samples were freeze-dried in the same conditions, meaning that all differences that were visible on the SEM micrographs were caused by the marinating temperature, and not by the freeze-drying process. The results of scanning electron microscopy showed that the ISP extracts from meat contained oblong microstructures similar to myofibrils and showed the effect of the marinating temperature on differences in the three-dimensional microstructure. As shown in Figure 3a, the myofibrils ISP-extracted from the marinades at 2 °C were tightly connected, and arranged in parallel, and the spaces between the myofibrils that were observed indicated a well-organized structure of the herring muscle. The SEM image of the ISP extract from the marinades-7 °C demonstrated shorter myofibrils, which formed small aggregates of fibers that began to appear, indicating that the well-organized structure of the muscle tissue was destroyed (Figure 3b). The results of SEM of ISP extract from the marinades at 12 °C showed a loss of myofibril structure, visibly dense aggregates, and a low population of small cavities, unevenly distributed (Figure 3c). It is tempting to suggest that a lower ripening temperature decreased cathepsins activity, which to a lesser degree destroyed the myofibrils' structure, that can be extracted from meat to water, despite the precipitation process. The ISP technique does not obtain 100% efficiency in protein precipitation, and short myofibrils treated by proteases probably precipitated at a lower degree. The SEM analysis of the ISP extracts from the brine samples indicated that neither myofibril structure was observed. The globular particles and their aggregates (Figure 3d–f) visible on SEM micrographs of the brine samples were larger than the myofibrils visible on the micrographs of the meat samples, suggesting that these globular particles were NaCl particles. The globular proteins extracted from the myofibrils (cut or dissolved) would have been lower than the fibers they were extracted from. Additionally, seven days of marinating in pH 4.3, close to the isoelectric point of herring proteins, probably precipitated most of the proteins diffused to the brine.

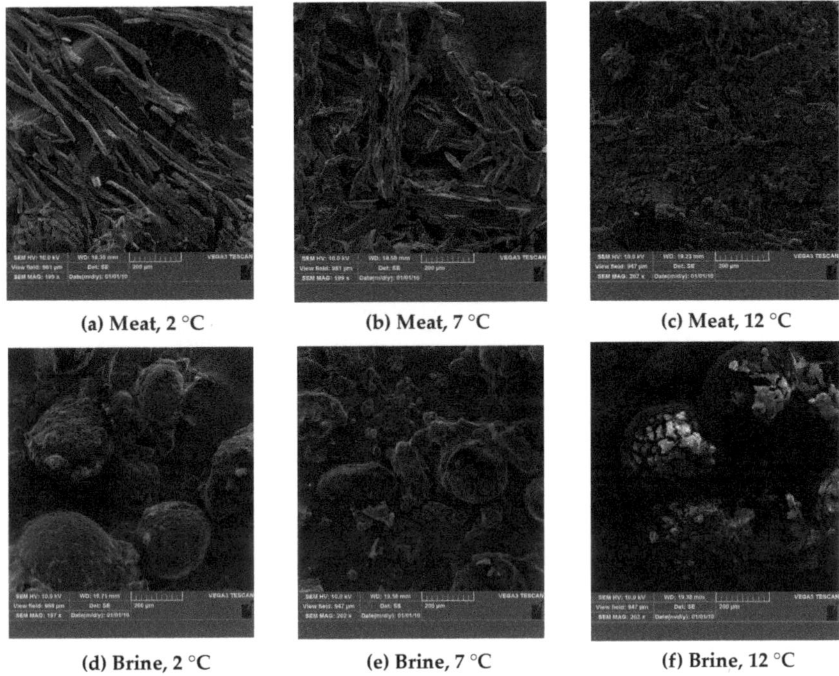

(a) Meat, 2 °C     (b) Meat, 7 °C     (c) Meat, 12 °C

(d) Brine, 2 °C     (e) Brine, 7 °C     (f) Brine, 12 °C

**Figure 3.** SEM images of lyophilized (**a**–**c**) meat and (**d**–**f**) brine ISP extracts.

## 4. Conclusions

The marinating temperature between 2 and 12 °C had a significant effect on the ripening of the marinades, and especially on the composition of the meat and the marinating brine. On the one hand, the use of the lowest temperature had advantages, such as the highest marinating yield, the lowest nitrogenous substance loss, the highest cathepsin L activity, the best surface appearance of the marinated fillets, and the highest antioxidant activity of the TEAC. On the other hand, the lowest temperature also had disadvantages, such as the slowest ripening dynamics, the lowest-rated texture of the marinades, the lowest cathepsin B and D activity, and the lowest content of protein hydrolysis products (PHP), especially free amino acids, which give the characteristic flavor of marinades.

The results show that the marinating brine contained high amounts of PHPs, which have strong antioxidant properties in vitro, as evidenced by their strong radical-scavenging and metal-binding activities. The fractionation of PHP released during the marinating of herring, on the basis of molecular weight, had a negative effect, by reducing antioxidant synergy with respect to DPPH in the meat, while it had a positive effect with respect to FRAP, ABTS, and DPPH activity in the brine. This shows that brine PHPs should be industrially fractionated and used separately in food technology. Chromatography and mass spectrometry may serve as tools for monitoring the effect of simultaneous proteolysis, and the extraction of peptides and proteins, from meat to brine. The RP-HPLC results correlate with the antioxidative activity of the extracts.

**Supplementary Materials:** The following supporting information can be downloaded at: https://www.mdpi.com/article/10.3390/app13127225/s1, Figure S1: Entire chromatogram of brine 2 °C ISP extract; Figure S2: Entire chromatogram of meat 2 °C ISP extract; Figure S3: Entire chromatogram of brine 7 °C ISP extract; Figure S4: Entire chromatogram of meat 7 °C ISP extract; Figure S5: Entire chromatogram of brine 12 °C ISP extract; Figure S6: Entire chromatogram of meat 12 °C ISP extract; Figure S7: Entire MALDI-ToF MS spectrum of brine 2 °C ISP extract; Figure S8: Entire MALDI-ToF MS spectrum of meat 2 °C ISP extract; Figure S9: Entire MALDI-ToF MS spectrum of brine 7 °C ISP extract; Figure S10: Entire MALDI-ToF MS spectrum of meat 7 °C ISP extract; Figure S11: Entire MALDI-ToF MS spectrum of brine 12 °C ISP extract; Figure S12: Entire MALDI-ToF MS spectrum of meat 12 °C ISP extract.

**Author Contributions:** Conceptualization, M.S. (Mariusz Szymczak), P.K., P.M. and M.M.; methodology, M.S. (Mariusz Szymczak), P.K., P.M., J.B., D.M. and M.M.; formal analysis, M.S. (Mariusz Szymczak), P.K., P.M., M.T., M.M. and M.S. (Magdalena Stobinska); investigation, resources and data curation, M.S. (Mariusz Szymczak), P.K., P.M., J.B., D.M., M.T. and M.M.; writing—original draft preparation, M.S. (Mariusz Szymczak), P.K., P.M., M.T. and M.M.; writing—review and editing, M.S. (Mariusz Szymczak), P.K. and P.M.; visualization, M.S. (Mariusz Szymczak), P.K., P.M., M.T. and M.M. All authors have read and agreed to the published version of the manuscript.

**Funding:** The authors are grateful to Krzysztof Formicki, the Dean of the Faculty of Food Science and Fisheries, for financial support toward chemicals for the antioxidant activity assays. The contributions of authors from UWM in Olsztyn were financially supported by the funds of the University of Warmia and Mazury in Olsztyn (Project No 17.610.014-110), and by the Minister of Education and Science, within the program entitled "Regional Initiative of Excellence" for the years 2019–2023 (Project No. 010/RID/2018/19), to the amount of 12,000,000 PLN.

**Institutional Review Board Statement:** Not applicable.

**Informed Consent Statement:** Not applicable.

**Data Availability Statement:** Any data or material supporting this study's findings can be made available by the corresponding author upon request.

**Acknowledgments:** We would like to thank Alicja Tarnowiecka-Kuca from CBIPM for her assistance in the analysis of antioxidant activity.

**Conflicts of Interest:** The authors declare no conflict of interest.

# References

1. Shenderyuk, V.I.; Bykowski, P.J. Salting and marinating of fish. In *Seafood: Resources, Nutritional Composition, and Preservation*; Sikorski, Z., Ed.; CRC Press Inc.: Boca Raton, FL, USA, 1990; Volume 9, pp. 147–162.
2. Domiszewski, Z.; Bienkiewicz, G.; Plust, D.; Kulasa, M. Quality of lipids in marinated herring. *Electron. J. Pol. Agric. Univ.* **2011**, *14*, 9. Available online: http://www.ejpau.media.pl/volume14/issue1/art-09.html (accessed on 1 May 2023).
3. Trafiałek, J.; Zwoliński, M.; Kolanowski, W. Assessing hygiene practices during fish selling in retail stores. *Br. Food J.* **2016**, *118*, 2053–2067. [CrossRef]
4. Szymczak, M. Effect of technological factors on the activity and losses of cathepsins B, D and L during marinating of Atlantic and Baltic herrings. *J. Sci. Food Agric.* **2017**, *97*, 1488–1496. [CrossRef] [PubMed]
5. Szymczak, M. Recovery of cathepsins from marinating brine waste. *Inter. J. Food Sci. Technol.* **2017**, *52*, 154–160. [CrossRef]
6. Szymczak, M.; Kołakowski, E. Losses of nitrogen fractions from herring to brine during marinating. *Food Chem.* **2012**, *132*, 237–243. [CrossRef]
7. Szymczak, M.; Tokarczyk, G.; Felisiak, K. Marinating and salting of herring, nitrogen compounds' changes in meat and brine. In *Processing and Impact on Active Components in Food*; Preedy, V., Ed.; Academic Press: Cambridge, MA, USA; Elsevier: Amsterdam, The Netherlands, 2015; Volume 53, pp. 439–445. [CrossRef]
8. Szymczak, M.; Felisiak, K.; Szymczak, B. Characteristics of herring marinated in re-used brines after microfiltration. *J. Food Sci. Technol.* **2018**, *55*, 4395–4405. [CrossRef] [PubMed]
9. Szymczak, M.; Felisiak, K.; Tokarczyk, G.; Szymczak, B. The reuse of brine to enhance the ripening of marine and freshwater fish resistant to marinating. *Int. J. Food Sci. Technol.* **2019**, *54*, 1151–1159. [CrossRef]

10. Gringer, N.; Hosseini, S.V.; Svendsen, T.; Undeland, I.; Christensen, M.L.; Baron, C.P. Recovery of biomolecules from marinated herring (*Clupea harengus*) brine using ultrafiltration through ceramic membranes. *LWT Food Sci. Technol.* **2015**, *63*, 423–429. [CrossRef]

11. Forghani, B.; Sørensen, A.-D.M.; Sloth, J.J.; Undeland, I. Liquid Side Streams from Mussel and Herring Processing as Sources of Potential Income. *ACS Omega* **2023**, *8*, 8355–8365. [CrossRef]

12. Iwaniak, A.; Darewicz, M.; Minkiewicz, P. Peptides Derived from Foods as Supportive Diet Components in the Prevention of Metabolic Syndrome. *Compr. Rev. Food Sci. Food Saf.* **2018**, *17*, 63–81. [CrossRef]

13. Barati, M.; Javanmardi, F.; Jazayeri, S.M.H.M.; Jabbari, M.; Rahmani, J.; Barati, F.; Nickho, H.; Davoodi, S.H.; Roshanravan, N.; Khaneghah, A.M. Techniques, perspectives, and challenges of bioactive peptide generation: A comprehensive systematic review. *Compr. Rev. Food Sci. Food Saf.* **2020**, *19*, 1488–1520. [CrossRef]

14. Szymczak, M.; Kołakowski, E.; Felisiak, K. Effect of addition of different acetic acid concentrations on the quality of marinated herring. *J. Aquat. Food Prod. Technol.* **2015**, *24*, 566–581. [CrossRef]

15. Chan, S.S.; Roth, B.; Jessen, F.; Løvdal, T.; Jakobsen, A.N.; Lerfall, J. A comparative study of Atlantic salmon chilled in refrigerated seawater versus on ice: From whole fish to cold-smoked fillets. *Sci. Rep.* **2020**, *10*, 17160. [CrossRef] [PubMed]

16. Boziaris, I.S.; Stamatiou, A.P.; Nychas, G.-J.E. Microbiological aspects and shelf life of processed seafood products. *J. Sci. Food Agric.* **2013**, *93*, 1184–1190. [CrossRef] [PubMed]

17. Gringer, N.; Osman, A.; Nielsen, H.H.; Undeland, I.; Baron, C.P. Chemical Characterization, Antioxidant and Enzymatic Activity of Brines from Scandinavian Marinated Herring Products. *J. Food Process. Technol.* **2014**, *5*, 1000346. [CrossRef]

18. Kołakowski, E. Analysis of proteins, peptides, and amino acids in foods. In *Methods of Analysis of Food Components and Additives*; Ötles, S., Ed.; CRC, Taylor & Francis Group: Boca Raton, FL, USA, 2005; pp. 59–96.

19. Nagai, H.; Minatani, T.; Goto, K. Development of a method for crustacean allergens using liquid chromatography/tandem mass spectrometry. *J. AOAC Int.* **2015**, *98*, 1355–1365. [CrossRef]

20. Visser, S.; Slangen, C.J.; Rollema, H.S. Phenotyping of bovine milk proteins by reversed-phase high-performance liquid chromatography. *J. Chromatogr.* **1991**, *548*, 361–370. [CrossRef]

21. Dolan, J.W. Gradient elution, Part V: Baseline drift problems. *LC-GC North Am.* **2013**, *31*, 538–543. Available online: https://www.chromatographyonline.com/view/gradient-elution-part-v-baseline-drift-problems (accessed on 1 July 2013).

22. Kolanowski, W.; Karaman, A.D.; Akgul, F.Y.; Ługowska, K.; Trafiałek, J. Food Safety When Eating out—Perspectives of Young Adult Consumers in Poland and Turkey—A Pilot Study. *Int. J. Environ. Res. Public Health* **2021**, *18*, 1884. [CrossRef]

23. Rodger, G.; Hastings, R.; Cryne, C.; Bailey, J. Diffusion properties of salt and acetic acid into herring and their subsequent effect on the muscle tissue. *J. Food Sci.* **1984**, *49*, 717–720. [CrossRef]

24. Zamojski, J. Przemiany związków azotowych mięsa śledzia w kwaśnym środowisku zalewy octowej (Changes in nitrogenous compounds of herring meat in an acidic environment of acetic brine). *Przemysł Spożywczy* **1970**, *XXIV*, 191–194.

25. Kolodziejska, I.; Sikorski, Z.E. Muscle cathepsins of marine fish and invertebrates. *Pol. J. Food Nutr. Sci.* **1995**, *4*, 3–10.

26. Zhao, N.; Yang, X.; Li, Y.; Wu, H.; Chen, Y.; Gao, R.; Xiao, F.; Bai, F.; Wang, J.; Liu, Z.; et al. Effects of protein oxidation, cathepsins, and various freezing temperatures on the quality of superchilled sturgeon fillets. *Mar. Life Sci. Technol.* **2021**, *23*, 117–126. [CrossRef]

27. Iwaniak, A.; Hrynkiewicz, M.; Minkiewicz, P.; Bucholska, J.; Darewicz, M. Soybean (*Glycine max*) protein hydrolysates as sources of peptide bitter-tasting indicators: An analysis based on hybrid and fragmentomic approaches. *Appl. Sci.* **2020**, *10*, 2514. [CrossRef]

28. Iwaniak, A.; Minkiewicz, P.; Hrynkiewicz, M.; Bucholska, J.; Darewicz, M. Hybrid approach in the analysis of bovine milk protein hydrolysates as a source of peptides containing di- and tripeptide bitterness indicators. *Pol. J. Food Nutr. Sci.* **2020**, *70*, 139–150. [CrossRef]

29. Iwaniak, A.; Mogut, D.; Minkiewicz, P.; Żulewska, J.; Darewicz, M. Gouda cheese with modified content of β-casein as a source of peptides with ACE- and DPP-IV-inhibiting bioactivity: A study based on in silico and in vitro protocol. *Int. J. Mol. Sci.* **2021**, *22*, 2949. [CrossRef]

30. Fornstedt, T.; Forssén, P.; Westerlund, D. System peaks and their impact in liquid chromatography. *Trends Anal. Chem.* **2016**, *81*, 42–50. [CrossRef]

31. Krokhin, O.V.; Craig, R.; Spicer, V.; Ens, W.; Standing, K.G.; Beavis, R.C.; Wilkins, J.A. An improved model for prediction of retention times of tryptic peptides in ion pair reversed-phase HPLC. *Mol. Cell. Proteom.* **2004**, *3*, 908–919. [CrossRef] [PubMed]

32. Krokhin, O.V. Sequence Specific Retention Calculator—A novel algorithm for peptide retention prediction in ion-pair RP-HPLC: Application to 300 Å and 100 Å pore size C18 sorbents. *Anal. Chem.* **2006**, *78*, 7785–7795. [CrossRef]

33. Bucholska, J.; Minkiewicz, P. The use of peptide markers of carp and herring allergens as an example of detection of sequenced and non-sequenced proteins. *Food Technol. Biotechnol.* **2016**, *54*, 266–274. [CrossRef]

34. Orsini Delgado, M.C.; Nardo, A.; Pavlovic, M.; Rogniaux, H.; Añón, M.C.; Tironi, V.A. Identification and characterization of antioxidant peptides obtained by gastrointestinal digestion of amaranth proteins. *Food Chem.* **2016**, *197*, 1160–1167. [CrossRef]

35. Baugreet, S.; Gomez, C.; Auty, M.A.E.; Kerry, J.P.; Hamill, R.M.; Brodkorb, A. In vitro digestion of protein-enriched restructured beef steaks with pea protein isolate, rice protein and lentil flour following sous vide processing. *Innov. Food Sci. Emerg. Technol.* **2019**, *54*, 152–161. [CrossRef]

36. Kürzl, C.; Wohlschläger, H.; Schiffer, S.; Kulozik, U. Concentration, purification and quantification of milk protein residues following cleaning processes using a combination of SPE and RP-HPLC. *MethodsX* **2022**, *9*, 101695. [CrossRef]

37. Shelke, P.A.; Sabikhi, L.; Khetra, Y.; Ganguly, S.; Baig, D. Effect of skim milk addition and heat treatment on characteristics of cow milk Ricotta cheese manufactured from Cheddar cheese whey. *LWT* **2022**, *162*, 113405. [CrossRef]

38. Duncan, M.W.; Roder, H.; Hunsucker, S.W. Quantitative matrix-assisted laser desorption/ionization mass spectrometry. *Brief. Funct. Genom. Proteom.* **2008**, *7*, 355–370. [CrossRef] [PubMed]

39. Šebela, M. Biomolecular profiling by MALDI-TOF mass spectrometry in food and beverage analyses. *Int. J. Mol. Sci.* **2022**, *23*, 13631. [CrossRef] [PubMed]

40. Cozzolino, R.; Passalacqua, S.; Salemi, S.; Malvagna, P.; Spina, E.; Garozzo, D. Identification of adulteration in milk by matrix-assisted laser desorption/ionization time-of-flight mass spectrometry. *J. Mass Spectrom.* **2001**, *36*, 1031–1037. [CrossRef] [PubMed]

41. Flensburg, J.; Haid, D.; Blomberg, J.; Bielawski, J.; Ivansson, D. Applications and performance of a MALDI-ToF mass spectrometer with quadratic field reflectron technology. *J. Biochem. Biophys. Methods* **2004**, *60*, 319–334. [CrossRef]

42. Keller, B.O.; Li, L. Discerning matrix-cluster peaks in Matrix-Assisted Laser Desorption/Ionization Time-of-Flight mass spectra of dilute peptide Mixtures. *J. Am. Soc. Mass Spectrom.* **2000**, *11*, 88–93. [CrossRef]

43. Treu, A.; Römpp, A. Matrix ions as internal standard for high mass accuracy matrix-assisted laser desorption/ionization mass spectrometry imaging. *Rapid Commun. Mass Spectrom.* **2021**, *35*, e9110. [CrossRef]

44. Søtoft, L.F.; Lizarazu, J.M.; Parjikolaei, B.R.; Karring, H.; Christensen, K.V. Membrane fractionation of herring marinade for separation and recovery of fats, proteins, amino acids, salt, acetic acid and water. *J. Food Eng.* **2015**, *158*, 39–47. [CrossRef]

45. Portmann, R.; Jiménez-Barrios, P.; Jardin, J.; Abbühl, L.; Barile, D.; Danielsen, M. A multi-centre peptidomics investigation of food digesta: Current state of the art in mass spectrometry analysis and data visualisation. *Food Res. Int.* **2023**, *169*, 112887. [CrossRef] [PubMed]

46. Girgih, A.T.; He, R.; Hasan, F.M.; Udenigwe, C.C.; Gill, T.A.; Aluko, R.E. Evaluation of the in vitro antioxidant properties of a cod (*Gadus morhua*) protein hydrolysate and peptide fractions. *Food Chem.* **2015**, *173*, 652–659. [CrossRef] [PubMed]

*Article*

# Effects of Primary Processing and Pre-Salting of Baltic Herring on Contribution of Digestive Proteases in Marinating Process

Patryk Kamiński [1,*], Mariusz Szymczak [1] and Barbara Szymczak [2]

1  Department of Toxicology, Dairy Technology and Food Storage, Faculty of Food Sciences and Fisheries, West Pomeranian University of Technology in Szczecin, Papieza Pawla VI 3, 71-459 Szczecin, Poland
2  Department of Microbiology, Faculty of Food Sciences and Fisheries, West Pomeranian University of Technology in Szczecin, Papieza Pawla VI 3, 71-459 Szczecin, Poland
*  Correspondence: patryk-kaminski@zut.edu.pl; Tel.: +48-91-449-65-01

Citation: Kamiński, P.; Szymczak, M.; Szymczak, B. Effects of Primary Processing and Pre-Salting of Baltic Herring on Contribution of Digestive Proteases in Marinating Process. *Appl. Sci.* **2022**, *12*, 10877. https://doi.org/10.3390/app122110877

Academic Editor: Anna Lante

Received: 3 October 2022
Accepted: 24 October 2022
Published: 26 October 2022

**Publisher's Note:** MDPI stays neutral with regard to jurisdictional claims in published maps and institutional affiliations.

**Abstract:** The low-technological quality of herring caught during the feeding season makes it impossible to achieve full ripeness of the meat in marinades. One solution may be to assist ripening using herring digestive tract proteases. Therefore, whole herring, headed herring and fillets were marinated for 2–14 days using the German (direct) and Danish (pre-salted) methods. The results showed that the mass of marinades from fillets was lower than from herring with intestines and correlated strongly with salt concentration in the Danish method, in contrast to the German method. Marinades from whole and headed herring had significantly higher trypsin, chymotrypsin, carboxypeptidase-A and cathepsin activities than marinated fillets. The herring marinated with viscera had 2–3 times higher non-protein nitrogen, peptide and amino acid fractions, as well as ripened 3 days faster than the marinated fillets. After 2 weeks of marinating, the fillets did not achieve full ripeness of the meat, unlike marinades made from whole and headed herring. The pre-salting stage in the Danish method significantly reduced cathepsin D activity by the tenth day of marinating, which was compensated by digestive proteases only in the case of whole or headed herring. The digestive proteases activity in the fillets was too low to achieve the same effect. Sensory evaluation of texture and hardness-TPA correlated strongly with several proteases in whole herring marinades, in contrast to a weak correlation with only one protease when marinating fillets. Marinating with intestines makes it possible to produce marinades faster, more efficiently and with higher sensory quality from herring of low-technological quality.

**Keywords:** herring; marinades; primary processing; pre-salting; cathepsins; intestines proteases

## 1. Introduction

The high degree of ripeness of the meat of herring marinades promotes high sensory quality and nutritional value, and facilitates digestion, especially in the elderly or those with certain gastrointestinal diseases. Cold-ripened marinades are produced mainly from herring, the technological quality of which varies seasonally [1]. Due to the high price of the raw material, the industry tries to marinate cheaper herring caught at the beginning of the feeding season, resulting in low ripeness of marinated meat. The ripening of marinades is mainly related to cathepsins D, L and B activity [2,3]. Herring caught during heavy feeding season have low cathepsins activity, and therefore, are worse suited for marinades production. Higher cathepsins activity causes better ripeness of meat and the formation of a higher content of peptides and free amino acids, which provides sensory and nutritional characteristics to marinades. Protein hydrolysis products (PHP) released from marinated herring proteins also show biological activity [4,5]. These peptides may have antihypertensive, anticoagulant, antioxidant or even immunomodulatory effects [6].

Kamiński et al. [7] showed that cold storage of herring prior to marinating allows the use of digestive proteases to improve the low technological quality of the raw material.

During the 7 days in cold storage, trypsin, chymotrypsin and carboxypeptidases diffused from the viscera into the herring muscle, which started proteolysis of the meat and increased cathepsin activity during marinating. The result was an increase in the sensory evaluation of marinades for taste, aroma and texture. The disadvantages of this method are the requirement of additional time before marinating and a lower yield of marinade mass. Currently, Baltic herring after primary processing (heading, filleting) are marinated in two ways. The longest known method, called German or classic or direct method, consists of marinating of fillets in acid-salty brine for 7 days. This method has a 75–85% mass yield of the marinating process [8]. The second method, called the Danish method, has a 90–100% mass yield and is based on wet salting of fillets for 2 days, followed by marinating in an acid-salty brine [9]. Despite the fact that the Danish method has been known in the industry for many years, there is a lack of research regarding the effect of pre-salting on protease activity in the process of marinating herring.

Currently, cold marinades are mainly made from herring fillets, but marinated sprats in northern and central Europe and marinated sardines in southern Europe and western Asia are also available on the market. Baltic herring caught at the beginning of the feeding season are characterized by a small size, similar to that of a large sprat [10]. Large sprats are subjected to nobbing before marinating, which involves separating the head along with the digestive tract, without opening the abdominal cavity and without removing the pyloric caceae and gonads [11]. The pyloric caceae contain very active proteases [7,12]. Practice shows that mechanical nobbing is not always effective and some of the sprat is marinated with whole intestines. There are no results in the literature for the marinating method of whole or nobbed herring, especially in combination with the Danish marinating method, in which pre-salting may change the activation of digestive proteases, as occurs during traditional salting of herring [13,14].

Protease activity is crucial in the ripening process of marinated herring meat, but so far, protease activity has only been studied during the classic marination of fillets [2,15]. The effect of pre-salting on protease activity in marinades is unknown, while the contribution of digestive proteases during marination has been described in only one publication [7]. Therefore, the aim of this study is to determine the effect of primary processing of Baltic herring (whole, headed, fillets) with and without pre-salting stage on contribution of digestive and cathepsins proteases during ripening of marinades using Danish and German methods.

## 2. Materials and Methods

### 2.1. Baltic Herring and Marinating Methods

Baltic herring (*Clupea harengus membras*) were caught in December (FAO 27IIId), transported within 24 h in a polystyrene ice box to the laboratory, where Quality Index Method (QIM) and morphometry of whole herring were assessed [7,16]. The fish condition and fullness of the herring's stomachs indicated foraging. The gonads had a mass of $11.3 \pm 4.0$ g and maturity stage III on the Maier scale, while the weight of the digestive tract was $6.5 \pm 1.8$ g. The herring were $22.3 \pm 0.8$ cm long and weighed $113.0 \pm 28.2$ g, with meat containing $11.7 \pm 0.1\%$ lipids, $71.9 \pm 0.1\%$ water and $15.9 \pm 0.3\%$ protein.

The herring were divided into three raw material groups: whole herring, headed herring and fillets (control sample), which were marinated by the (G-) German or (D-) Danish method obtaining marinated samples: G-whole, G-headed, G-fillets, and D-whole, D-headed, D-fillets (Figure 1). The German method was performed by marinating each group of herring separately in 5% vinegar and 6% rock salt solution for 14 days [2]. The Danish method was performed using salting for 2 days in 10% brine, followed by marinating in 5% vinegar and 1.8% rock salt solution for 12 days [9]. All herring samples were marinated at 7 °C; fish to brine ratio was 1:1 (*w:w*). After 2, 4, 7, 10 and 14 days of marinating, the fish were transferred to a sieve and herrings were weighed.

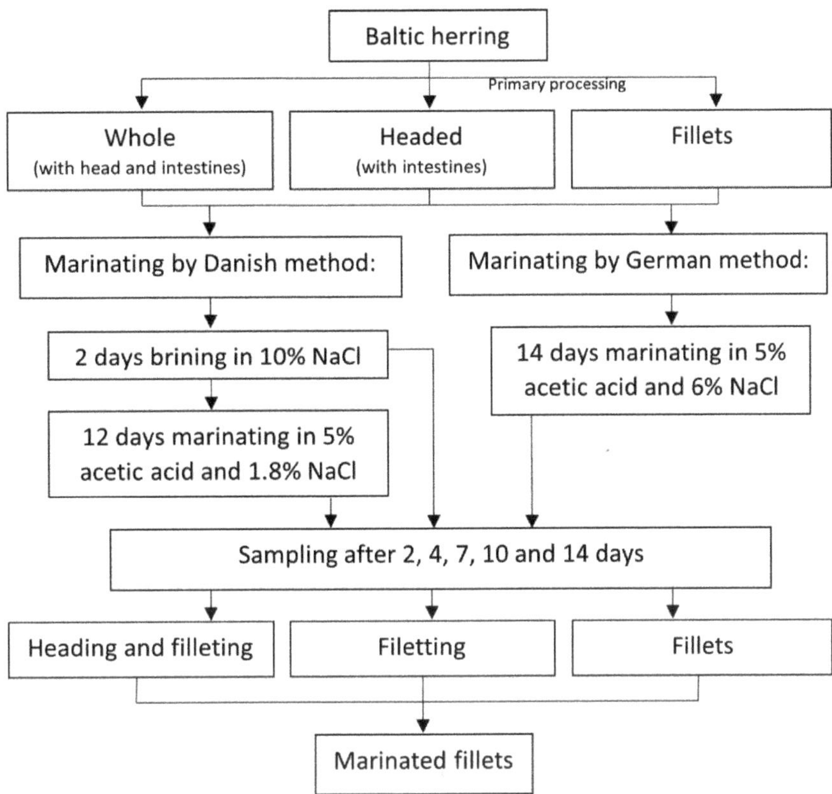


**Figure 1.** Diagram showing the performance of marinades using the German and Danish methods.

*2.2. Total Nitrogen, Lipids, Salt, pH, Acidity and Meat Moisture Analysis*

The marinated herring fillets were skinned and minced. The pH value in the meat with distilled water homogenate (1:10, *w:v*) was determined using a digital pH-meter. Total nitrogen Kjeldahl's method, total lipids Soxhlet method, moisture, salt contents and total acidity were determined using standard AOAC analytical techniques no. 940.25, 960.39, 950.46B and 976.09, respectively.

*2.3. Extractions and Activity Assay of Proteases*

Digestive protease extracts were made from fresh (raw material) and marinated herring meat using water (1:10, *v:w*) and TrisHCl 150 mM buffer at pH 7.8 (to buffer acetic acid in marinades), respectively [7]. Cathepsins from raw material and marinated meat were extracted with ultra-pure water. Samples were homogenized for 30 s at 34,000 rpm and centrifuged after 30 min (10 min, $9000 \times g$, 4 °C). Obtained supernatant was centrifuged at $20,000 \times g$ (10 min, 4 °C) to obtain the crude protease extract, where digestive proteases and cathepsins activity were determined. Amidase trypsin (Am-Tr) was measured at pH 8.2 against N-benzoyl-DL-arginine (BApNA) [17]; Esterase trypsin (Es-Tr) at pH 8.2 against N$\alpha$-*p*-Tosyl-L arginine methyl ester hydrochloride (TAME) and Chymotrypsin at pH 7.8 against N-Benzoyl-L-tyrosine ethyl ester (BTEE) [18]; Carboxypeptidase-A (Cp-A) at pH 7.5 against 4-Methoxyphenylazoformyl-Phe (AAFP) [19]. The activity of cathepsin B against Z-RR-MCA, cathepsin D against Mca-GKPILFFRLK(Dnp)-r-NH2 with and without pepstatin-A, and cathepsin L against Z-FR-MCA were determined according to [2]. Protease activity assays were performed at 37 °C using continuous spectrophotometric rate determination, and units of enzyme activity were converted per 1 g of tissue. Chemicals from PeptaNova (Concord, CA, USA) and ultrapure water were used for analysis.

### 2.4. Non-Protein Nitrogen Fraction Analysis

Determinations of (a) non-protein nitrogen by Kjeldahl's method (AOAC no. 940.25), (b) peptides [PHP(R)] and amino acids [PHP(A)] fractions by Lowry's methods as modified by [20], were carried out in 50 g·kg$^{-1}$ trichloroacetic acid (TCA) extract.

### 2.5. Texture Profile Analysis (TPA)

Hardness-TPA was determined in four fillets from each sample with a TA-XTplus Texture Analyzer (Stable Micro Systems, Godalming, UK). TPA tests consisted of twice compressing samples in the same spot with a cylindrical probe P10 (10 mm) up to 50% of fillet height at the speed of 5 mm·s$^{-1}$, and were conducted in three replications for each fillet separately (12 replications in each sample), and only on the dorsal muscle in an area from 2/10 to 6/10 fillets length measured from the head side. Their courses were recorded as curves representing changes of force in time and expressed in newtons, N.

### 2.6. Sensory Profiling

Marinated meat was analyzed by sensory profiling performed by a trained sensory panel, using a five-point unstructured scale with a 0.5-point accuracy anchored at their extremes with minimum and maximum degrees of acceptance [21]. A higher note signified better sensory attributes (0 points being the worst sensory/extremely disliked; 5 points being the best sensory/extremely liked). Four skinned fillets from each sample were served in porcelain trays. The assessors used water and flat bread to clean their palates between samples. Sensory attributes were: texture, flavor, odor and appearance. The sum of individual scores gave a total score that represented the overall sensory evaluation of the marinated fillets. The seven panelists participating in the sensory evaluation of marinades in this study had many years of experience in the subject of marinating fish. In addition, before starting the research, the panelists were trained on the sensory assessment methods described above.

### 2.7. Microbiological Analysis

Three fillets from marinated herring were taken for analysis using disinfected (peracetic acid, 30% $H_2O_2$) knives and plastic boards. The fillets were held with sterile tweezers and rinsed for 1–2 s with cold tap water to remove any remaining viscera and blood. The preparation of the test samples was performed according to the ISO standard [22]. Samples in triplicate were weighed at 20 g each into sterile stomacher bags and 180 mL of sterile dilution fluid (P-0061, BTL, Warsaw, Poland) was added. Samples were homogenized (BagMixer 400P, Interscience, Saint Nom la Bretêche, France), and then dilutions and surface cultures were performed on nutrient agar medium (P-0075, BTL, Poland) for total psychrophilic bacteria [23], and on Sabouraud medium for total yeast and mold [24]. In addition, flooded cultures were performed on VRBGLA medium (BT5158.02, Biomaxima, Lublin, Poland) for *Enterobacteriacea* [25] and on nutrient agar (01140, Scharlab, Barcelona, Spain) for total mesophilic bacteria [26].

### 2.8. Statistical Analysis

Results were statistically analyzed using one-way analysis of variance (ANOVA) with Statistica 13.3 software (Statsoft, Tulsa, OK, USA). The differences between treatments were examined using analysis of variance (ANOVA) and the post hoc Tukey's test of honestly significant differences ($p < 0.05$) [27]. Correlations between marinating time, proteases activities, proteolysis indicators and sensory analyses for each marinating method and raw materials were separately investigated using Principal Component Analyses (PCA) [28,29]. All analyses were performed in triplicate (n = 3), except TPA (n = 12), and results were presented as mean and standard deviation.

## 3. Results and Discussion

### 3.1. Mass Yield, pH and Salt Concentration

The mass of herring marinated by the German method (G-) decreased by day 14 (Figure 2A), which was caused by a decrease in meat pH close to the isoelectric point and in the resulting loss of water (Figure 2B). The mass of German marinades significantly correlated with marinating time, most strongly for G-whole r = −0.987 (Figure 3A). In the case of the Danish (D-) method, the yield increased after 2 days due to protein salting and moisture increase, but when acetic acid was added, fish mass and moisture significantly decreased (Figure 2A,B). The mass of Danish marinates also significantly correlated with marinating time, but the correlation was lower than that of German marinates. After 4–7 days of marinating, the mass yield of G-fillets marinades averaged 78.7%, while G-headed and G-whole marinades had higher yields of 4.1 and 15.0 percentage points, respectively (Figure 2A). At the same time, D-fillets marinades had an average mass yield of 83.3%, while D-headed and D-whole marinades were 4.7 and 9.1 percentage points higher, respectively. Thus, the advantage of higher mass yield of the Danish method over the German method when marinating fillets [9] was not present in the case of marinating whole herring.

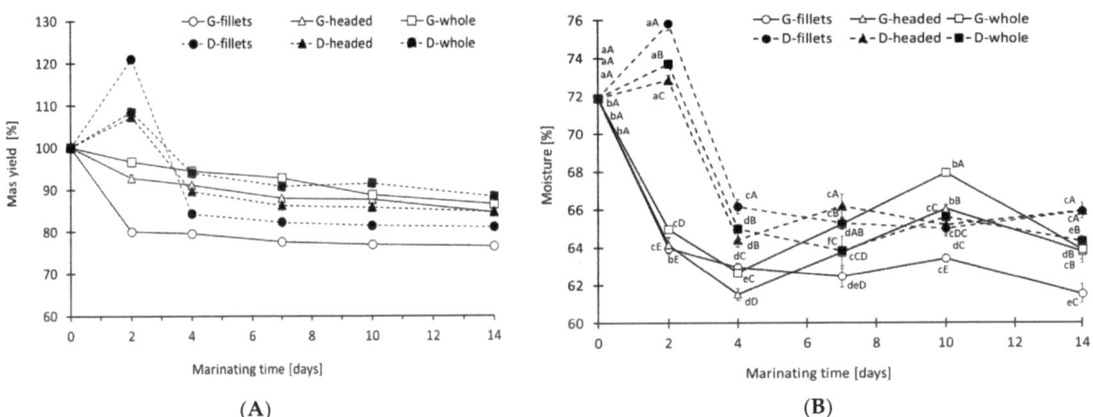

**Figure 2.** Effect of marinating method and degree of primary processing of Baltic herring (whole, headed, fillets) on (**A**) mass yield and (**B**) moisture during marinating. G—German method, D—Danish method. a,b,c Results marked with the same lower case letter do not statistically differ by the influence of marinating time; A,B,C Results marked with the same capital letter do not statistically differ by the influence of the marinating method.

G-fillets marinades after 4–7 days of marinating had an average meat pH value of 3.91, while G-headed and G-whole had 4.16 and 4.26, respectively (Figure 4A). The pH value significantly correlated with the decreasing mass of German marinades: G-whole (0.973), G-fillets (0.891) and G-headed (0.871) (Figure 3A,C,E). The D-fillets marinades had a meat pH of 3.88, while the D-headed and D-whole marinades had a pH of 4.22 and 4.38, respectively. The pH values of D-fillets, D-headed and D-whole marinades significantly correlated with a mass yield of 0.991, 0.877 and 0.750, respectively (Figure 3B,D,F). The content of table salt after 4–7 days of marinating in the meat of G-fillets marinades was 2.96% and was 0.07–0.17% higher than in G-whole and G-headed marinades (Figure 4B). For marinating using the Danish method, the differences in salt content between D-fillets marinades (2.80%) and D-headed (2.38%) and D-whole (2.16%) marinades were higher than that of the German method (Figure 4B). PCA analysis showed that the higher the degree of primary processing of the herring (whole < headed < fillets), salt concentration correlated less strongly with the mass yield of marinades performed by the German method (G-whole −0.820, G-headed −0.801, G-fillets −0.469) (Figure 3A,C,E), in contrast to Danish method (D-whole 0.750, D-headed 0.877 and D-fillets 0.991) (Figure 3B,D,F). Therefore, in marinades made of whole herring, the skin reduced the loss of water and protein from the meat [30], and also reduced the diffusion of acetic acid and salt into the meat (Figure 4A,B), which synergistically reduced the water holding capacity of the marinated meat [31].

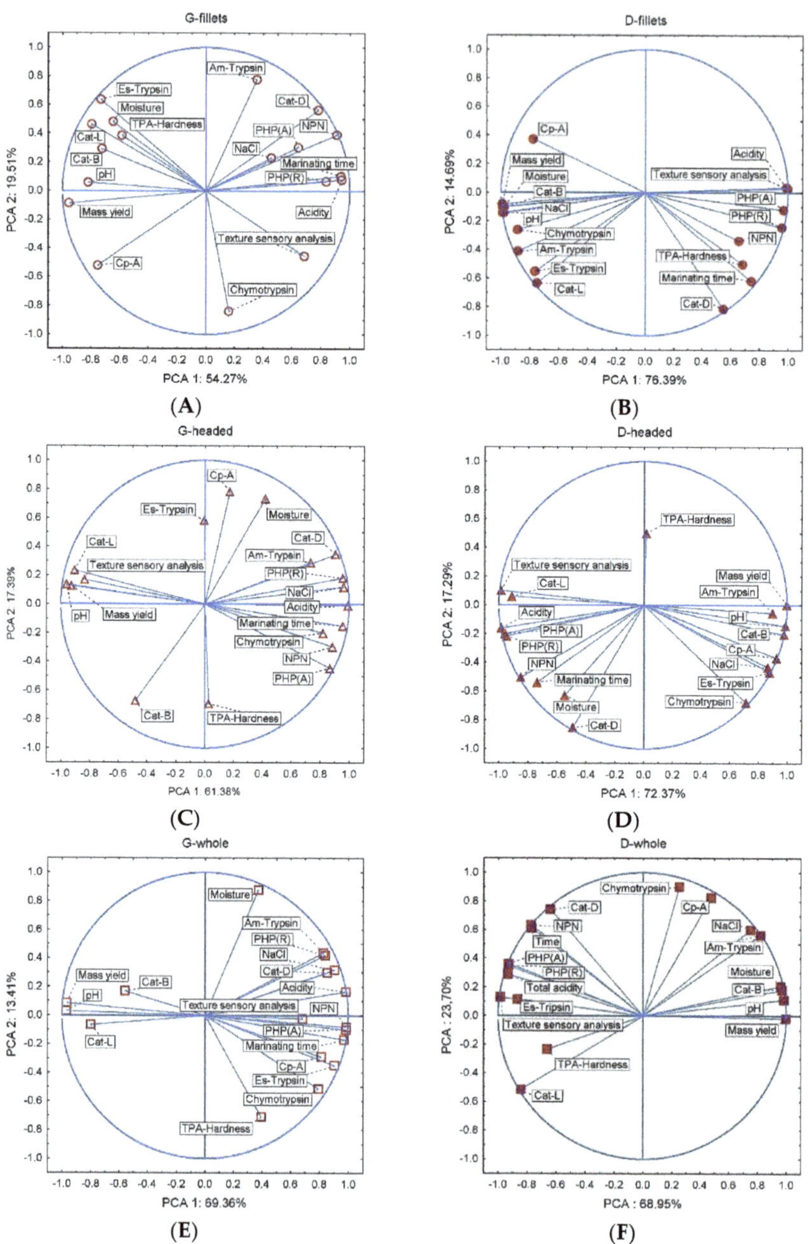

**Figure 3.** PCA biplot for correlations between protease activity, physicochemical parameters and sensory analyses of marinades depending on marinating methods: (**A,C,E**) German method (G-), (**B,D,F**) Danish method (D-) and depending on degree of primary processing of Baltic herring: (**A,B**), fillets of herring, (**C,D**) headed herring and (**E,F**) whole herring.

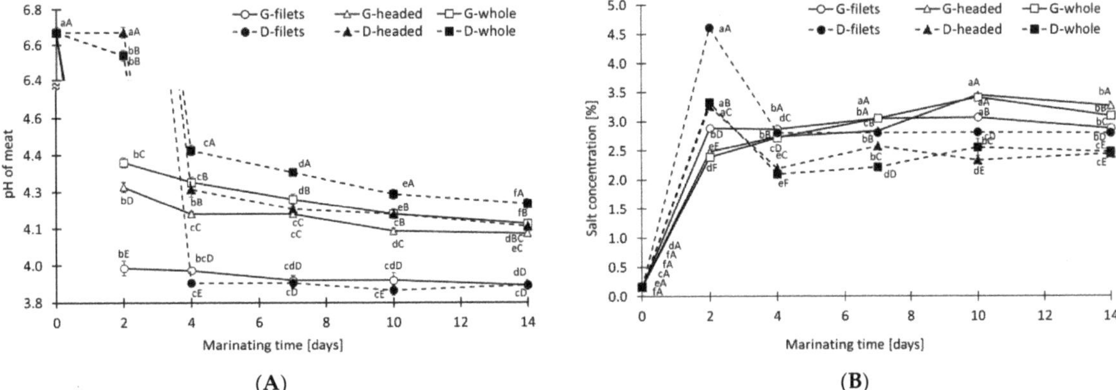

**Figure 4.** Effect of marinating method (G—German, D—Denmark) and degree of primary processing of Baltic herring (whole, headed, fillets) on (**A**) mass yield and (**B**) moisture during marinating. [a,b,c] Results marked with the same lower case letter do not statistically differ by the influence of marinating time; [A,B,C] Results marked with the same capital letter do not statistically differ by the influence of the marinating method. For a better presentation of the results, a break ($\approx$) in the Y axis was used.

### 3.2. Activity of Digestive Proteases and Cathepsins

During marinating, mainly muscle proteases—cathepsins—were active, while during marinating whole and headed herring, digestive proteases such as trypsin, chymotrypsin and carboxypeptidases were also diffused into the meat. The results showed that digestive proteases were already active in fresh herring meat (0 day of marinating) and their activity decreased during marinating (Figure 5), mainly due to a decrease in pH [7]. After the fourth and tenth days of marinating, trypsin esterase activity was higher in the marinades made using the German method than the Danish method, and G-whole marinades had the highest activity after 4–14 days of marinating (11.95–20.12 U) (Figure 5A). Chymotrypsin activity was lower in marinades from fillets than from whole or headed herring, with the exception of day 4 (Figure 5C). Chymotrypsin activity decreased the most after 2 days of marinating using the German method, or after 4 days of marinating using the Danish method. On subsequent days, chymotrypsin activity increased, especially in marinades made from whole and headed herring, and more so in Danish than German marinades. Additionally, carboxypeptidase-A activity decreased by 2–4 days of marinating, except for fillets, where activity decreased by 10–14 days (Figure 5D). Between 4 and 14 days of marinating, an increase in carboxypeptidase-A activity was noted in marinades made using both methods from whole herring. D-whole and G-whole marinades from days 7 to 14 had the significantly highest carboxypeptidase-A activity, 9.22–9.54 U and 6.89–8.95 U, respectively. Marinating time significantly and positively correlated with trypsin esterase activity in G-whole (0.956) and D-whole (0.726) (Figure 3E,F). There was also a significant positive correlation between marinating time and all digestive proteases in G-whole marinades (Figure 3E), except for Cp-A activity, which negatively correlated with marinating time in fillets (G-fillets −0.842, D-fillets −0.739) (Figure 3E,F). Salt concentration and pH values in marinades significantly correlated with all proteases in G-whole and D-whole marinades (Figure 3E,F), and in D-fillets and D-headed marinades (Figure 3B,D). It can be noted that the marinating of whole herring with intestines contributed to greater digestive protease activity in meat than the marinating of fillets obtained by cold-stored whole herring [7].

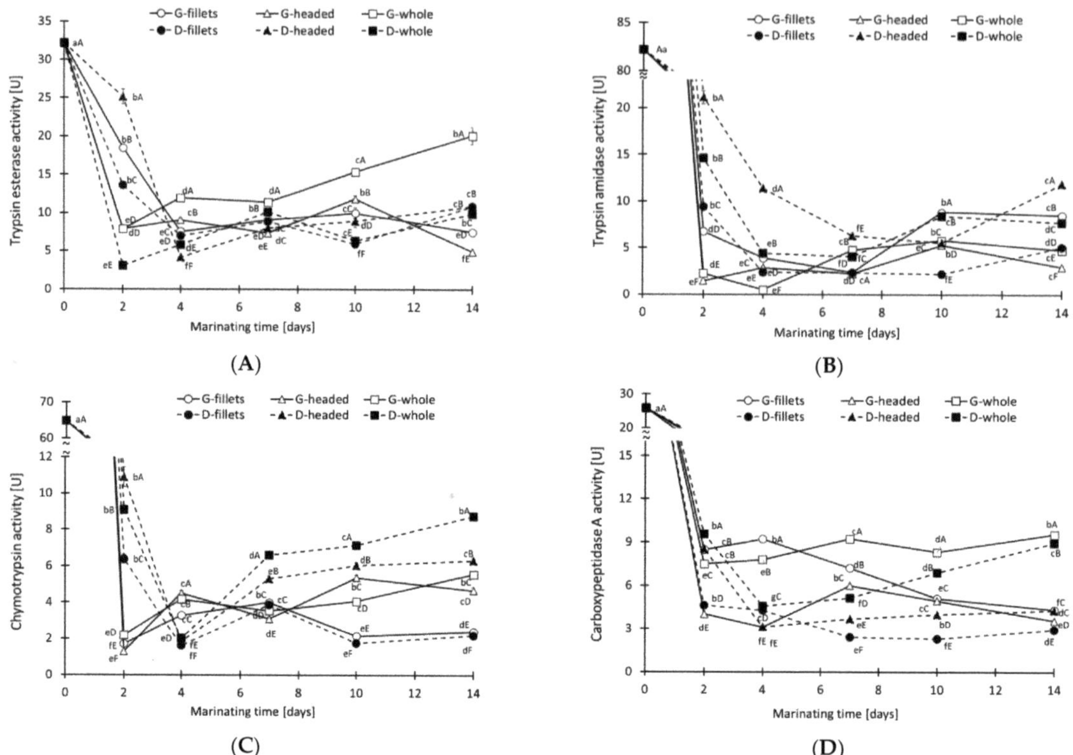

**Figure 5.** Effect of marinating method (G—German, D—Denmark) and degree of primary processing of Baltic herring (whole, headed, fillets) on activity of (**A**) trypsin esterase, (**B**) trypsin amidase, (**C**) chymotrypsin and (**D**) carboxypeptidase-A during marinating. [a,b,c] Results marked with the same lower case letter do not statistically differ by the influence of marinating time; [A,B,C] Results marked with the same capital letter do not statistically differ by the influence of the marinating method. For a better presentation of the results, a break ($\approx$) in the Y axis was used.

Cathepsin D activity in marinades made by the German method increased by 3–4 times after 4–7 days of marinating; while in marinades made by the Danish method, it increased by 2 to 3 times (Figure 6A). Therefore, cathepsin D activity significantly and positively correlated with marinating times in all samples (Figure 3). German marinades had maximum cathepsin D activity (6987–8051 U) after 10 days, while in Danish marinades, its activity increased until the fourteenth day, reaching 6913–7867 U. Low pH, along with the lowest possible salt concentration, are required to activate cathepsin D [2]. In G-headed and G-whole samples, cathepsin D activity increased with decreasing pH, as confirmed by a significant correlation, $-0.793$ and $-0.861$, respectively. In Danish marinades, after 2 days, the salt concentration was about 1.5 times higher than in German marinades, while the pH was close to neutral compared to strongly acidic in German marinades. Therefore, it is likely that the lowest cathepsin D activity in Danish marinades was due to the inhibitory effect of salt and the lack of activating acetic acid during the first two days of brining [32]. Cathepsin D activity is important for the ripening of marinated meat, because this endoprotease prepares substrates for the other cathepsins [33].

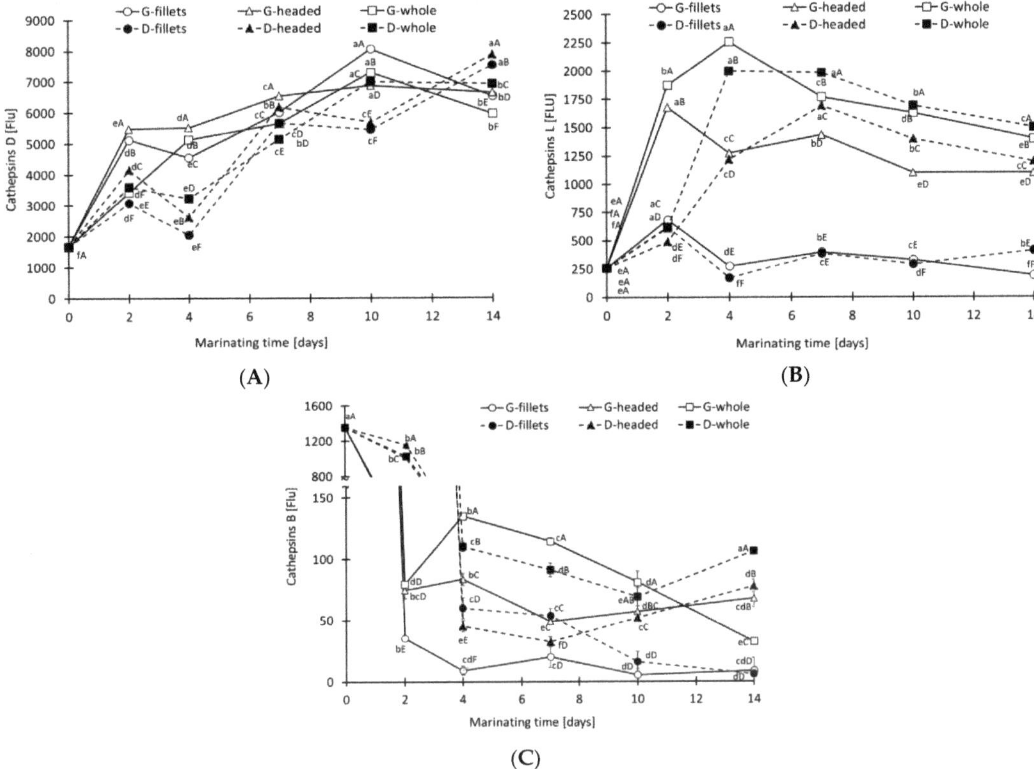

**(A)**

**(B)**

**(C)**

**Figure 6.** Effect of marinating method (G—German, D—Denmark) and degree of primary processing of Baltic herring (whole, headed, fillet) on activity of (**A**) cathepsin D, (**B**) cathepsin L and (**C**) cathepsin B during marinating. [a,b,c] Results marked with the same lower case letter do not statistically differ by the influence of marinating time; [A,B,C] Results marked with the same capital letter do not statistically differ by the influence of the marinating method. For a better presentation of the results, a break ($\approx$) in the Y axis was used.

Cathepsin L activity, after 2 days of marinating the fillets, increased twice to 612–683 U, after which it decreased to 170–380 U (Figure 6B). In marinades from headed herring, cathepsin L activity increased 6.5 times after 2 days by the German method (1679 U) or after 7 days by the Danish method (1692 U). In marinades from whole herring, cathepsin L activity increased as much as 8 and 9 times after 4 days in the Danish (2000 U) and German (2260 U) methods, respectively, and then decreased, reaching 1400–1499 U after 14 days of marinating. Cathepsin L, like cathepsin D, is also an acidic endopeptidase, but requires the presence of salt and a higher pH for optimal activity [2]. Therefore, cathepsin L activity positively correlated with the pH value in only G-fillets (0.593), G-headed (0.964) and G-whole (0.736) marinades. In turn, salt concentration significantly correlated with cathepsin L activity for all marinades, except G-fillets.

In the case of cathepsin B, the activity of this endopeptidase decreased 20–40 times after 2–7 days of marinating; the fastest in G-fillets marinades and the slowest in D-whole marinades (Figure 5C). The cause was the strong inactivation of cathepsin B by salt and acetic acid [2]. Thus, the results showed that the activity of digestive proteases and cathepsins was higher in marinades made from whole and headed herring than from fillets, especially after 2–7 days of marinating. The higher cathepsin activity in marinades from ungutted herring was possibly related to (i) the release of cathepsins from lysosomes by digestive proteases [6,34], (ii) the reduction of cathepsin loss from meat to brine by the herring skin [2] and (iii) the slower diffusion of acetic acid into meat [32], which inhibits the activity of alkaline digestive proteases. Moreover, higher cathepsin D activity may have promoted the release of intracellular proteases [35].

*3.3. Concentration of Protein Hydrolysis Products (PHP)*

The process of enzymatic ripening of herring meat leads to the formation of TCA-soluble proteins, peptides and amino acids, the quantitative composition of which allows assessment of the dynamics of the ripening process of marinades [3,36]. The average NPN content of fillets marinated for 2–14 days was significantly lower by 66–91% than in marinades from whole or headed herring (Figure 7A). NPN content in fillets after 2 days decreased to 156 and 117 mg in the German and Danish methods, respectively. The reason for the decrease in NPN content in the German method was the diffusion of NPN from the meat into the marinating brine [30], while in the Danish method, there was an additional increase in moisture during the salting stage. As a result, fillets marinated for 2 days contained 16–18% less NPN than the raw material. This phenomenon did not occur in whole or headed herring, where diffusion was limited by the skin, causing these marinades to contain 344–356 mg of NPN after 2 weeks, 60–65% more than the raw material (Figure 7A). The higher NPN content was also due to higher proteases activity in marinades form whole and headed herring.

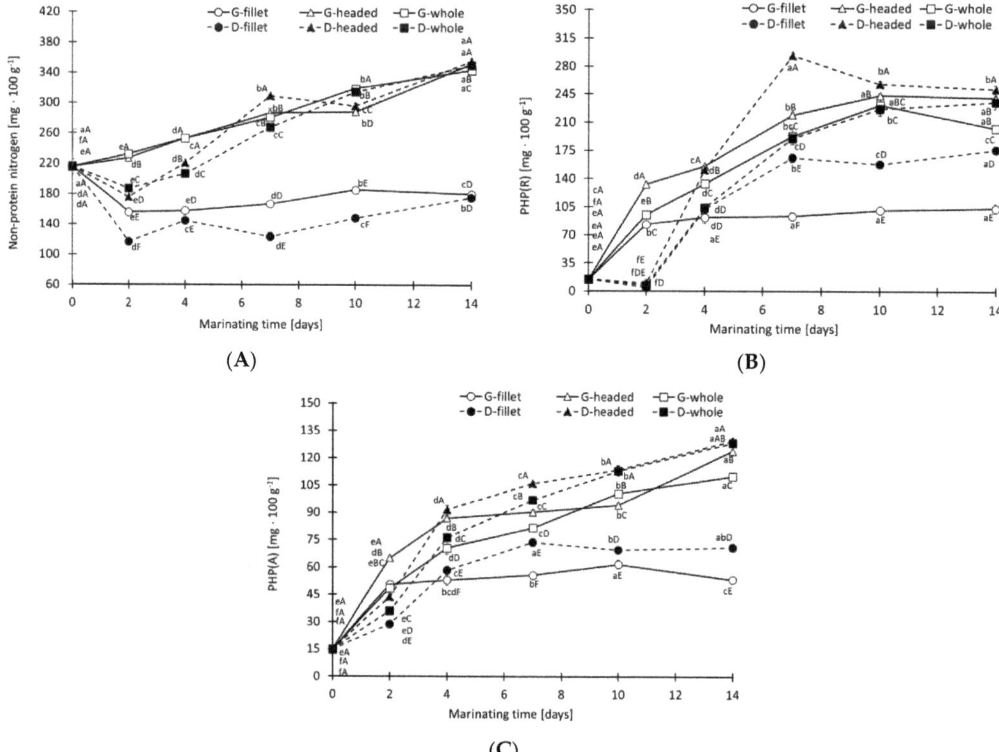

**Figure 7.** Effect of marinating method (G—German, D—Denmark) and degree of primary processing of Baltic herring (whole, headed, fillet) on concentration of (**A**) NPN, (**B**) PHP(R) and (**C**) PHP(A) during marinating. [a,b,c] Results marked with the same lower case letter do not statistically differ by the influence of marinating time; [A,B,C] Results marked with the same capital letter do not statistically differ by the influence of the marinating method.

The peptide fraction content of G-fillets marinades significantly increased only up to the fourth day, reaching 92–103 mg (Figure 7B), which is characteristic for the German method [32]. On the other hand, during marinating of whole and headed herring, the peptide content increased up to the tenth day, reaching 232–234 mg PHP(R). In the Danish method, the peptide content decreased by a few mg after 2 days due to an increase in the moisture of the salted meat, after which it sharply increased to 7 days in fillets and headed herring or to 14 days in whole herring. After 7–10 days, marinades from fillets and headed herring obtained by the Danish method contained more peptides compared to the German method (Figure 7B). Peptide content was not significantly different after 7–10 days for whole herring marinated using both methods. This shows that the inhibition of cathepsin D by salt, and the lower dynamics of mainly peptide formation in the Danish method than in the German method, were balanced by high digestive protease activity. Additionally, the content of the amino acid fraction changed during marinating, similar to the content of NPN and peptides. After 7 days of marinating, G-fillets marinades contained the least PHP(A) (55.8 mg), followed by D-fillets (73.7 mg), G-whole (81.6 mg), G-headed (90.3 mg), D-whole (96.9 mg), and the most, D-headed (106.1 mg) (Figure 7C). For the amino acid fraction, the increase in content was characteristic up to 7–10 days in fillets, while up to 14 days in marinades from ungutted herrings.

Statistical analysis showed that NPN, peptides and amino acid content mainly correlated in G-fillets marinades with cathepsin D (0.637–0.931) and Am-Tp (0.633) activity, in G-headed marinades with the addition of chymotrypsin activity (0.610–0.705) and in G-whole marinades with the addition of Es-Tp and Cp-A activity (0.580–0.946) (Figure 3A,C,E). For the Danish method, PHP content mainly correlated in D-fillets marinades with the addition of cathepsin D activity (0.565–0.746), in D-headed marinades with the addition of cathepsin L (0.704–0.935), and in D-whole marinades with Es-Tp (0.725–0.815) (Figure 3B,D,F). Thus, the lower degree of preliminary processing of herring promoted more significant correlations of digestive proteases with PHP and with higher cathepsin D activity. The results showed a synergistic effect of digestive and muscle enzymes in the proteolysis of marinated herring proteins (Figure 3), confirming the results of Kamiński et al. [7].

### 3.4. Hardness-TPA and Sensory Assessment of Marinated Meat

For two weeks of marinating, the highest texture rating in sensory analysis was regularly attributed to marinades made using the German method; with the exception of the 14th day in the case of marinades made from headed herring (Figure 8A). G-whole marinades scored highest in texture, with scores ranging from 4.79 to 4.93 after 4–14 days of marinating. D-fillets received the lowest marks of 4.20–4.43 for texture. German marinades needed only 4–7 days to achieve maximum marks for texture, while Danish marinades needed at least twice as long. Texture profile analysis (TPA) showed that the hardness-TPA of the meat of G-whole and G-headed marinades averaged 6.5 N, while the hardness of Gs marinades was significantly higher, at 9.6 N (Figure 8B). On the other hand, in the case of the Danish method, the hardness-TPA of the D-headed and D-whole marinades increased to from the fourth and decreased by the tenth day, obtaining higher values than the German G-headed and G-whole marinades (except D-headed after the seventh day). This could have been due to the lower activity of cathepsins L and D in the Danish marinades than German marinades. Cathepsin L hydrolyzes collagen, which has a greater impact on meat hardness than muscle proteins [37], which are mainly hydrolyzed by cathepsin D [38]. It was observed that cathepsin L activity positively correlated with texture in marinades obtained from herring with intestines (G-headed 0.847, D-headed 0.861, D-whole 0.779), contrary to a negative correlation in marinades from fillets (G-fillets −0.709, D-fillets −0.772). Although both marinating methods had a positive effect of marinating with digestive proteases on improving meat texture, significant hardness-TPA correlations with texture in sensory analysis were only for G-whole (0.620) and D-whole (0.568) marinades (Figure 3E,F).

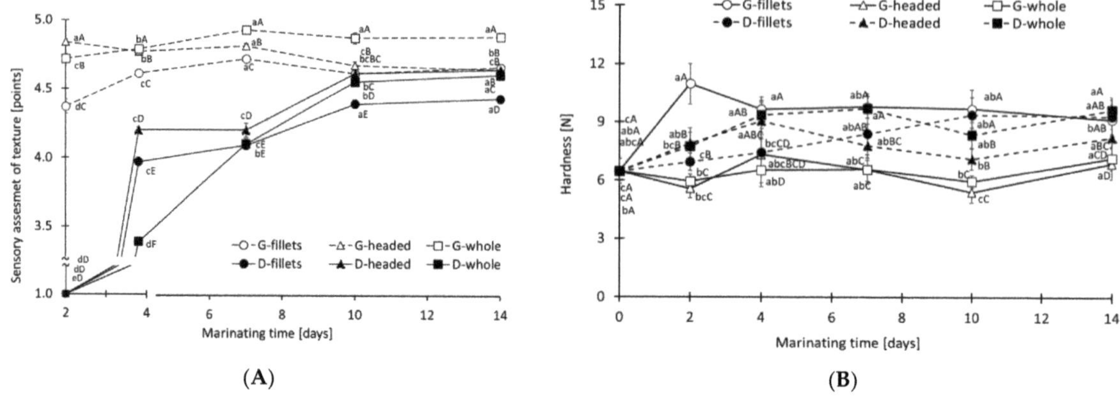

**Figure 8.** Effect of marinating method (G—German, D—Denmark) and degree of primary processing of Baltic herring (whole, headed, fillets) on (**A**) texture sensory evaluated and (**B**) hardness evaluated by TPA. [a,b,c] Results marked with the same lower case letter do not statistically differ by the influence of marinating time; [A,B,C] Results marked with the same capital letter do not statistically differ by the influence of the marinating method. For a better presentation of the results, a break ($\approx$) in the Y axis was used.

Additionally, the results of sensory analysis showed that a bitter taste appeared in the D-whole and D-headed marinades after 4 days of ripening, the intensity of which increased up to the tenth day of marinating from 1.0 to 4.0 points and from 0.5 to 2.0 points, respectively. The bitter taste was not present in D-fillets marinades, indicating that it was caused by the intestines of the fish. In the case of the German method, the bitter taste also appeared, but only after 10–14 days of marinating in G-whole and G-headed marinades, obtaining 1.0 and 0.5 points, respectively. The bitter taste could have come from blood and/or from the gastrointestinal tract and/or from bitter peptides, which are mainly released by chymotrypsin [39]. Chymotrypsin activity after 10–14 days of marinating was highest in only the D-whole and D-headed marinades, then was significantly lower in G-whole and G-headed marinades, and was lowest in fillets marinades (Figure 5C). In turn, carboxypeptidase-A, which is known to reduce bitter taste [40], was most active in G-whole marinades (Figure 5D), which likely protected these marinades from the rapid appearance of bitter taste.

### 3.5. Microbiota of Herring

Contamination of fresh herring meat with psychrophilic bacteria was 6.60 log cfu/g, mesophilic bacteria < 0.1 log cfu/g, moulds and yeasts < 0.1 log cfu/g and bacteria Enterobacteriaceae 3.00 log cfu/g (data not shown). After marinating, microbiological contamination of the four groups of bacteria was below 0.1 log cfu/g, except for psychrophile contamination after 2 days in Danish marinades: D-fillets 5.46, D-headed 6.30 i D-whole 5.28 log cfu/g. Microbiological analysis showed that the addition of acetic acid completely inhibited the proliferation of microorganisms in marinated meat, including Enterobacteriaceae from the digestive tract of the herring. Acetic acid, among other organic acids used in marinades, has the strongest bacteriostatic effect [41]. Kamiński et al. [7] showed that Enterobacteriaceae bacteria were the most resistant to marinating, but did not pose a threat to the microbiological quality of the marinades.

### 4. Conclusions

Studies have shown that marinating whole herring using the German method increased the mass yield of marinated fillets by several percent (10–15% after 4–14 days of marinating), which was also possible using the Danish method. However, the Danish method increased the salt content by 0.7–1.7 percent after 2 days of marinating, while in congtrast, the German method inhibited cathepsin D by 32–55% and promoted a bitter taste. The technology of marinating with viscera made it possible to reduce the ripening time of marinades by 3 days, from 7–10 days to 4–7 days. Diffusion of viscera proteases into the meat increased viscera proteases activity and cathepsin activity, which made it possible to obtain well-ripened meat as early as 4–7 days. Fillets from the same herring had mainly cathepsin activity, and did not achieve full and proper meat ripeness even after 14 days. In addition, marinating of ungutted herring promoted a longer and greater increase in the content of protein hydrolysis products (PHP), which give marinades their characteristic flavor. Therefore, especially important was the twice higher carboxypeptidase A activity in G-whole marinades (8.3–9.5 U) compared to G-headed and G-fillets (4.3–5 U). The 2–3 times higher PHP content for whole and headed samples, compared to fillets samples, was due to two phenomena: higher proteolytic activity in the meat and reduced PHP loss from the meat to the brine.

The results showed that the contribution of cathepsin D to meat ripening can be supplemented or even largely replaced by digestive proteases, but the right conditions for their activity (faster or previous diffusion of proteases than acetic acid and/or faster or previous diffusion of salt than acetic acid) must be applied. This is crucial knowledge in the field of marinating herring of low technological quality or non-herring fish resistant to marinating, whose meat does not reach full ripeness.

**Author Contributions:** Conceptualization, methodology, formal analysis, investigation writing—review and editing P.K., M.S. and B.S.; visualization P.K.; All authors have read and agreed to the published version of the manuscript.

**Funding:** This work was supported by Rector of the West Pomeranian University of Technology in Szczecin for PhD students of the Doctoral School, grant number: 540/080—Patryk Kamiński.

**Institutional Review Board Statement:** Not applicable.

**Informed Consent Statement:** Not applicable.

**Data Availability Statement:** Not applicable.

**Conflicts of Interest:** The authors declare no conflict of interest.

# References


1. Hyldig, G.; Jørgensen, B.M.; Undeland, I.; Olsen, R.E.; Jónsson, A.; Nielsen, H.H. Sensory Properties of Frozen Herring (*Clupea harengus*) from Different Catch Seasons and Locations. *J. Food Sci.* **2012**, *77*, S288–S293. [CrossRef] [PubMed]
2. Szymczak, M. Effect of technological factors on the activity and losses of cathepsins B, D and L during the marinating of Atlantic and Baltic herrings. *J. Sci. Food Agric.* **2017**, *97*, 1488–1496. [CrossRef] [PubMed]
3. Szymczak, M.; Kamiński, P.; Felisiak, K.; Szymczak, B.; Dmytrów, I.; Sawicki, T. Effect of constant and fluctuating temperatures during frozen storage on quality of marinated fillets from Atlantic and Baltic herrings (*Clupea harengus*). *LWT-Food Sci. Technol.* **2020**, *133*, 109961. [CrossRef]
4. Gringer, N.; Osman, A.; Nielsen, H.H.; Undeland, I.; Baron, C.P. Chemical Characterization, Antioxidant and Enzymatic Activity of Brines from Scandinavian Marinated Herring Products. *J. Food Process. Technol.* **2014**, *5*, 346. [CrossRef]
5. Szymczak, M.; Felisiak, K.; Tokarczyk, G.; Szymczak, B. The reuse of brine to enhance the ripening of marine and freshwater fish resistant to marinating. *Int. J. Food Sci. Technol.* **2019**, *54*, 1151–1159. [CrossRef]
6. Benjakul, S.; Yarnpakdee, S.; Senphan, T.; Halldorsdottir, S.M.; Kristinsson, G. *Fish Protein Hydrolysates: Production, Bioactivities, and Applications*; John Wiley & Sons, Ltd.: Hoboken, NJ, USA, 2014.
7. Kamiński, P.; Szymczak, M.; Szymczak, B. The effect of previous storage condition on the diffusion and contribution of digestive proteases in the ripening of marinated Baltic herring (*Clupea harengus*). *LWT-Food Sci. Technol.* **2022**, *165*, 113723. [CrossRef]
8. Szymczak, M. Comparison of physicochemical and sensory changes in fresh and frozen herring (*Clupea harrengus* L.) during marinating. *J. Sci. Food Agric.* **2011**, *91*, 68–74. [CrossRef]
9. Laub-Ekgreen, M.H.; Martinez-Lopez, B.; Frosch, S.; Jessen, F. The influence of processing conditions on the weight change of single herring (*Clupea harengus*) fillets during marinating. *Food Res. Int.* **2018**, *108*, 331–338. [CrossRef]
10. Babikova, J.; Hoeche, U.; Boyd, J.; Noci, F. Nutritional, physical, microbiological, and sensory properties of marinated Irish sprat. *Int. J. Gastron. Food Sci.* **2020**, *22*, 100277. [CrossRef]

11. Mansur, M.A.; Uddin, M.N.; Rahman, S.; Horner, W.F.A.; Uga, S. Northern Europe Processing Technique, Nutritional Composition, Quality and Safety, Flavour Compounds, Bio-Chemical Change During Pickle Curing of North Sea Herring (*Clupea harengus*) of Britain. *Oceanogr. Fish. Open Access J.* **2018**, *6*, 555679. [CrossRef]

12. Felberg, H.S.; Hagen, L.; Slupphaug, G.; Batista, I.; Nunes, M.L.; Olsen, R.L.; Martinez, I. Partial characterisation of gelatinolytic activities in herring (*Clupea harengus*) and sardine (*Sardina pilchardus*) possibly involved in post-mortem autolysis of ventral muscle. *Food Chem.* **2010**, *119*, 675–683. [CrossRef]

13. Engvang, K.; Nielsen, H.H. In situ activity of chymotrypsin in sugar-salted herring during cold storage. *J. Sci. Food Agric.* **2000**, *80*, 1277–1283. [CrossRef]

14. Gudmundur, S.; Nielsen, H.H.; Torstein, S.; Reinhard, S.; Jörg, O.; Joop, L.; Gudný, G. Frozen herring as raw material for spice-salting. *J. Sci. Food Agric.* **2000**, *80*, 1319–1324.

15. Kilinc, B.; Cakli, S. Chemical, enzymatical and textural changes during marination and storage period of sardine (*Sardina pilchardus*) marinades. *Eur. Food Res. Technol.* **2005**, *221*, 821–827. [CrossRef]

16. Nielsen, D.; Hyldig, G. Influence of handling procedures and biological factors on the QIM evaluation of whole herring (*Clupea harengus* L.). *Food Res. Int.* **2004**, *37*, 975–983. [CrossRef]

17. Erlanger, B.F.; Kokowsky, N.; Cohen, W. The preparation and properties of two new chromogenic substrates of trypsin. *Arch. Biochem. Biophys.* **1961**, *95*, 271–278. [CrossRef]

18. Hummel, B.C.W. A modified spectrophotometric determination of chymotrypsin, trypsin and thrombin. *Can. J. Biochem. Physiol.* **1959**, *37*, 1393–1399. [CrossRef]

19. Mock, W.L.; Liu, Y.; Stanford, D.J. Arazoformyl Peptide Surrogates as Spectrophotometric Kinetic Assay Substrates for Carboxypeptidase A. *Anal. Biochem.* **1996**, *239*, 218–222. [CrossRef]

20. Kołakowski, E. Analysis of proteins, peptides, and amino acids in foods. In *Methods of Analysis of Food Components and Additives*; Ötles, S., Ed.; CRC, Taylor & Francis Group: Boca Raton, FL, USA, 2005; pp. 59–96.

21. Felisiak, K.; Szymczak, M. Use of Rapid Capillary Zone Electrophoresis to Determine Amino Acids Indicators of Herring Ripening during Salting. *Foods* **2021**, *10*, 2518. [CrossRef]

22. *ISO 6887-1:2017-05*; Microbiology of the Food Chain—Preparation of Test Samples, Starting Suspension and Decimal Dilutions for Microbiological Tests—Part 1: General Rules for the Preparation of Starting Suspension and Decimal Dilutions. International Organization for Standardization: Geneva, Switzerland, 2017.

23. *ISO 4833-2:2013-12*; Microbiology of the Food Chain. Horizontal Method for the Determination of the Number of Microorganisms. Part 2. Determination of the Number by Surface Culture at 20 °C. International Organization for Standardization: Geneva, Switzerland, 2013.

24. *ISO 21527-1:2009*; Microbiology of Food and Animal Feeding Stuffs—Horizontal Method for the Enumeration of Yeasts and Moulds—Part 1: Colony Count Technique in Products with Water Activity Greater than 0,95. International Organization for Standardization: Geneva, Switzerland, 2008.

25. *ISO 21528-1:2017-08*; Microbiology of the Food Chain. Horizontal Method for Detection and Enumeration of Enterobacteriaceae. Part 1: Detection of Enterobacteriaceae. International Organization for Standardization: Geneva, Switzerland, 2017.

26. *ISO 4833-1:2013*; Microbiology of the Food Chain—Horizontal Method for the Enumeration of Microorganisms—Part 1: Colony Count at 30 Degrees C by the Pour Plate Technique. International Organization for Standardization: Geneva, Switzerland, 2013.

27. Pielak, M.; Czarniecka-Skubina, E.; Trafiałek, J.; Głuchowski, A. Contemporary Trends and Habits in the Consumption of Sugar and Sweeteners—A Questionnaire Survey among Poles. *Int. J. Environ. Res. Public Health* **2019**, *16*, 1164. [CrossRef]

28. Kolanowski, W.; Trafialek, J.; Drosinos, E.H.; Tzamalis, P. Polish and Greek young adults' experience of low quality meals when eating out. *Food Control* **2020**, *109*, 106901. [CrossRef]

29. Trafiałek, J.; Zwolinski, M.; Kolanowski, W. Assessing hygiene practices during fish selling in retail stores. *Br. Food J.* **2016**, *118*, 2053–2067. [CrossRef]

30. Szymczak, M.; Kołakowski, E. Losses of nitrogen fractions from herring to brine during marinating. *Food Chem.* **2012**, *132*, 237–243. [CrossRef] [PubMed]

31. Sharedeh, D.; Gatellier, P.; Astruc, T.; Daudin, J.-D. Effects of pH and NaCl levels in a beef marinade on physicochemical states of lipids and proteins and on tissue microstructure. *Meat Sci.* **2015**, *110*, 24–31. [CrossRef]

32. Szymczak, M.; Kołakowski, E.; Felisiak, K. Influence of salt concentration on properties of marinated meat from fresh and frozen herring (*Clupea harengus* L.). *Int. J. Food Sci. Technol.* **2012**, *47*, 282–289. [CrossRef]

33. Kołodziejska, I.; Sikorski, Z.E. Muscle cathepsins of marine fish and invertebrates. *Pol. J. Food Nutr. Sci.* **1995**, *45*, 3–10.

34. Talukdar, R.; Sareen, A.; Zhu, H.; Yuan, Z.; Dixit, A.; Cheema, H.; George, J.; Barlass, U.; Sah, R.; Garg, S.K.; et al. Release of Cathepsin B in Cytosol Causes Cell Death in Acute Pancreatitis. *Gastroenterology* **2016**, *151*, 747–758.e5. [CrossRef]

35. Zaidi, N.; Kalbacher, H. Cathepsin E: A mini review. *Biochem. Biophys. Res. Commun.* **2008**, *367*, 517–522. [CrossRef]

36. Özden, Ö. Changes in amino acid and fatty acid composition during shelf-life of marinated fish. *J. Sci. Food Agric.* **2005**, *85*, 2015–2020. [CrossRef]

37. Felberg, H.S.; Slizyté, R.; Mozuraityte, R.; Dahle, S.W.; Olsen, R.L.; Martinez, I. Proteolytic activities of ventral muscle and intestinal content of North Sea herring (*Clupea harengus*) with full and emptied stomachs. *Food Chem.* **2009**, *116*, 40–46. [CrossRef]

38. Johnston, I.A.; Solberg, C.; Johnston, I.A. Activity of aspargate (cathepsin D), cysteine proteases (cathepsins B, B + L, and H), and matrix metallopeptidase (collagenase) and their influence on protein and water-holding capacity of muscle in commercially farmed Atlantic halibut (*Hippoglossus hippoglossuss* L.). *J. Agric. Food Chem.* **2008**, *56*, 5953–5959.

39. Umetsu, H.; Ichishima, E. Mechanism of digestion of bitter peptide from a fish protein concentrate by wheat carboxypeptidase. *J. Food Sci. Technol.* **1985**, *32*, 281–287. [CrossRef]

40. Broncano, J.; Timón, M.; Parra, V.; Andrés, A.; Petrón, M. Use of proteases to improve oxidative stability of fermented sausages by increasing low molecular weight compounds with antioxidant activity. *Food Res. Int.* **2011**, *44*, 2655–2659. [CrossRef]

41. Logrén, N.; Hiidenhovi, J.; Kakko, T.; Välimaa, A.-L.; Mäkinen, S.; Rintala, N.; Mattila, P.; Yang, B.; Hopia, A. Effects of Weak Acids on the Microbiological, Nutritional and Sensory Quality of Baltic Herring (*Clupea harengus membras*). *Foods* **2022**, *11*, 1717. [CrossRef] [PubMed]

MDPI

St. Alban-Anlage 66

4052 Basel

Switzerland

www.mdpi.com

*Applied Sciences* Editorial Office

E-mail: applsci@mdpi.com

www.mdpi.com/journal/applsci